Macro-Econophysics

Recent availability of big data on heterogeneity and synchronization of economic networks provides a good opportunity to study macro-economy from micro and mesoscopic perspectives in terms of heterogeneous interacting agents. Through collaborations between physicists and economists, we find that the key to understand macro-economy is economic networks, on which a large-number of economic agents are interacting with others in heterogeneity, and synchronization, that is, spatial or temporal regularities that emerge from the heterogeneity and interaction. In tandem with the growing necessity of studying macro-economy through such a new perspective and methodology, there is a surge of interest among researchers and students in the subject. This book is tailor-made to meet the needs of a wide range of readers in economics, complex systems, complex network science and related disciplines of research.

This book discusses economic networks and synchronization from the perspectives of statistical physics and complex networks. It aims at discussing application of big data in macro-economy, specifically, business cycles, systemic risks, inflation and deflation, productivity dispersion and innovation, and related topics. It offers detailed discussion on previous studies in macroeconomics and new insights found by using new methodologies. Each topic is elaborated by exploiting recently available big data and by employing new tools. It covers most of the recent research that is based on empirical and large-scale data in Japan as well as in Europe and the US carried out using tools and concepts in non-equilibrium statistical physics, complex networks and statistical science. The text offers new methods based on complex network and statistical physics to macroeconomics, especially for the understanding of interactions and aggregate dynamics in economic systems.

Hideaki Aoyama is Professor of Theoretical Physics at the Graduate School of Science, Kyoto University and also is Faculty Fellow of the Research Institute of Economy, Trade and Industry (RIETI, Tokyo) and Senior Researcher of Asia Pacific Institute of Research (APIR, Osaka). His research interests are macro-econophysics, network science, linguistics and theoretical physics applied to broad range of subjects in general.

Yoshi Fujiwara is Professor at the Graduate School of Simulation Studies, University of Hyogo and also is Senior Visiting Scientist of RIKEN, Advanced Institute of Computational Science, in Kobe. His research interests are Pareto-Zipf law, firms growth and failure; large-scale economic networks including production network, debtor-creditor relationship among banks and firms; high performance computing and visualization.

Yuichi Ikeda is Professor of Physics at Graduate School of Advanced Integrated Studies in Human Survivabity, Kyoto University. His current research interest includes econophysics, network science, data science, and computational science.

Hiroshi Iyetomi is Professor of Mathematics at Faculty of Science, Niigata University. His research interests include econophysics with an emphasis on collective motion in economic systems, multivariate time series analysis, complex networks, strongly-coupled plasma physics, and computational material science.

Wataru Souma is Associate Professor of Physics at College of Science and Technology, Nihon University. His main fields of research are econophysics, network science, scientometrics, functional renormalization group, and quantum gravity.

Hiroshi Yoshikawa is Emeritus Professor of the University of Tokyo and Professor at Faculty of Economics, Rissho University. His research interests are macroeconomics and macro-econophsyics.

Physics of Society: Econophysics and Sociophysics

This book series is aimed at introducing readers to the recent developments in physics inspired modelling of economic and social systems. Socio-economic systems are increasingly being identified as 'interacting many-body dynamical systems' very much similar to the physical systems, studied over several centuries now. Econophysics and sociophysics as interdisciplinary subjects view the dynamics of markets and society in general as those of physical systems. This will be a series of books written by eminent academicians, researchers and subject experts in the field of physics, mathematics, finance, sociology, management and economics.

This new series brings out research monographs and course books useful for the students and researchers across disciplines, both from physical and social science disciplines, including economics.

Series Editors:

Bikas K. Chakrabarti
Professor
Saha Institute of Nuclear Physics, Kolkata
India

Mauro Gallegati
Professor of Economics
Polytechnic University of Marche
Italy

Alan Kirman
Professor Emeritus of Economics
University of AixMarseille III, Marseille
France

H. Eugene Stanley
William Fairfield Warren Distinguished Professor
Boston University, Boston
USA

Editorial Board Members:

Frédéric Abergel
Professor of Mathematics
Centrale Supélec
Chatenay Malabry
France

Hideaki Aoyama
Professor of Physics
Kyoto University, Kyoto
Japan

Anirban Chakraborti
Professor of Physics
School of Computational and Integrative Sciences
Jawaharlal Nehru University, New Delhi
India

Satya Ranjan Chakravarty
Professor of Economics
Indian Statistical Institute, Kolkata
India

Arnab Chatterjee
TCS Innovation Labs, Delhi
India

Shu-Heng Chen
Professor of Economics and Computer Science
Director, AIECON Research Center
National Chengchi University, Taipei
Taiwan

Domenico Delli Gatti
Professor of Economics
Catholic University, Milan
Italy

Kausik Gangopadhyay
Professor of Economics
Indian Institute of Management, Kozhikode
India

Cars Hommes
Professor of Economics
Amsterdam School of Economics
University of Amsterdam
Director, Center for Nonlinear Dynamics in
Economics and Finance (CeNDEF), Amsterdam
Netherlands

Giulia Iori
Professor of Economics
School of Social Science
City University, London
United Kingdom

Teisei Kaizoji
Professor of Economics
Department of Economics and Business
International Christian University, Tokyo
Japan

Kimmo Kaski
Professor of Physics
Dean, School of Science
Aalto University, Espoo
Finland

János Kertész
Professor of Physics
Center for Network Science
Central European University, Budapest
Hungary

Akira Namatame
Professor of Computer Science and Economics
Department of Computer Science
National Defense Academy, Yokosuka
Japan

Parongama Sen
Professor of Physics
University of Calcutta, Kolkata
India

Sitabhra Sinha
Professor of Physics
Institute of Mathematical Science, Chennai
India

Victor Yakovenko
Professor of Physics
University of Maryland, College Park
USA

Physics of Society: Published Titles

- *Limit Order Books* by Frédéric Abergel, Marouane Anane, Anirban Chakraborti, Aymen Jedidi and Ioane Muni Toke

- *Interactive Macroeconomics: Stochastic Aggregate Dynamics with Heterogeneous and Interacting Agents* by Corrado Di Guilmi, Simone Landini and Mauro Gallegati

Physics of Society: Forthcoming Titles

- *A Statistical Physics Perspective on Socio Economic Inequalities* by Arnab Chatterjee and Victor Yakovenko

Physics of Society: Econophysics and Sociophysics

Macro-Econophysics
New Studies on Economic Networks and Synchronization

Hideaki AOYAMA
Yoshi FUJIWARA
Yuichi IKEDA
Hiroshi IYETOMI
Wataru SOUMA
Hiroshi YOSHIKAWA

CAMBRIDGE
UNIVERSITY PRESS

University Printing House, Cambridge CB2 8BS, United Kingdom
One Liberty Plaza, 20th Floor, New York, NY 10006, USA
477 Williamstown Road, Port Melbourne, vic 3207, Australia
4843/24, 2nd Floor, Ansari Road, Daryaganj, Delhi - 110002, India
79 Anson Road, #06–04/06, Singapore 079906

Cambridge University Press is part of the University of Cambridge.

It furthers the University's mission by disseminating knowledge in the pursuit of education, learning and research at the highest international levels of excellence.

www.cambridge.org
Information on this title: www.cambridge.org/9781107198951

© Authors 2017

This publication is in copyright. Subject to statutory exception and to the provisions of relevant collective licensing agreements, no reproduction of any part may take place without the written permission of Cambridge University Press.

First published 2017

Printed in India by Nutech Print Services, New Delhi

A catalogue record for this publication is available from the British Library

ISBN 978-1-107-19895-1 Hardback

Additional resources for this publication at www.cambridge.org/9781107198951

Cambridge University Press has no responsibility for the persistence or accuracy of URLs for external or third-party internet websites referred to in this publication, and does not guarantee that any content on such websites is, or will remain, accurate or appropriate.

We are at the very beginning of time for the human race. It is not unreasonable that we grapple with problems. But there are tens of thousands of years in the future. Our responsibility is to do what we can, learn what we can, improve the solutions, and pass them on.

– Richard Feynman

Contents

Figures	*xv*
Tables	*xxiii*
Foreword	*xxv*
Acknowledgements	*xxvii*
Prologue	*xxix*

1 Introduction: Reconstructing Macroeconomics
- 1.1 Background and Motivation ... 1
- 1.2 Outline ... 6

2 Basic Concepts in Statistical Physics and Stochastic Models
- 2.1 Stochastic Models and Fat Tails ... 11
  - 2.1.1 Yule and Simon's models and multiplicative processes ... 11
  - 2.1.2 Marsili–Zhang model and interaction of agents ... 18
- 2.2 Entropy ... 24
- 2.3 Statistical Equilibrium ... 26
- 2.4 Stationarity in Time Series ... 27
  - 2.4.1 Stationary and non-stationary time series ... 27
  - 2.4.2 Methods to obtain stationary time series ... 30
- 2.5 Long-Term Memory in Time Series ... 35
  - 2.5.1 Autocorrelation function and average mutual information ... 35

2.5.2 Volatility of return time series ... 36
2.6 Basic Model of Time Series ... 37
  2.6.1 Autoregression model ... 37
  2.6.2 Time dependence of forecast error ... 39
2.7 Stochastic Model of Non-Stationary Time Series ... 41
  2.7.1 Standard brownian motion ... 41
  2.7.2 Fractal brownian motion ... 42
  2.7.3 Fractal dimension ... 43
2.8 Advanced Model of Time Series ... 46
  2.8.1 Autoregressive fractionally integrated moving average model ... 46
  2.8.2 ARCH model and GARCH model ... 48
2.9 Exercise ... 51

3 Income and Firm-size Distributions

3.1 Distributions with Fat-Tails ... 54
  3.1.1 Pearson system ... 55
  3.1.2 Burr and dagum system ... 57
  3.1.3 The Beta-type distributions ... 61
3.2 Large Values and their Shares ... 69
  3.2.1 Maximum value and its share ... 69
  3.2.2 The second largest value and its share ... 78
  3.2.3 All about shares ... 80
3.3 Personal Income ... 83
  3.3.1 Short history ... 83
  3.3.2 Distributions and fluctuations ... 86
  3.3.3 Two-factor model ... 90
3.4 Firm Size ... 94
3.5 Excercise ... 95

4 Productivity Distribution and Related Topics

4.1 Production Function and Production Copula ... 97
  4.1.1 Value added and labor productivity ... 97
  4.1.2 Production function ... 99
  4.1.3 Copula theory ... 103
  4.1.4 Construction of a production copula ... 108
4.2 Stochastic Macro-equilibrium ... 116
  4.2.1 The basic idea ... 116

		4.2.2 Empirical distribution of productivity	120
		4.2.3 Universal statistics	123
		4.2.4 Fitting of the model to the Japanese data	129
	4.3	Exercise	132

5 Multivariate Time Series Analysis
	5.1	Principal Component Analysis (PCA)	134
	5.2	Complex Hilbert PCA (CHPCA)	137
		5.2.1 Continuous time	137
		5.2.2 Discrete time series in a finite time range	138
		5.2.3 Absolute values and phases of the complexified time series	143
		5.2.4 Complex correlation matrix	145
		5.2.5 Examples	148
	5.3	Identification of Comovements	150
		5.3.1 Random matrix theory	150
		5.3.2 Autocorrelation effects	152
		5.3.3 Rotational random shuffling	153
	5.4	Case Study: Equity and Currency in the World	155
	5.5	Exercise	161

6 Business Cycles
	6.1	What Causes Business Cycles	163
		6.1.1 Indices of industrial production	165
		6.1.2 Power spectrum	168
		6.1.3 Identification of dominant factors	169
		6.1.4 Dominant eigenmodes and the business cycles	172
		6.1.5 Production and shipments	173
		6.1.6 The 2008–09 economic crisis	176
		6.1.7 Summary	179
	6.2	Synchronization in Japanese Business Cycles	180
		6.2.1 Background	180
		6.2.2 Time series of the production indices	180
		6.2.3 Phase time series	182
		6.2.4 Frequency entrainment	184
		6.2.5 Phase locking	186
		6.2.6 Common shock and individual shock	187
		6.2.7 Summary	190
	6.3	Synchronization in International Business Cycles	191
		6.3.1 Background	191

	6.3.2 GDP time series	191
	6.3.3 Phase time series	195
	6.3.4 Frequency entrainment	197
	6.3.5 Phase locking	198
	6.3.6 Common shocks versus individual shocks	199
	6.3.7 Summary	202
6.4	Coupled Limit-Cycle Oscillator Model	202

7 Price Dynamics and Inflation/Deflation

7.1	Individual Price Data	210
	7.1.1 CHPCA+RRS results	213
	7.1.2 A case study	214
7.2	Correlation between Mode-signals and Macroeconomic Indices	216
7.3	Collective Motion of Prices and Business Cycles	221
7.4	Summary	223

8 Complex Networks, Community Analysis, Visualization

8.1	Basic Tools	226
	8.1.1 Graph search and simple applications	226
	8.1.2 Random graphs	236
	8.1.3 Basic properties of network structures	239
8.2	How to Identify Communities	247
8.3	Visualization	250
8.4	Nationwide Production Network	253
	8.4.1 Large-scale structure of the production network	254
	8.4.2 Visualization of directed graph of production network	258
8.5	Stock Correlation Network	260
	8.5.1 Group correlations	260
	8.5.2 Community detection	262
	8.5.3 Frustrated correlation structure	263
8.6	Globally-coupled Financial Networks	266
	8.6.1 Lead–lag relations	267
	8.6.2 Synchronization networks	268
8.7	Community Analysis of Trade and Production Networks	272
	8.7.1 Background	272
	8.7.2 Data	273
	8.7.3 Community analysis	277
	8.7.4 Community structure of trade network	280

		8.7.5 Linked communities of trade network	282

        8.7.5   Linked communities of trade network   282
        8.7.6   Synchronization of international business cycles   287
        8.7.7   Significant modes in economic crisis   289
        8.7.8   Community structure of production network   290
        8.7.9   Linked communities of production networks   292
        8.7.10 Communities in multiplex production networks   294
  8.8   Controllability of Production Network   295
        8.8.1   Theory of structural controllability   295
        8.8.2   Illustrative example for identifying driver nodes   296
        8.8.3   Controllability of production networks   298

9 Systemic Risks

  9.1   Nation-wide Production Network   307
        9.1.1   DebtRank method   308
        9.1.2   Bowtie structure and up/down streams of production   310
        9.1.3   Calculation of DebtRank using the K computer   313
        9.1.4   Summary   320
  9.2   Bank–Firm Credit Network   321
        9.2.1   Japanese bank–firm data   321
        9.2.2   'Too big to fail?' and other questions   327
        9.2.3   Vulnerability   331
        9.2.4   Summary   335

A Computer Programs for Beginners

  A.1   *Mathematica* Codes for Finance   337
  A.2   Tools for Network Analysis   338
  A.3   Python Codes for Basic Graph Algorithms   339
  A.4   *Mathematica* Codes for Network Analysis   345

Solution to Exercises   347

Epilogue   356

Bibliography   358

Index   379

Color Plates   387

Figures

2.1 Genus-size distribution constructed from Yule (1925)   12
2.2 Empirical distribution for citations of papers   12
2.3 Daily change of stock indices   28
2.4 Unit root test   30
2.5 Stationary time series calculated from Dow Jones Industrial Average   32
2.6 Stationary time series calculated from NIKKEI-225 Stock Average   34
2.7 Time series x_n and Lag τ 36
2.8 Daily change of volatility and its surrogate variable   36
2.9 Autocorrelation function for log return and volatility of Nikkei-225   37
2.10 Daily temporal change of log-return calculated by ARMA(1,2) model   39
2.11 Cantor set   44
2.12 Period τ and normalized range S/V 46
2.13 Daily change of return calculated by ARFIMA(1,2)   48
2.14 Daily change of return calculated by GARCH(1,2)   49
2.15 Autocorrelation function of return volatility   51
3.1 Beta-type distribution tree   61
3.2 The behavior of $f_N^{(\max)}(\tilde{x})$ 71
3.3 $H_N(t)$ at $\mu = 0.8$ (left) and $\mu = 1.2$ (right)   73

3.4 $\langle R_N^{(\max)} \rangle$ for $\mu > 1$ $N \to \infty$ — 75

3.5 $\langle R_N^{(\max)} \rangle$ at $\mu = 1$ — 77

3.6 The share of the maximum value $\langle R_N^{(\max)} \rangle$ — 79

3.7 The share of the second largest value $\langle R_N^{(2)} \rangle$ — 81

3.8 Share of the top values (3.148) — 82

3.9 Income distribution in Great Britain and Ireland during 1893–94 — 84

3.10 Distributions of personal income in Japan from 1987 to 2000 — 87

3.11 Scatterplot for personal income tax for two consecutive years — 89

3.12 Gibrat's law and breakdown of Gibrat's law — 89

3.13 Income sources of high-income earners — 91

3.14 Double logarithmic plot of the simulation results — 93

3.15 CDF of Japanese company size in 2002 — 93

3.16 Relation between firm sizes (sales) and variance of growth rate — 95

3.17 Relation between company-size (sales) and variance of growth-rate — 95

4.1 The distributions of labor productivity for electrical machinery sector — 98

4.2 The distributions of labor productivity for service sector — 99

4.3 Correlations for each pair of the three financial quantities — 100

4.4 Difference between the actual Y and the best-fit CD function $\Phi(L, K)$ — 101

4.5 Complementary CDFs of $Y/\Phi(L, K)$ — 102

4.6 Contour plots of the bivariate copula densities — 106

4.7 Modeling of the pair correlations in Fig. 4.3 in terms of the Gumbel copula — 109

4.8 Contour plots of the copula densities corresponding to the fitted copulas in Fig. 4.7 — 109

4.9 Contour plots of the copula cumulant — 112

4.10 Typical cross sections in u_L-u_K-u_Y space — 113

4.11 Comparison of the results for $\Omega(x, x, x)$ between the copula model and real data — 113

4.12 A simulated result for the production activity of listed Japanese firms — 114

4.13 The CCDFs of $Y/\Phi(L, K)$ calculated in Model (IV) — 115

4.14 Relationship between entropy S and aggregate demand D — 119

4.15 PDF of $\log c$ for firms and workers in 2012 — 121

4.16	Dependence of the average number of workers of individual firms on c in 2012	122
4.17	Elementary binary process and its reverse for workers' job change	124
4.18	A model of worker limitation $L(n,c)$ in Eq. (4.66)	126
4.19	Ceiling function $f(n;g)$ given by Eq. (4.81)	129
4.20	The best fit to the empirical data in 2012	130
5.1	The integration contour for the integration in Eq. (5.20) on the complex z-plane	137
5.2	Rotating behavior of Eq. (5.41)	141
5.3	The sample time series of sin and cos defined in Eq. (5.98)	148
5.4	The time series $\tilde{r}_{1,2}(t)$ complexified as defined in Eq. (5.42)	149
5.5	PDF of the eigenvalues for a completely random correlation matrix	151
5.6	PDF of the PCA eigenvalues for the TSE market during 1996-2006	151
5.7	Same as Fig. 5.5, but for the correlation matrices based on the AR(1) model	153
5.8	A combination dial lock	154
5.9	PDF of the PCA eigenvalues for real data preprocessed with the RRS	154
5.10	CHPCA eigenvalues for the world currency and equity market	157
5.11	Comparison between the PCA and CHPCA eigenvalues	158
5.12	The first eigenvector of the FX and equity markets	158
5.13	The sixth eigenvector components of the world equity and foreign exchange markets	159
5.14	Significance of the sixth eigenmode	160
6.1	Demonstration of Slutsky's effect	164
6.2	Input-output relationship in industry as measured by IIP in Japan	166
6.3	Averaged IIP data S_α for production, shipments, and inventory	167
6.4	Averaged power spectrum $p(\omega)$ of the normalized growth rate	169
6.5	PDF of the eigenvalues compared with the RMT prediction	170
6.6	Components of the two dominant eigenvectors, $V^{(1)}$ and $V^{(2)}$	171
6.7	Contributions to the total power spectrum of the two dominant eigenmodes	173
6.8	Comparison between the extracted business cycles and the original data	175
6.9	Decomposition of volatility of the extended IIP data	176
6.10	Relative contributions of the dominant modes, corresponding to Fig. 6.9	178

6.11	Indices of exports and industrial production	178
6.12	The time series of log returns of the production indices	181
6.13	The time series of log returns of the production indices in the complex plane	183
6.14	The phase time series of log returns of the production indices	184
6.15	The angular frequencies for the 16 sectors	185
6.16	The indicators of the phase locking $\sigma(t)$	187
6.17	The average amplitudes $\langle A(t)\rangle$ for the 16 oscillators	188
6.18	The average phases $\langle \cos\theta(t)\rangle$ for the 16 oscillators	189
6.19	The individual shocks for the 16 oscillators	190
6.20	The growth rate of GDP for the six counties	192
6.21	The band-pass filtering of growth rate of GDP for the six counties	193
6.22	The filtered change in inventory stock for the four countries	194
6.23	The growth rate of GDP vs the change in inventory stock for the four countries	195
6.24	The time series of growth rate of GDP in the complex plane for the six countries	196
6.25	The time series of phase obtained using Hilbert transform for the six countries	197
6.26	The estimated angular frequencies for the six countries	198
6.27	The estimated indicator of phase locking $\sigma(t)$.	199
6.28	The average amplitudes $\langle A(t)\rangle$ and the average phases $\langle \cos\theta(t)\rangle$ for the 6 oscillators	199
6.29	The time series of common and individual shocks	201
6.30	Trade linkage structure of the coupled limit-cycle oscillator model	203
6.31	Temporal changes of the amount of international exports and imports	204
6.32	Temporal changes of the interaction strengths	207
6.33	Network structure of the coupled limit-cycle oscillator model	208
6.34	Synchronization of international business cycle and the interaction strength	209
7.1	Bird's eye view of the monthly price changes	212
7.2	Eigenvalues obtained from the CHPCA+RRS analysis	213
7.3	The cumulative eigenvalues S_n	214
7.4	The actual time series and the cleansed time series	215
7.5	The 1st eigenvector components in the complex plane	216

7.6	The eight macroeconomic indices	218
7.7	The absolute values of the correlation coefficient $\mathcal{A}_{j,1}$	219
7.8	The absolute values of the correlation coefficient $|\mathcal{A}_{j,n}|$ for $n = 2$–5	220
7.9	Collective motion of individual prices in the first mode of the CHPCA	222
7.10	Collective motion of individual prices in the second mode of the CHPCA	222
7.11	Contributions of the first and second eigenmodes to the coincident index	223
8.1	Undirected graph, adjacency list and matrix	228
8.2	Directed graph, adjacency list and matrix	228
8.3	Depth-first search for a directed graph	230
8.4	DFS trees	231
8.5	Decomposition into strongly connected component	233
8.6	An undirected graph for BFS	234
8.7	Breath-first search for an undirected graph	235
8.8	A BFS tree	236
8.9	Schematic picture of the modularity landscape	248
8.10	Community structure of the Karate club network	250
8.11	Spring-charge model for network visualization	251
8.12	Optimized layout of the network in Fig. 8.11	251
8.13	Hierarchical domain decomposition	252
8.14	Coarse-graining of a network by the hierarchical domain decomposition	253
8.15	Application of the coarse-graining procedure to the power grid network	253
8.16	Visualization of directed network of production network in Japan	259
8.17	Selected industrial sectors visualized in the same visualisation	259
8.18	Frustrated stock group structure in the TSE	263
8.19	Triangular networks with friendly or unfriendly bilateral relations	264
8.20	Three-dimensional correlation state vectors of the stocks in the TSE	265
8.21	Same as Fig. 8.18, but for the S&P 500	265
8.22	Same as Fig. 8.20, but for the S&P 500	266
8.23	Community structure of the financial network on a world map	269
8.24	Adjacency matrices of the financial network for the four periods	270
8.25	Optimized layouts of the financial network corresponding to Fig. 8.24	271
8.26	Trade and business cycles in USA	274

8.27	Community structure and temporal change of modularity	281
8.28	Variation of information: (a) 1995 and (b) 2011	282
8.29	Jaccard index: (a) between 1995 and 1996, and (b) between 2010 and 2011	283
8.30	Polar plot of the 2009 phase: (a) all sectors, (b)–(d) community $c_{1,2,3}$	287
8.31	Temporal change of amplitude for the order parameter $r(t)$	288
8.32	Eigenvalues of significant modes for production indices in G7 countries	290
8.33	Community structure: (a) 2004, (b) 2007, (c) 2010, and (d) 2013	291
8.34	Jaccard index: (a) 2003–2004, (b) 2006–2007, (c) 2009–2010, and (d) 2012–2013	293
8.35	Comparison of communities between multiplex network and single layer network	295
8.36	The first example for identifying driver nodes	297
8.37	The second example for identifying driver nodes	298
8.38	Matchability distribution: (a) 2004, (b) 2007, (c) 2010, and (d) 2013	299
8.39	Temporal changes of the number of driver nodes	299
8.40	Degree distributions: (a) 2004, (b) 2007, (c) 2010, and (d) 2013	304
9.1	Schematic diagram of economic networks for banks, firms and credit relationships	307
9.2	A photograph of the K computer system used for DebtRank calculation	310
9.3	Bowtie structure of production network to identify upstream and downstream	312
9.4	Cumulative distribution for the DebtRank of a million firms in Japan	315
9.5	Relation between the DebtRank and firm-size of sales for a million firms in Japan	315
9.6	Relation between the DebtRank and firm-size of number of employees	316
9.7	Relation between the DebtRank and size of industrial sectors in Japan	317
9.8	Vulnerability of each sector owing to another sector's financial distress	319
9.9	Number of banks and firms in the database	322
9.10	Bank-firm credit network	323
9.11	MST of banks	325
9.12	$C_{\beta f}$ from bank β to firm f	326
9.13	CDF of C_β (left) and C_f (right)	327
9.14	The DebtRanks of the banks in 2010	328
9.15	Total asset vs total DebtRank in 2010	329

9.16 The importance-ratio function $R(x)$	330
9.17 Vulncrability of banks	331
9.18 Vulnerability of banks to distress in automobile sector and the construction sector	332
9.19 Vulnerability of the banks caused by each sector	333
9.20 Vulnerability of sectors by other sectors	335
A.1 Diagram obtained by dividing a line segment into four and removing every other line segment	350
A.2 CDF for $f(x) = \delta(x-a)$	352

Tables

2.1 AIC of the ARIMA $(p, 1, q)$ model for stock price   38

2.2 AIC of the ARIMA $(p, 0, q)$ model for log return   38

2.3 Hurst index H 46

2.4 Comparrison of Hurst index H 50

3.1 Pearson distributions   57

3.2 Burr distributions   58

3.3 Dagum distributions   60

3.4 Share of the top 10, 50, 100 firms in % given by Eq. (3.148)   82

3.5 The number of individuals of 1987 − 2000 Japanese income   86

4.1 Maximized log-likelihood ℓ in fitting bivariate Archimedean copulas   108

4.2 Degree of asymmetry in the pair correlations in Fig. 4.3   110

4.3 Maximum likelihood optimization in the models for the production copula   111

4.4 Estimated parameters in Eq. (4.67) and the peak position c_{p} 132

5.1 List of 48 countries with their stock market indices   156

5.2 List of CHPCA eigenvalues and RRS results   157

6.1 Classifications of 21 goods in IIP   166

6.2 Eigenvalues and eigenvectors   200

6.3 Estimation of model parameters for Australia   204

6.4	Estimation of model parameters for Canada	205
6.5	Estimation of model parameters for France	205
6.6	Estimation of model parameters for UK	205
6.7	Estimation of model parameters for Italy	206
6.8	Estimation of model parameters for USA	206
8.1	Assortative mixing in sexual partnerships, a survey in epidemiology	240
8.2	Performance of typical modularity optimization methods	249
8.3	Polarization of stocks at sector classification level in $V^{(2)}$, $V^{(3)}$ and $V^{(4)}$	260
8.4	G7 Global Production Data1	275
8.5	G7 Global Production Data2	276
8.6	Linked communities	284
8.7	Linked community 1 (Total = US\$ 24.360 trillion)	284
8.8	Linked community 4 (Total = US\$ 20.447 trillion)	284
8.9	Linked community 3 (Total = US\$ 17.227 trillion)	285
8.10	Linked community 2 (Total = US\$ 16.520 trillion)	285
8.11	Linked community 5 (Total = US\$ 15.096 trillion)	285
8.12	Linked community 6 (Total = US\$ 7.900 trillion)	285
8.13	Linked communities before the crisis	293
8.14	Linked communities during the crisis	293
8.15	Linked communities after the crisis	294
8.16	Driver nodes with $m \leq 0.2$ in 2004	300
8.17	Driver nodes with $m \leq 0.2$ in 2007	301
8.18	Driver nodes with $m \leq 0.2$ in 2010	302
8.19	Driver nodes with $m \leq 0.2$ in 2013	303
9.1	Classification of a million firms in Japan into industrial sectors	311
9.2	Bowtie structure: Sizes of different components for a million firms in Japan	312
9.3	All 33 sectors specified in the Nikkei database	334

Foreword

Economics has emerged as a major discipline today of interest to all because of its impact on our day-to-day life. What has been achieved so far has been truly impressive, although the discipline is not as successful as one would expect. Notwithstanding what mainstream economics does or strives for, it does not really meet the criteria to be called a natural science yet. This book is an attempt to steer it in that direction.

Although natural sciences, such as physics, chemistry, biology, or geology employ logic and mathematics (as a condensed form of logic), it is never the sole ingredient. Stepwise observations are organized in a logical fashion, often with the help of tentative or approximate hypotheses, and both the existing observations and predicted outcomes are carefully compared. The understanding of the next level or of similar but different systems grows progressively, based on the successful ideas or an understanding developed earlier. Naturally, there is interdependence in natural sciences as a consequence of this kind of development. In general, precise knowledge, successful ideas, or techniques developed in one area of the natural sciences become easily translated into another.

This interdependent structure of research in the natural sciences also gets reflected in the graduate level course structure for students in their respective disciplines. Students of one major discipline of the natural sciences have to learn the basic and established concepts in other disciplines: Physics majors learn concepts of chemistry, biology, or geology; biology majors learn basic concepts of physics and chemistry, along with others. This practice is somehow not there yet for the social sciences; the graduate students here do learn mathematics and statistics but not the basic concepts of physics, chemistry, or biology. Personal interests are, of course, exceptions and are not counted here!

To many, this is the main reason why economics, which also started becoming formalized much later compared with most other branches of the natural sciences, could not boast of the spectacular successes achieved by other disciplines. Among others, econophysicists believe in the need for a similar mutation of ideas from economics and physics, for the healthy evolutionary growth in both.

MACRO-ECONOPHYSICS is an attempt by an internationally renowned group of (econo-) physicists and economists to recast macroeconomics in the mold of physics. The subject deals with collective or evolving economic or financial dynamics of a cluster of companies, firms, banking and other networks, where the healthy or sick status of the dynamics of individual agents or companies may not imply the same for the collective society or the nation. Successful ideas, models, and techniques developed in statistical physics over the past century or so can indeed lead to a very satisfactory understanding in macroeconomics and is shown in this book. This attempt is indeed pioneering and balanced. Most of its authors, sometimes with other collaborators, had earlier indicated similar possibilities in their well-known books published earlier by CUP. In that sense, this book details the latest developments in this attempt and in a very comprehensive way. The book should be of immense value to graduate students and researchers in economics and physics interested in exploring the natural science frontier of (macro) economics.

Bikas K. Chakrabarti
Condensed Matter Physics Division, Saha Institute of Nuclear Physics
Economic Research Unit, Indian Statistical Institute

Acknowledgments

We wish to express our sincere thanks to Bikas Chakrabarti for giving us an opportunity to publish this work. We would like to thank our collaborators and other researchers in this and other related fields. With them, we enjoyed discussions over many years of study that led to this book. We cannot list them all, but some of them are: Masanao Aoki, Yuta Arai, Yoshiyuki Arata, Yuji Aruka, Stefano Battiston, Guido Caldarelli, Giulia De Masi, Corrado Di Guilmi, Yuji Fujita, Yudai Fujiwara, Mauro Gallegati, Bruce Greenwald, Takashi Iino, Hiroyasu Inoue, the late Jun-ichi Inoue, Taisei Kaizoji, Yuichi Kichikawa, Eliza Olivia Lungu, Thomas Lux, Luca Marotta, Rasario Mantegna, Salvatore Miccichè, Kazuo Minami, Takayuki Mizuno, Makoto Nirei, Takaaki Ohnishi, Bertrand M. Roehner, Yohei Sakamoto, Yukiko Saito, Katunori Shimohara, Fumiyoshi Shoji, Robert M. Solow, Didier Sornette, Gene Stanley, Joe Stiglitz, Masaaki Terai, Ken-ichi Ueda, Irena Vodenska, Tsutomu Watanabe, Yoshihiro Yajima, Hiwon Yoon, Takeo Yoshikawa, Kikuo Yuta, and Leon Suematsu Yutaka.

This manuscript is written by researchers who are members of a research project supported by the Research Institute of Economy, Trade and Industry (RIETI), Tokyo. We would like to thank RIETI for support in preparing this book.

This work is also partially supported by *Grant-in-Aid for Scientific Research (KAKENHI) Grant Numbers 22300080, 24243027, 25282094, 25400393, and 26350422* by JSPS, *the Kyoto University Supporting Program for Interaction-based Initiative Team Studies: SPIRITS*, as part of the Program for Promoting the Enhancement of Research Universities, the Ministry of Education, Culture, Sports, Science and Technology (MEXT), Japan, *Nihon University College of Science and Technology Grants-in-Aid*,

Japan Center of Economic Research, and ISHII Memorial Securities Research Promotion Foundation Research Fund (ISHII-2015-329).

The authors thank the Yukawa Institute for Theoretical Physics at Kyoto University. Discussions during the YITP workshop YITP-W-15-15 on "Econophysics 2015" were useful to complete this work.

Part of the results are obtained by using the K computer at the RIKEN Advanced Institute for Computational Science (AICS), Kobe, partially supported by *The Post K Computer Exploratory Challenges* under the Flagship 2020 initiated by MEXT.

Finally, we are grateful to a number of organizations and individuals, who have understood the value of our endeavors and advised or provided us with crucial data for our evidence-based scientific research of economics: The Credit Risk Database Association for small-to-medium firm data in Japan, Tokyo Shoko Research Ltd., K. Itoh, and N. Shinozaki (both at NHK — Japan Broadcasting Corporation).

Prologue

> All truths are easy to understand once they are discovered; the point is to discover them.
>
> — Galileo Galilei

SALVIATI: Greetings, Sagredo, and Simplicio, my good friends. Yesterday, we resolved to meet today and discuss as clearly and in as much detail as possible the character and the efficacy of those laws of macro economics, which up to the present, have been put forth by the books of Aoki and Yoshikawa (2006) and Aoyama et al. (2010a), the very same authors of this book.

SAGREDO: Indeed, I am truly glad and honored to meet you and Simplicio on this occasion of the completion of this book.

SIMPLICIO: Indeed, indeed, (with a touch of doubt on his face) what is "macro-econophysics"?

SALV.: You must be familiar with "econophysics". It is evidence-based economics as a science. You may recall that it has the word "physics" in it as many physicists have devoted their research to this area, guided by the concepts and ideals of physics in their heart.

SAGR.: I see that most of the authors are physicists, except for Prof. Yoshikawa, who I heard is a macro-economist.

SALV.: They both have put forth the same ideals of revolutionizing the way real economy is studied in their respective books before. Now they have joined forces to introduce the term **Macro-econophysics**.

SAGR.: I have heard that it shares its ideals with "agent-based modelling", which is yet another great approach.

SIMP.: That is good. But isn't this book a mere collection of the respective topics from each of the authors?

SALV.: Absolutely not. They have been working together for the last few years, combining the best of physics and economics and publishing papers. They have spent many days and nights discussing all things big and small included in this book.

SAGR.: And look... they have Professor Richard Feynman's very hopeful words on the front cover.

SALV.: This book is one of the latest efforts to construct a science of economics, which forms a part of the basis of this new development and will be improved and passed on to the next generation of academics.

SAGR.: And look at this photograph taken by the first author (who, by the way, is the first author because of the names being listed in alphabetical order) on December 1, 1979, at UC Irvine, California. He and his famous van!

SIMP.: These words must have been uttered in the 1960s or 1970s, when the human race was very positive and hopeful about its future. Now, with singularity facing them, it is no longer appropriate. Besides, this van was burned down by *Sheldon Cooper* and his company.

SAGR.: No, no, no! Come to your senses, Simplicio. I am more hopeful of the future of the human race than ever and you know well that the van was burned down only on *The Big Bang Theory*!

SALV.: Speaking of the Big Bang, they say this is the age of the **Information Big Bang**.

SIMP.: That much is true.

SAGR.: Agreed. And we need to make use of it as well for our future.

SALV.: Indeed. Just as in physics where physicists have been successful in delving into the depths of vast experimental and observational data to obtain surprisingly precise details of the inner workings of Nature, we need to work on the vast economic data available to understand its true nature.

SIMP.: Are you implying that this book is successful in presenting the necessary tools and some results?

SALV.: That is up to you two to read and figure out. At the least, they have covered the basics that everyone needs to know and laid out the set of tools that they have been developing in the past few years together with the latest results.

SAGR.: As I quickly look at this, I see that they even used the **K computer** to obtain results on economic networks. Moreover, the authors told me that they are starting a project with an eye on "Post-K", the next–generation supercomputer.

SIMP.: Oh, that is the world's top supercomputer in the GRAPH 500 benchmark in June 2016! Even I know that! They are serious about the science of economics as a data-intensive, evidence-based enterprise.

SALV.: I am glad that I have successfully stimulated your appetite. Oh, a last word before we depart. They told me that they will provide the readers with supporting materials on:

> http://www.econophysics.jp/book_macroeconophysics

Let us then read through this book and meet again soon.

1

Introduction: Reconstructing Macroeconomics

We are no river. But we are not made of clay.

Miroslav Penkov

1.1 Background and Motivation

Macroeconomics is aimed at understanding the behavior of the economy as a whole—business cycles, economic growth, employment/unemployment, inflation/deflation, and inequality of income. A macroeconomy, such as that of the U.S. and Japan, consists of more than 100 million consumers and one million firms. Evidently, the behavior of the macroeconomy is the outcome of the aggregation and interactions of a large number of micro units such as consumers and firms. To understand such interactions beneath the surface of the macroeconomy, one must resort to new methods that are different from those of standard microeconomics. We propose that this new method be called *Macro-econophysics*. It provides one not only with novel theoretical insights but also with new empirical methods. Following the spirit of physics and other natural sciences, macro-econophysics puts equal importance on theory and on empirical findings. This book is concerned with both macroeconomic theory and important findings that advanced the study of macroeconomics.

Macroeconomics was born in the first half of the 20th century. Above all, it was expected to make a good diagnosis of, and hopefully provide a prescription for depression and high unemployment. Keynes' *General Theory of Employment, Interest, and Money*, published in 1936, was a landmark work, and had such a profound impact on the discipline that after the war, macroeconomics became synonymous with Keynesian economics

(Keynes, 1936). Paul Samuelson once proposed "the neoclassical synthesis," wherein it is held that the achievement of full employment requires Keynesian intervention but neoclassical theory is valid and useful for efficient resource allocation when full employment is achieved. To many, it was then obvious that the economy occasionally lapses into a recession where labor and other resources are left unemployed. The usefulness of Keynesian economics was taken for granted. Thus, during the 1950s and 1960s, the profession accepted the neoclassical synthesis, and held that economics stood on two pillars, namely, neoclassical microeconomics and Keynesian macroeconomics.

However, Arrow (1967), in his review of the *Collected Scientific Papers of Paul Samuelson*, pointed out that the neoclassical synthesis was actually nothing but a common sense argument and lacked rigor. The relation between neoclassical microeconomics and Keynesian macroeconomics, which Arrow called "one of the major scandals of price theory," was indeed a great challenge, and attracted the profession's research interest with gathering momentum. While the Walrasian general equilibrium was transparent, Keynesian macroeconomics was suspect because it allegedly lacked sound microeconomic foundations.

To many, providing microeconomic foundations for macroeconomics meant analyzing the optimization of economic agents in detail in macro model. The rational expectations model of Lucas Jr. (1972, 1987) and Sargent (2015), the real business cycle theory (RBC) of Kydland and Prescott (1982), the equilibrium labor search theory of Mortensen (2011) and Pissarides (2011), and the dynamic stochastic general equilibrium (DSGE) models, all sought to analyze the micro behaviors of economic agents. Mainstream economists take it that this "micro-founded macroeconomics" is superior to the old Keynesian macroeconomics. This is how macroeconomics has changed drastically in the past 30 years.

Lucas Jr. (1987) declared the victory of micro-founded macroeconomics over the old Keynesian economics thus:

> "The most interesting recent developments in macroeconomic theory seem to me describable as the reincorporation of aggregative problems such as inflation and the business cycle within the general framework of 'microeconomic' theory. If these developments succeed, the term 'macroeconomic' will simply disappear from use and the modifier 'micro' will become superfluous. We will simply speak, as do Smith, Ricardo, Marshall and Walras, of *economic* theory. If we are honest, we will have to face the fact that at any given time there will be phenomena that are well-understood from the point of view of the economic theory we have, and other phenomena that are not. We will be tempted, I am sure, to relieve the discomfort induced by discrepancies between theory and facts by saying that the ill-understood facts are the province of some other, different kind of economic theory. Keynesian 'macroeconomics' was, I think, a surrender (under great duress) to this temptation. It led to the abandonment, for a class of problems of great importance, of the use of the only 'engine for the

discovery of truth' that we have in economics. Now we are once again putting this engine of Marshall's to work on the problems of aggregate dynamics." Lucas Jr., 1987, pp.107–108

Based on the economic theory that he developed, Lucas Jr. (2003), in his presidential address to the American Economic Association dismissed the role of Keynesian stabilization policies. In this address, he repeated his own estimation of possible welfare gains from stabilization for the post-war U.S. economy (Lucas Jr., 1987). There, based on the representative agent assumption, he argues that if aggregate consumption fluctuations of the magnitude experienced since World War II were eliminated, it would raise the level of utility by only $8.50 per person! Thus, he concludes that "economic instability at the level we have experienced since the Second World War is a minor problem." Fair (1989) in his review of Lucas Jr. (1987) pointed out that quite a different picture emerges when one considers a model in which business cycles are caused by demand failures. Referring to the 1980–82 recession, he estimates that had real GNP grown at an annual rate of 3.0% from 1979, approximately $560 billion more in output would have been produced in the three years. He thus concludes that "this is a large lunch for everyone, approximately $2400 per person." Moreover, we must recall that the burden of a severe recession is not equally shared in society; for example, only a small fraction of people become unemployed. That is why business cycles are a major economic problem. However, the representative agent assumption on which modern micro-founded macroeconomics is built provides one with a very different picture of the macroeconomy!

Only a few years later, history disproved Lucas' verdict. In September 2008, the Lehman Brothers went bankrupt, and in the subsequent financial crisis, the world fell into the Great Recession. Across the world, no policy maker or government questioned the need for major stabilization policies. This historical event was surely a big embarrassment for the modern micro-founded macroeconomics. It was such a big embarrassment that the *Economist*, in the July 18–24, 2009 issue, featured a cover article titled "Modern economic theory – Where it went wrong, and how the crisis is changing it." Earlier, Paul Krugman in his Lionel Robbins lectures delivered at the London School of Economics on June 10, 2009, feared that most macroeconomics of the past 30 years was "spectacularly useless at best, and positively harmful at worst" (Krugman et al., 1998). Joseph Stiglitz also criticized mainstream macroeconomics as follows (Stiglitz, 2010):

> "The blame game continues over who is responsible for the worst recession since the Great Depression – the financiers who did such a bad job of managing risk or the regulators who failed to stop them. But the economics profession bears more than a little culpability. ...
>
> It is hard for non-economists to understand how peculiar the predominant macroeconomic models were. Many assumed demand had to equal supply – and that meant there could be no unemployment. (Right now a lot of people are just enjoying an extra dose of leisure; why they are unhappy is a matter

for psychiatry, not economics). Many used 'representative agent models' – all individuals were assumed to be identical, and this meant there could be no meaningful financial markets (who would be lending money to whom?). ...

Changing paradigm is not easy. Too many have invested too much in the wrong models. Like the Ptolemaic attempts to preserve earth-centric views of the universe, there will be heroic efforts to add complexities and refinements to the standard paradigm. The resulting models will be an improvement and policies based on them may do better, but they too are likely to fail. Nothing less than a paradigm shift will do." (*Financial Times*, August 20, 2010)

Many feel uncomfortable with modern micro-founded macroeconomics. Stiglitz's column was titled "Needed: a new economic paradigm." The problem is that the profession is not sure what a new promising paradigm is. We believe that the needed new paradigm is *macroeconophysics*.

Yes, as Stiglitz argued, too many have invested too much in the wrong models. What precisely is wrong with modern micro-founded macroeconomics? The fundamental problem is the assumption of a representative consumer/firm (Kirman, 1992). Unlike RBC, which is built literally on representative agent assumptions, other models, such as Lucas' rational expectations models and Mortensen's (2011) and Pissaride's (2011) equilibrium labor search models, actually emphasize heterogeneity of microeconomic agents. Thus, more explanation is necessary.

Most modern micro-founded macroeconomic models are stochastic. The fundamental assumption is that all the microeconomic agents, such as consumers and firms, share a common stationary distribution of the economic variable of interest. On this assumption, optimization of the micro agent is analyzed in great detail. In this sense, modern micro-founded macroeconomics, effectively, presumes a representative agent. These models differ only to the extent that realizations of stochastic variables differ.

This is not an accurate characterization of heterogeneity of economic agents in the actual macro-economy. Most economists, of course, agree that consumers and firms are different. In standard models, representative agent assumptions are justified because the cross-sectional variance over means approaches zero as the number of economic agents becomes large; namely, the system is presumed to be self-averaging.

This analytical framework is taken for granted, but is actually based on extremely unrealistic assumptions that all the economic agents face the same probability distribution, and that micro shocks are independent. Aoki and Yoshikawa (2012) demonstrated that once these unrealistic assumptions are dropped, non-self-averaging behavior emerges naturally. Non-self-averaging is not pathological. Rather, it arises naturally in a wide range of both natural and social phenomena (Sornette, 2006). Garibaldi and Scalas (2010, pp.101–102) also demonstrated how it emerges in the well-known Pólya process. One should, therefore, expect that non-self-averaging is typical rather than exceptional in economics. Because non-self-averaging makes representative agents meaningless, mainstream micro-founded macroeconomics based on representative agent assumptions actually has no foundations.

To analyze a system, such as a macroeconomy comprising a large number of interacting micro units, it is useless to pursue the behavior of a representative micro unit in detail. Instead, we must resort to methods of statistical physics. The whole purpose of this book is to show how fruitfully such methods can be applied for analyzing a wide range of problems in macroeconomics.

However, economists are skeptical of the application of the methods of statistical physics in their field. The most important reason for their reservation is that movements of particles are mechanical whereas economic actions are outcomes of purposeful sophisticated dynamic optimization by the micro agents. Every student of economics knows that the behavior of dynamically optimizing economic agents, such as the Ramsey consumer, is described by the Euler equation for a problem of calculus of variation, namely the problem of maximizing utility over time. On the surface, such sophisticated economic behavior may look far removed from the "mechanical" movements of inorganic particles that only satisfy the law of motion. However, every student of physics knows that the Newtonian law of motion is actually nothing but the Euler equation for a certain variational problem. It is called the *principle of least action*; see Chapter 19 of Feynman et al. (1964)'s *Lectures on Physics*. Therefore, behaviors of dynamically optimizing economic agents and motions of inorganic particles are on par to the extent that they both satisfy the Euler equations for respective variational problems. The method of statistical physics can be usefully applied not because the motions of micro units are "mechanical," but because the object/system under investigation consists of many micro units, individual movements of which one knows nothing. The macroeconomy consisting of more than 100 million consumers and one milion firms is such a system to which methods of statistical physics can be usefully applied.

After seminal works such as those by Montroll (1981, 1987), the application of statistical physics to economics has been done mainly by physicists rather than by economists. Most notably, statistical physics has been successfully applied to empirical analysis of financial markets such as stocks and foreign exchange. *Econo-physics*, the name for this emerging interdisciplinary research area, was coined in 1995 (Mantegna and Stanley, 1999a). The availability of good high frequency data for stock prices and exchange rates, and the rapid technological progress of the modern computer certainly facilitated active research in finance. The research in this area not only seeks to establish empirical regularity but also clarifies why markets occasionally become so unstable (Sornette, 2003). In light of the 2008-09 global financial crisis, the importance of such research requires no explanation.

This book shows the reader that the promising application of statistical physics is not confined to finance. We apply *methods of statistical physics* to a wide range of problems in macroeconomics, and demonstrate that they produce fruitful results.

To carry out empirical investigations, we take full advantage of high quality micro data in Japan; such quality is rarely available in other countries. The data we use include

- Personal income on the largest ever scale, covering 10,000 *individual* high-income persons and their growth; the same is true for firms (Chapter 3);
- Productivity for an exhaustive list of 1,000 listed firms (Chapter 4);
- The Credit Risk Database (CRD), which covers a large fraction of small and medium enterprises (SMEs), with detailed data of financial statements as well as qualitative data of one million firms for more than 15 years (Chapter 4);
- Prices for 1,000 individual goods and services with comovements in price changes (Chapter 7);
- A nationwide production network comprising one million firms and several millions of supplier–customer relationships among them for several years; a credit network of lending and borrowing among all banks and large firms over more than 20 years (Chapter 8 and 9).

These rich micro data enable us to test macroeconomic theories. Many of the results we obtain have fundamental implications for macroeconomics. Mainstream micro-founded macroeconomics must make way for macro-econophysics. It turns out that the new approach actually enriches the old macroeconomics of the 1950s and 1960s.

1.2 Outline

In Chapter 2, we explain the basic concepts and methods of statistical physics. We introduce the concepts of statistical equilibrium and probabilistic description as indispensable in understanding macroeconomic phenomena, including the methods of entropy and statistical equilibrium. This will help explain the distribution of labor productivity as detailed in Chapter 4. Since a lot of time series data in the economy is characterized by non-stationarity, the methods explained here will be used at many places in later chapters. Finally, to study fat-tailed distributions frequently encountered in economic systems, as we do in Chapter 3, we explain a class of stochastic processes of "the rich get richer" and the role of externalities.

In Chapter 3, we analyze personal income and firm-size distributions. Income inequality is a major problem facing advanced countries today. Among others, Piketty (2014) eloquently reminded us of the importance of the problem of income distribution. In fact, as early as the end of the 19th century, Pareto (1897) discovered that the distribution of income is so skewed that it is actually what is now called the Pareto distribution, namely, a power law for high incomes. The mechanism that generates such a distribution was subsequently analyzed by Gibrat (1931) and Champernowne (1953, 1973). Their model is one of stochastic processes. Following their lead, Chapter 3 presents modern treatments of the subject and our own new findings. A similar method is applied to firm-size distributions, which is another topic of Chapter 3.

Chapter 4 presents our analyses of productivity. As Yoshikawa (2003, 2014) explains, the problem of productivity dispersion is closely related to the micro-foundations of Keynesian macroeconomics.

The core of Keynesian macroeconomics is the recognition that an underemployment of resources, such as labor, is almost always present in the macroeconomy, and that real aggregate demand determines the level of utilization of such resources. The difficulty lies in the fact that the determination of utilization of labor is not mechanical because both workers' job searches and firms' job offers are economic decisions. The equilibrium labor search theory is aimed at describing the labor market (Mortensen, 2011; Pissarides, 2011). Although this theory is widely accepted by the profession, it is actually based on untenable representative agent assumptions. To be specific, in this model, all the workers and firms are assumed to share a common distribution of wages. It is well known, however, that in reality, the labor market is segmented by region, industry, sex, age, educational level, and so forth. The equilibrium labor search theory apparently presumes heterogeneous agents, but like other micro-founded macro models, it is built, in fact, essentially on representative agent assumptions.

We present a model of *stochastic macro-equilibrium*. In this model, we maximize entropy under two macro constraints, namely, labor endowment and effective demand. The entropy maximization should be standard for physicists. The novel feature is the negative temperature. It may sound odd, but the notion of negative temperature is perfectly legitimate even in physics; see Appendix E of Kittel and Kroemer (1980). The difference between physics and economics arises because in physics, particles tend to achieve the lowest energy level possible, whereas in economics, workers strive for the highest productivity possible (Iyetomi, 2012).

In this model, based on the methods of statistical physics, we can show, quantitatively, how labor is mobilized when aggregate demand rises. The level of aggregate demand is the ultimate factor affecting the outcome of the random matching of workers and monopolistically competitive firms. By so doing, it changes not only unemployment but also the distribution of productivity and, as a consequence, the level of aggregate output. This is the market mechanism underlying Keynes' principle of effective demand. Contrary to what many economists believe, the old principle of effective demand has a solid micro-foundation. The market mechanism underlying Keynes' principle of effective demand is the general equilibrium of monopolistic competition coupled with the search by workers and firms under friction and uncertainty. Keynesian economics, in effect, claims that in the short run, aggregate demand ultimately affects the matching of workers and firms, thereby determining the utilization of labor and the level of output in the macroeconomy. The logic here does not depend on details of the micro behavior of economic agents. Without the help of statistical physics, this insight would not be possible.

In Chapter 5, we introduce a new method of extracting comovements hidden in a number of time series data, which we call *the complex Hilbert principal component analysis* (CHPCA). The principal component analysis (PCA) or factor analysis is routinely used in many research areas, including economics, but it fails when significant leads and lags are present. A simple example is sine and cosine waves with an obvious lead or lag. The CHPCA takes care of such leads and lags. However, as this method is not known in economics, it is explained fully in this chapter. This method will be used frequently in this book to extract comovements out of a large number of time series.

The topic of Chapter 6 is business cycles. The Great Recession of 2008–10 underlined that contrary to the assertion by Lucas Jr. (2003), business cycles remain the major problem and the top policy concern in advanced countries. Yet, even after so many years of research, economists have not reached a good consensus on many important issues, such as whether the fundamental cause of business cycles is real or monetary, and if real, which is more important, productivity or real aggregate demand. To explore these problems, we resort to CHPCA. Our analysis based on this new analytical method demonstrates that business cycles are basically caused by real aggregate demand in the line with the old Keynesian argument. Chapter 6 also presents an analysis of the synchronization of international business cycles.

In Chapter 7, we analyze price dynamics and inflation/deflation. In the 21st century, Japan was the first to be affected by deflation, followed by the European Union. Even the U.S. was driven to the brink of deflation. The interest rates were lowered worldwide to unprecedentedly low levels, and finally to zero and even negative values. At first, mainstream economists argued that deflation can be stopped by simply increasing the money supply. An early and influential contribution to this line of thinking was done by Krugman et al. (1998). Many mainstream economists followed suit. In their models, representative consumers and firms are supposed to react to such an increase in money by raising prices. Alas! There are no such representative firms or consumers in the actual economy. Prices are set by heterogeneous firms, and this gives rise to non-trivial price dynamics. Now, deflation and inflation refer to changes in the average level of many prices of individual goods and services. Thus, to understand deflation/inflation, one must fully understand the dynamics of micro prices. Chapter 7 explores this price dynamics, and shows how the average price changes. One of the important results we obtained is that the average price does not respond to money supply as mainstream economists argued. This finding casts a serious doubt on aggressive quantitative easing (QE) pursued by many central banks fighting deflation.

Chapter 8 explains the complex network present in a macroeconomy and community analysis. The macroeconomy consists of many micro agents who are not scattered randomly but form networks. They interact with each other through such networks, and these interactions generate the dynamics of the macroeconomy. This chapter explains the basic concepts and techniques necessary for exploring complex networks.

Real networks are often divided into densely connected subsets called communities. Given that the firm is a basic micro unit, an industry would be such a community. Chapter 8 introduces methods for identifying communities. It also presents our analysis of international business cycles based on the world input–output database. One of our major findings is that the community structure in the world economy changed drastically after the 2008 global financial crisis.

Chapter 9 explores systemic risks. The global financial crisis after the bankruptcy of Lehman Brothers in September 2008 and the subsequent Great Recession once again reminded us of the fact that contrary to Lucas's claim, business cycles remain a major problem today. In particular, a financial crisis, if it happens, significantly worsens a downturn of the real economy. Obviously, we need to avoid or, at least, mitigate financial troubles.

The world of finance is a typical network. Financial institutions, such as banks, interact with each other in inter-bank networks while banks supply credit to firms who, in turn, interact with each other in supplier/customer networks. The stability of the financial system depends crucially on the health of the real economy, namely, the production sector. Trouble facing one micro agent propagates to others through networks. For example, a firm that incurs significant losses may not be able to repay its debt to the bank. This adversely affects the bank's profitability. Conversely, a bank that is not confident enough of its financial position may curtail credit to firms. This, in turn, affects the firms' business operations. The stability of the financial system as a whole must be analyzed based on this complex network structure. Our approach is to use the DebtRank method. Chapter 9 explains this method in detail, and presents the result of our analysis.

As Stiglitz observed, mainstream macroeconomics is a dead end. Most economists are still confident, of course, but to say that "economic instability at the level we have experienced since the Second World War is a minor problem," and that the best we can do is to let the economy be, is just embarrassing. In November 2008, Queen Elizabeth II visited the London School of Economics and asked why nobody had noticed that a credit crunch was on its way. The Queen looks at the economy more squarely than mainstream economists because modern micro-founded macroeconomics preaches that the macroeconomy is stable and works well.

Evidently, we need a new economic paradigm. We believe that this new paradigm is macro-econophysics. First, we must recognize that to understand a macroeconomy consisting of many micro agents, it is useless to pursue the behaviors of those micro agents in detail. Instead, we must resort to methods of statistical physics and network analysis. They provide us with the right micro foundations for macroeconomics.

On the whole, our results resurrect old Keynesian macroeconomics (Tobin, 1993). The macro-economy, a complex network, occasionally becomes trapped in troubles where aggregate real demand plays a central role. Our analyses also suggest a way to mitigate, if not completely remove, macroeconomic problems.

2

Basic Concepts in Statistical Physics and Stochastic Models

> The whole is simpler than its parts.
>
> Josiah Willard Gibbs

This chapter explains basic concepts and analytical methods extensively used in subsequent chapters. Readers begin with learning fundamental stochastic modeling to reproduce distributions with power-law tails. Such skewed distributions are discussed extensively in Chapter 3, with important examples of income and firm-size distributions. Next, we explain entropy. Readers must have already recognized that the concept of *statistical equilibrium* and *probabilistic description*, developed in physics, would be indispensable for understanding complex macroeconomic phenomena. *Mechanical equilibrium* cannot adequately accommodate the diversity of an economic system. Entropy, which measures the degree of randomness or disorder in a system, is a major player in thermal and statistical physics. The statistical equilibrium of a system is regarded as a manifestation of the maximum entropy principle. We first reiterate the basics of entropy especially for readers who have no physics background; this serves as a prelude to Chapter 4 that deals with the distribution of labor productivity. Finally, we provide a brief account of the stochastic modelling of stationary and non-stationary time series with a flavor of nonlinear physics. This part is expanded in Chapters 5, 6, 7, and 8 on multivariate time series analysis, business cycles, collective motion of prices, and correlation and synchronizing networks, respectively.

2.1 Stochastic Models and Fat Tails

Many phenomena in socio and economic systems exhibit fat-tailed and skewed distributions. We have observed and will observe this fact for personal income, firm-size, productivity dispersion, economic networks, and so forth in this book. It is often called "A few giants and many dwarfs". Presence of a few giants and comovement of many dwarfs has important consequences to a macroeconomic system as we shall explain at places in this book.

Natural science has also witnessed many phenomena, notably in the study of fractals and scaling, and attracted many researchers who found models and scenarios specific to each domain, and even speculated "universal" explanations for the origin of such distributions, especially of power-laws that possess mathematical properties of scale-free.

In fact, one of the oldest and most widely known models is the so-called Yule's model to explain distribution of biological species. This kind of model has been "discovered" over and over again in various contexts under different names including the following:

- Process of "The rich get richer" (anecdotal, e.g. see the Bible, Matthew 13:12)
- Law of proportional effect or Gibrat's law (Gibrat, 1931; Sutton, 1997)
- Cumulative advantage (de S. Price, 1965)
- Matthew effect (Merton, 1968)
- Multiplicative process, mathematical theory (Kesten, 1973)
- Preferential attachment (Barabsi and Albert, 1999)

Let us make a compact description of the Udny Yule's model, and the closely related class of models by Herbert Simon, and introduce elementary mathematics of stochastic processes. Then we will turn our attention to city-size and related phenomena to illustrate simple applications of these stochastic processes.

2.1.1 Yule and Simon's models and multiplicative processes

Biological species are classified in a hierarchy of *order, family, genus (pl. genera)*, and *species*, from large taxonomic ranks to small ones. G. U. Yule found that the numbers of species belonging to each genus in a family, namely genus-size, have a fat-tailed distribution (Yule, 1925). Figure 2.1 shows two examples of such distributions constructed from Yule's tabulated data collected originally by J. C. Willis. One can observe two distinct features; that a few genera exists, each having a large number of species, and also that there are many genera, each having only a small number of species. To explain the two features, he invented a probabilistic model which is based on how species diversify (speciation) due to geographical separation of mating population, genetic drift and other effects. We shall call it the **Yule 1925 model**.

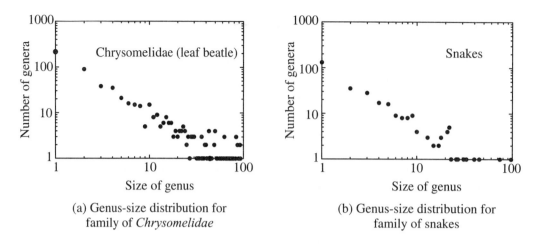

Fig. 2.1 Genus-size distribution constructed from Yule (1925)

Later, H. A. Simon, using the Yule's model, proposed a possible explanation for a wide range of phenomena such as word frequencies in a document, number of papers written by a researcher, city-size, personal income, biological abundance, and so forth by calling those distributions as skewed distributions and inventing a class of probabilistic models (Simon, 1955). A similar model could be applied to citations of papers. Figure 2.2 is such an empirical distribution, in the forms of histogram and cumulative probability distribution, constructed from a modern data in Redner (1998). Let us call the probabilistic model the **Simon 1955 model**.

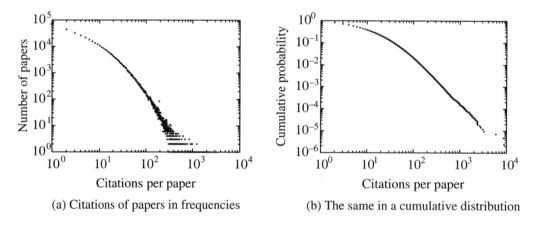

Fig. 2.2 Empirical distribution for citations of papers

These models by Yule and Simon have been reinvented in a variety of contexts for many phenomena. It would be valuable to understand them in a fundamental framework in this section. Then we shall point out that they lack interaction of agents in a context such as city-size formation, and proceed to include it in the next section.

Yule 1925 model

Yule's idea to consider speciation is that species change in time and a large deviation brings about a new species. Suppose, initially there is a single species. Consider the following two elementary processes.

Y1 *In a time interval $(t, t+dt)$, there is a probability $\sigma\, dt$ such that a species bifurcates into two species, both of which are assumed to belong to a same genus.*

Y2 *In the same interval, independently of Y1, there is a probability $\gamma\, dt$ for an existing genus such that a new species is generated as a genus of size 1.*

The existence of large-sized genera is explained from **Y1**: the larger genera get larger. Small-sized genera exist due to **Y2** which generates a new genus with size 1. The process of **Y1** is a pure birth process in the framework of stochastic process, called a Yule process (Feller, 1968, [XVII])

Denote by $p_n(t)$ the probability to find a genus with size n at time t. This can be easily calculated as follows. Let $t = 0$ be the initial time. Because $p_1(t)$ is the probability that the process **Y1** does not take place during time t

$$p_1(t) = \exp(-\sigma t). \tag{2.1}$$

Suppose at time s ($0 < s < t$), a genus has size n. In a time interval $(s, s+ds)$, if one of the n species bifurcates to have $n+1$ ones, and no further speciation takes place, then the state becomes precisely $n+1$. One can then express $p_{n+1}(t)$ as an integral of the probability over s:

$$p_{n+1}(t) = \int_0^t ds\, n\sigma \cdot p_n(s) \cdot e^{-(n+1)\sigma(t-s)}. \tag{2.2}$$

By mathematical induction from $p_1(t)$ and Eq. (2.2), one has

$$p_n(t) = e^{-\sigma t}(1 - e^{-\sigma t})^{n-1} \quad (n = 1, 2, \ldots). \tag{2.3}$$

On the other hand, due to the process **Y2**, the number of genera grows exponentially as $\exp(\gamma t)$. If a genus is randomly chosen, the probability to find its lifetime to be equal to t, denoted by $p(t)$, is

$$p(t) = \gamma e^{-\gamma t}. \tag{2.4}$$

The probability p_n that a genus has size n is therefore given by

$$p_n = \int_0^\infty dt\, p_n(t)\, p(t) = \int_0^\infty dt\, \gamma e^{-\gamma t} \cdot e^{-\sigma t}(1 - e^{-\sigma t})^{n-1}$$

$$= \frac{\gamma}{\sigma} B\left(n, 1 + \frac{\gamma}{\sigma}\right). \tag{2.5}$$

This is called a Yule distribution. Here $B(x, y)$ is the beta function:

$$B(x, y) := \int_0^1 du \, u^{x-1}(1-u)^{y-1} = \frac{\Gamma(x)\Gamma(y)}{\Gamma(x+y)} = B(y, x), \qquad (2.6)$$

and $\Gamma(x)$ is the gamma function:

$$\Gamma(x) := \int_0^\infty du \, e^{-u} u^{x-1}. \qquad (2.7)$$

By using the Stirling formula

$$\Gamma(x+1) \simeq \sqrt{2\pi x} \, (x/e)^x, \qquad (2.8)$$

one can easily prove that for a fixed parameter a and for a sufficiently large number of x

$$\frac{\Gamma(x)}{\Gamma(x+a)} \propto x^{-a}. \qquad (2.9)$$

Employing this relation in the distribution in Eq. (2.5), one finds the power-law

$$p_n \propto n^{-(1+\gamma/\sigma)}, \qquad (2.10)$$

in the region for a large number of n. In the study of biological species, the exponent $1 + \gamma/\sigma$ was close to 1.5 (Yule, 1925).

One can modify the original model slightly. Namely, instead of **Y2**, consider

Y2' *Independently of **Y1**, there is a probability $\gamma \, dt$ for an existing species such that a new species is generated as a genus of size 1.*

Let us call this model the modified Yule model. From **Y2'**, the number of genera grows exponentially as $\exp((\sigma + \gamma)t)$. Then a randomly chosen genus has the lifetime t with probability

$$p(t) = (\sigma + \gamma)e^{-(\sigma+\gamma)t}. \qquad (2.11)$$

A similar calculation yields

$$p_n = \left(1 + \frac{\gamma}{\sigma}\right) B\left(n, 2 + \frac{\gamma}{\sigma}\right) \sim n^{-(2+\gamma/\sigma)}, \qquad (2.12)$$

for large n. Note that the power-law exponent is larger than 2. This modification is no longer appropriate as a model of speciation, in which the exponent was found to be less than 2. Nevertheless, the modified Yule model is equivalent to Simon's model as we shall see.

Simon 1955 model

A good example to explain Simon's model would be the word frequency, which exhibits a power-law, the well-known Zipf's law. Let us consider the following two elementary processes.

S1 *When the number of words is N at a time of writing, choose a word to be added next among the already appeared words with probability $1 - \alpha$.*

S2 *And choose a new word to be added next with probability α.*

At time step s, denoting the number of words with frequency k by $n_k(s)$, the number of words, $N(s)$, and the number of vocabularies (different words), $W(s)$, can be respectively expressed as

$$N(s) := \sum_k k\, n_k(s), \tag{2.13}$$

$$W(s) := \sum_k n_k(s). \tag{2.14}$$

Because choosing an already existing word with frequency k at the next time step is given by

$$\text{Choosing prob.} = \frac{k\, n_k(s)}{\sum_k k\, n_k(s)} = \frac{k\, n_k(s)}{N(s)}, \tag{2.15}$$

the two elementary processes of **S1** and **S2** obey *in average*

$$n_k(s+1) - n_k(s) = (1-\alpha)\left[\frac{(k-1)\, n_{k-1}(s)}{N(s)} - \frac{k\, n_k(s)}{N(s)}\right], \tag{2.16}$$

$$n_1(s+1) - n_1(s) = \alpha - (1-\alpha)\frac{n_1(s)}{N(s)}. \tag{2.17}$$

The equations are quite often called master equations, because most statistical quantities of interest can be derived starting from them.

Note that after summing Eq. (2.16) over $k \geq 2$ and adding Eq. (2.17) to the sum, it follows that $W(s+1) - W(s) = \alpha$. On the other hand, after summing the product of k and Eq. (2.16) over $k \geq 2$ and adding Eq. (2.17) to it, one finds that $N(s+1) - N(s) = 1$ as it should be. After a sufficient time, one has

$$W(s) = \alpha\, N(s). \tag{2.18}$$

In a stationary state, let us assume that for any k, $n_k(N)$ grows proportionally to the system size N, namely

$$\frac{n_k(N+1)}{n_k(N)} = \frac{N+1}{N}, \tag{2.19}$$

for all k. Because the ratio, $n_k(N)/n_{k-1}(N)$, does not depend on N due to Eq. (2.19), the relative frequencies of words or the frequency distribution defined by

$$p_k(s) := \frac{n_k(s)}{W(s)} \tag{2.20}$$

would be independent of time. Actually, by using Eq. (2.18), the stationary distribution Eq. (2.20) satisfies

$$p_k = \frac{k-1}{k+(1-\alpha)^{-1}} p_{k-1} \quad (k \geq 2), \tag{2.21}$$

$$p_1 = \frac{1}{2-\alpha}, \tag{2.22}$$

from Eq. (2.16) and Eq. (2.17). We can conclude that the frequency distribution in a stationary state reads

$$p_k = (1-\alpha)^{-1} B(k, 1+(1-\alpha)^{-1}), \tag{2.23}$$

which obeys a power-law

$$p_k \sim k^{-(1+(1-\alpha)^{-1})}, \tag{2.24}$$

in a region of large k.

If the Yule model is expressed in terms of Simon's model, a genus is a word; and the size of a genus is the frequency of a word. Denoting the number of genera by N_g and the total number of species by N_s, the probability of adding a new word in Simon's model corresponds to the probability of generating a new genus in the Yule model. For the case of the modified Yule model with the process of **Y2'**, one finds that

$$\alpha = \frac{\gamma N_s}{\sigma N_s + \gamma N_s} = \frac{\gamma}{\sigma + \gamma} \tag{2.25}$$

is a constant; the explanation of Simon's model applies here exactly. In fact, by inserting Eq. (2.25), Eq. (2.23) is equivalent to Eq. (2.12). In addition, while the number of words is incremented by $dN = 1$, the transition probabilities for word frequency and vocabularies can be calculated by

$$\text{prob. } \{k \to k+1\} = (1-\alpha)(k/N) \, dN = k \cdot (1-\alpha) \, a \, dt, \tag{2.26}$$

$$\text{prob. } \{W \to W+1\} = \alpha \, dN = N \cdot \alpha \, a \, dt. \tag{2.27}$$

Here t is measured by the logarithm of the total number of words, namely $a \, dt = d(\ln N)$. If the constants of α, a are changed into variables of σ, γ by using Eq. (2.25) and we consider the definition, $a := \sigma + \gamma$, Eq. (2.26) becomes $k \cdot \sigma \, dt$, while Eq. (2.27) equals $N \cdot \gamma \, dt$. Since they are precisely the transition probabilities for the processes **Y1** and **Y2'**, it is proved that Simon's model is equivalent to the modified Yule model.

Some applications of Simon's model

One can apply Simon's model to various phenomena. Corresponding to a word and its frequency, one could consider a city and its size, a paper and its citation, a web page and hyperlinks to it by others, a human and its friends, and so forth. See Newman (2005); Mitzenmacher (2004); Simkin and Roychowdhury (2006), for examples.

For the applications, at least two points should be additionally taken into account.

- For the case of city-size, a new city may be better assumed to start with a small size of more than two people with a parameter $k_0 > 1$.
- For the case of paper citation, it would be necessary to assume that a new paper has initially zero citations, yet has a non-zero probability to be cited by with a parameter c.

If one changes the probability of selection given by Eq. (2.15) in the following way, one can take them into account.

$$\text{Choosing prob.} = \frac{(k+c)\,n_k(s)}{\sum_{k \geq k_0}(k+c)\,n_k(s)} = \frac{(k+c)\,n_k(s)}{N(s) + c\,W(s)}. \tag{2.28}$$

Writing the master equation by using this probability, the stationary state can be easily calculated. Instead of Eq. (2.18), one can derive

$$\{1 + \alpha(k_0 - 1)\}\,W(s) = \alpha\,N(s), \tag{2.29}$$

from the master equation. This is left as an exercise. See also Newman (2005).

The resulting frequency distribution in the stationary state is obtained as

$$p_k = p_{k_0}\frac{B(k+c,\mu)}{B(k_0+c,\mu)}, \tag{2.30}$$

where

$$\mu := 2 + (k_0 + c)\,\alpha(1-\alpha)^{-1}. \tag{2.31}$$

For large k, $p_k \sim k^{-\mu}$. Note that when $k_0 = 1$ and $c = 0$, one has $\mu = 1 + (1-\alpha)^{-1}$ and we can recover Eq. (2.23) from Eq. (2.30).

It should be remarked that according to Eq. (2.31), one can construct models for different power-law exponents μ. How can one select a model? There is no panacea for a general solution; it would be important to observe directly the properties of fluctuations which lead to the distribution. This important point is studied in the next chapter. Additionally, the reader is referred to (Ijiri and Simon, 1977a, Chap.6) for an interesting viewpoint on this problem.

Going back to the example of city formation, the process of Eq. (2.15) or Eq. (2.28) does not include any interaction between people in a city. Each people moves to a city independently of others so that the transition probability for the city growth is expressed as what we have seen so far. How does interaction change the story? This important question is addressed in the next section.

2.1.2 Marsili–Zhang model and interaction of agents

When Simon's model is applied to city formation, one immediately comes up with at least two difficulties:

- There is no such process that people move from a city to another or that people can "exit" out of cities
- Rates of entry and exit per capita do not depend on the city size.

The first point refers to the fact that a city can become smaller; for example, a dweller can move to another city, may become a drifter who does not belong to any city ("reservoir" of drifters), or is prone to death. The second point is related to **externality**, an important aspect of interacting agents.

We shall explain in this section

- how externality is related to fat-tailed distribution, and
- how to describe externatlity by transition probabilities

by using an example of city-size distribution in the studies of Zanette and Manrubia (1997); Marsili and Zhang (1998a).

Master equation for a jump Markov process

Let us first review the mathematical framework of a jump Markov process in a concise way. Feller (1968); Gardiner (2009) are good references. A birth and death process for a population x is a stochastic process in which a single event causes a finite number of births and deaths with the corresponding transition probabilities being independent of time. A stochastic process is specified, if one assumes the transition probabilities per unit time as

$$w_+(x) \quad \text{for } x \to x+1,$$

$$w_-(x) \quad \text{for } x \to x-1.$$

Denoting by $P(x,t|x',t')$ the conditional probability that the population is x at time t under the condition that it is x' at previous time t', the master equation for it is essentially the conservation of probabilities, namely

$$\frac{\partial P(x,t)}{\partial t} = w_+(x-1)\, P(x-1,t) + w_-(x+1)\, P(x+1,t)$$

$$- \{w_+(x) + w_-(x)\}\, P(x,t). \tag{2.32}$$

Here the conditional part of $|x',t'\rangle$ was omitted for simplicity. Stationary state is a solution for the master equation, which does not depend on time, and is denoted by $P(x)$.

Equation (2.32) can be interpreted as the fact that the change of probability in the state of x is equal to the net flow of probabilities. In fact, the net flow at the "boundary" between the states x and $x - 1$ can be defined by

$$J(x) := w_-(x)\, P(x) - w_+(x - 1)\, P(x - 1). \tag{2.33}$$

Under the assumption of stationary states, Eq. (2.32) can be written by the net flow as

$$0 = J(x + 1) - J(x), \tag{2.34}$$

from which it follows that

$$\sum_{y=0}^{x-1} \{J(y+1) - J(y)\} = J(x) - J(0) = 0. \tag{2.35}$$

At the boundary of $x = 0$, the population does not decrease any more so that $w_-(0) = 0$. And $P(x) = 0$ for $x < 0$. The boundary condition can be expressed from these two facts by $J(0) = 0$. From Eq. (2.34), one has

$$J(x) = 0. \tag{2.36}$$

Solving this equation iteratively, one obtains the following stationary solution:

$$P(x) = \frac{w_+(x-1)}{w_-(x)} P(x-1) = P(0) \prod_{y=1}^{x} \frac{w_+(y-1)}{w_-(y)}. \tag{2.37}$$

Equation (2.36) which is valid under the stationary state can be expressed, due to the definition of Eq. (2.33), by the equation:

$$P(x, dt\,|\,x', 0) \cdot P(x') = P(x', dt\,|\,x, 0) \cdot P(x), \tag{2.38}$$

for $x' = x \pm 1$ in the limit of $dt \to 0$. This means that the condition of detailed balance holds.

Marsili–Zhang model

Zanette and Manrubia (1997) and Marsili and Zhang (1998a) studied a model for city-size distribution. Let us explain the model by using the Marsili–Zhang model.

Suppose that at time t there are Q cities, and that the population of city i is k_i. The growth and shrink of a city is stochastic, which we describe as a probabilistic model by specifying the transition probabilities of growth and shrink. Let us express the transition by

$$w_+(k_i)\, dt \qquad \text{for } k_i \to k_i + 1, \tag{2.39}$$

$$w_-(k_i)\, dt \qquad \text{for } k_i \to k_i - 1, \tag{2.40}$$

during the time interval $(t, t + dt)$. In addition, we assume that there exists a small probability such that a city with $k_i = 1$ can be

$$\text{born with probability } p \, dt \, . \tag{2.41}$$

The model does not specify whether the people exiting from a city by the transition Eq. (2.40), move to other cities or become drifters or dead. Also it is not specified whether the people entering a city by Eq. (2.39) enter from other cities or are drifters, or are born. So the total number of people who are staying in cities is generally not a constant. Instead, the whole population consists of people in cities and in a "reservoir" of this system.

Denote by $q(k,t)$ the number of cities with size k at time t. The master equation in Eq. (2.32) can be written for $q(k,t)$ as

$$\frac{\partial q(k,t)}{\partial t} = w_+(k-1)\, q(k-1,t) + w_-(k+1)\, q(k+1,t)$$
$$- \{w_+(k) + w_-(k)\}\, q(k,t) + p\, \delta_{k,1}. \tag{2.42}$$

Here $\delta_{k,1}$ is 1 for $k = 1$ and 0 otherwise. If the transition probabilities $w_+(k), w_-(k)$ and the probability p are given, a stochastic process is specified.

The total number of cities and the total population denoted by $Q(t)$ and $N(t)$ respectively are

$$Q(t) := \sum_k q(k,t), \tag{2.43}$$

$$N(t) := \sum_k k\, q(k,t). \tag{2.44}$$

Summing both sides of Eq. (2.42) or its multiple by k over k, one has

$$\frac{\partial Q(t)}{\partial t} = p - w_-(1)\, q(1,t), \tag{2.45}$$

$$\frac{\partial N(t)}{\partial t} = p - \sum_k [w_-(k) - w_+(k)]\, q(k,t). \tag{2.46}$$

The meanings of these equations are obvious.

The stationary state can be easily obtained by the method explained previously. Define the net flow between the states k and $k - 1$ by

$$J(k) := w_-(k)\, q(k) - w_+(k-1)\, q(k-1). \tag{2.47}$$

One can rewrite Eq. (2.42) under stationary state as

$$0 = J(k+1) - J(k) + p\, \delta_{k,1}. \tag{2.48}$$

Basic Concepts in Statistical Physics and Stochastic Models

Since the boundary condition is given by $q(k) = 0$ for $k < 0$, one has

$$J(1) = w_-(1) q(1) = p, \tag{2.49}$$

where we have used the fact that Eq. (2.45) is zero under the assumption of stationary state.

Summing Eq. (2.47) over k from 1 to $k - 1$, one has the condition of detailed balance, i.e. $J(k) = 0$ ($k > 1$). Solving this iteratively, one obtains the stationary state:

$$q(k) = \frac{w_+(k-1)}{w_-(k)} q(k-1) = q(1) \prod_{m=1}^{k-1} \frac{w_+(k-m)}{w_-(k-m+1)}. \tag{2.50}$$

Let us now specify the transition rates for the two cases without and with externality.

No externality

Consider the transition rates given by

$$w_-(k) = D \cdot k, \qquad w_+(k) = A \cdot k, \tag{2.51}$$

namely the *linear* case. This means that the probability of transition $k \to \pm 1$ depends on the population k, that is, individuals have mutually independent probabilities of entry and exit.

Inserting Eq. (2.51) into Eq. (2.50), the stationary state can be expressed by

$$q(k) = \left(\frac{A}{D}\right)^{k-1} \frac{q(1)}{k}. \tag{2.52}$$

To know the meanings of the parameters, D and A, write down the expressions of the number of cities and the total population, namely Q and N by inserting Eq. (2.52) into Eq. (2.45) and Eq. (2.46) to obtain

$$D = \frac{p}{q(1)}, \tag{2.53}$$

$$A = \frac{p}{q(1)} \left(1 - \frac{q(1)}{N}\right). \tag{2.54}$$

Note that $D > A$. The distribution for city-size, i.e. Eq. (2.52), is

$$q(k) = \frac{q(1)}{1 - q(1)/N} \frac{(1 - q(1)/N)^k}{k}, \tag{2.55}$$

Calculate Q to have

$$Q = \frac{q(1) \ln(N/q(1))}{1 - q(1)/N}. \tag{2.56}$$

We can see that D, A determine the system size through $q(1)$ and N in this equation.

Assuming that $p \ll 1$, the stationary distribution, Eq. (2.55), is approximated by

$$q(k) \simeq \frac{1}{k} e^{-k/k_*}. \tag{2.57}$$

Here the scale $k_* := N/q(1)$ satisfies the condition $k_* \gg 1$. The distribution, therefore, behaves as $q(k) \sim 1/k$ for a wide range of k. This implies that the cumulative probability of finding a city with size larger than k obeys $\bar{F}(k) \sim \ln(1/k)$. Let the rank of city with size k be r_k. Then the rank-size relation is given by $k \sim \exp(-r_k)$. One can conclude that the case without externality is inconsistent with the Zipf law, namely $k \times r_k = \text{const.}$, frequently observed in real data.

Externality

Consider a *non-linear* transition rate to include the effect of interaction among agents. A simple way would be to specify a quadratic terms in the transition rate:

$$w_-(k) = D \cdot k^2, \qquad w_+(k) = A \cdot k^2, \tag{2.58}$$

This means that the more a city is populated, the more likely people move into the city due to the increase of utility, opportunity and so forth. Conversely, a densely populated city may increase the chance for people to move out of it, because of heavy traffic, pollution, etc. These opposite effects can be regarded as externalities. The simplest case of it would be the aforementioned expression for the transition rate.

Inserting Eq. (2.58) into Eq. (2.50), the stationary state is written by

$$q(k) = \left(\frac{A}{D}\right)^{k-1} \frac{q(1)}{k^2}. \tag{2.59}$$

A similar calculation by inserting Eq. (2.59) into Eq. (2.45) and Eq. (2.46) leads us to Eq. (2.53) and

$$\frac{A}{D} = 1 - q(1)/N_2. \tag{2.60}$$

Here we defined

$$N_m := \sum_k k^m q(k). \tag{2.61}$$

In this case, the parameters D, A determine the system size through $q(1), N_2$.

The number of cities and the total population can be calculated from Eq. (2.59) as

$$N = \frac{q(1)}{1 - q(1)/N_2} \ln\left(\frac{q(1)}{N_2}\right), \tag{2.62}$$

$$Q \simeq \frac{\pi^2}{6} \frac{q(1)}{1 - q(1)/N_2}, \tag{2.63}$$

where we used the fact that

$$\sum_{k=1}^{\infty} k^{-2} = \frac{\pi^2}{6}. \tag{2.64}$$

The distribution, Eq. (2.59), is now given by

$$q(k) = \frac{q(1)}{1 - q(1)/N_2} \frac{(1 - q(1)/N_2)^k}{k^2} \simeq \frac{1}{k^2} e^{-k/k_*}. \tag{2.65}$$

Because the scale $k_* := N_2/q(1)$ satisfies $k_* \gg 1$, it follows that $q(k) \sim 1/k^2$ for a wide range of k. One can conclude that the rank-size relation becomes $k \propto 1/r_k$, that is, the Zipf law.

The quadratic transition Eq. (2.58) can also be interpreted in the following way. The transitions, $k \to k \pm 1$, are caused by elementary processes of entry and exit during a time-scale dt of days with probability proportional to k^2. Being accumulated for a longer time-scale $\nu \cdot dt$ of months and years, there would be a number of such transitions, proportional to $\nu \times k^2$. For a large number of ν, the net effect of such entries and exits should have transition of the order of $\sqrt{k^2} = k$ in an absolute value. In other words, during a sufficiently long time-scale, the process becomes a multiplicative process (Kesten, 1973). In the time-scale of months and years, the entry from and exit to other cities or the reservoir can be considered as a spatial diffusion. Thus the assumption of quadratic transition rate can be regarded at such a macroscopic time-scale as a combination of a set of multiplicative processes with a spatial diffusion (see Zanette and Manrubia (1997)).

It would be natural to consider in reality both the linear and non-linear effects. For this consideration, one can replace Eq. (2.66) with

$$w_-(k) = D\left(k^2 + d\,k\right), \qquad w_+(k) = A\left(k^2 + a\,k\right), \tag{2.66}$$

and perform a similar calculation. As a result, the stationary state is given by

$$q(k) = q(1) \cdot \frac{1}{k} \left(\frac{A}{D}\right)^{k-1} \frac{\Gamma(2+d)}{\Gamma(1+a)} \frac{\Gamma(k+a)}{\Gamma(k+1+d)}. \tag{2.67}$$

By taking a look at the meaning of the parameters D, A as we have done earlier, one has the condition that $k_* \gg 1$ by defining $A/D =: \exp(1/k_*)$. Because $\Gamma(x)/\Gamma(x+p) \propto x^{-p}$ for a fixed parameter p and a large value of x from the Stirling formula, one can conclude that

$$q(k) \simeq e^{-k/k_*} \cdot k^{-(2+d-a)}, \tag{2.68}$$

for the region of large k. This is a power-law with an exponential cut-off. Defining the power-law exponent by $q(k) \propto k^{-\mu-1}$, one has

$$\mu = 1 + d - a. \tag{2.69}$$

In the region of small and medium k, one can show that $q(k) \sim 1/k$ from Eq. (2.67). Thus, if the non-linear effect is dominant compared with the linear effect, and $|d - a| \ll 1$, the Zipf law can be a natural consequence of this generalized model.

We have seen in this section the role of externality in a solvable stochastic model of birth–death process. One can further examine how giant cities can emerge under certain conditions and in relation to monopoly and oligopoly. See Zanette and Manrubia (1997); Marsili and Zhang (1998a) for further details.

Readers may wonder how the argument based on a jump Markov process and assumptions of transition rates is related to the behaviors of economic agents. In other words, how can one bridge the gap between the microscopic behaviors of agents and the macroscopic description by a stochastic process? This is the problem of aggregation. We only refer to the pioneering works of Aoki (2002); Aoki and Yoshikawa (2006), in which the basic concept is what Masanao Aoki called the method of combinatorial stochastic processes.

2.2 Entropy

Entropy was first established by physicists to explain the irreversibility of natural processes such as the diffusion of heat. It is now known as the second law of thermodynamics. The total entropy of an isolated system never decreases over time. However, thermodynamics is solely a descriptive discipline of macroscopic thermal phenomena. Ludwig Boltzmann was the first physicist to make a connection between thermodynamic entropy and the probability of occurrence of a macroscopic state in a system by writing down

$$S = k_B \ln W, \tag{2.70}$$

where W is the number of possible microscopic states belonging to the macroscopic state and k_B is Boltzmann's constant named after him. Behind his seminal formula, Eq. (2.70), is the assumption of the equal a priori probability postulate, i.e., any possible microscopic state occurs with equal probability in thermal equilibrium.

Suppose that an *ideal* gas has n_k *distinguishable* particles belonging to microscopic states whose energy is ϵ_k ($k = 1, 2, \cdots, K$), where there are K energy levels arranged in the ascending order: $\epsilon_1 < \epsilon_2 < \cdots < \epsilon_K$. The total number N of particles and the total energy E in the system are then given by

$$N = \sum_{k=1}^{K} n_k, \tag{2.71}$$

and

$$E = \sum_{k=1}^{K} \epsilon_k n_k, \tag{2.72}$$

respectively. A vector $\boldsymbol{n} = (n_1, n_2, \cdots, n_K)$ represents a particular allocation of particles across microscopic states with different energies. The combinatorial number $W_{\boldsymbol{n}}$ of obtaining this allocation is equal to that of throwing N balls to K different boxes:

$$W_N(\boldsymbol{n}) = \frac{N!}{\prod_{k=1}^{K} n_k!} \qquad (2.73)$$

Applying the Stirling formula for large numbers,

$$\ln N! \simeq N \ln N - N \quad (N \gg 1), \qquad (2.74)$$

we can calculate the entropy, Eq. (2.70), as

$$S = k_{\rm B} \ln W_N(\boldsymbol{n}) = -N k_{\rm B} \sum_{k=1}^{K} p_k \ln p_k, \qquad (2.75)$$

where

$$p_k = \frac{n_k}{N}, \qquad (2.76)$$

with $\sum_{k=1}^{K} p_k = 1$. This equation captures the combinatorial aspect of the computation problem for entropy.

Entropy S is a non-negative definite function. For a completely ordered state ($p_i = 1$ and $p_j = 0$ otherwise), S exactly vanishes. For the completely random state ($p_i = 1/K$), on the other hand, S takes its maximum value, $S_{\max} = N k_{\rm B} \ln K$. Thus, one can use entropy as a measure of the diversity of a system.

Immediately after Boltzmann's work, Josiah Willard Gibbs defined his entropy as

$$S = -k_{\rm B} \sum_{\ell=1}^{W} p_\ell \ln p_\ell, \qquad (2.77)$$

with $p_\ell \geq 0$. Here p_ℓ is the occurrence probability of the ℓ th microscopic state. Although the functional form of Eq. (2.77) is essentially the same as that of the last form in Eq. (2.75), Gibbs' entropy is more general than Boltzmann's entropy in the sense that it is applicable to non-equilibrium states. If all possible microstates are equally likely, i.e.,

$$p_\ell = \frac{1}{W}, \qquad (2.78)$$

then Eq. (2.77) reduces to Eq. (2.70). Also Gibbs' entropy is more useful than Boltzmann's entropy with the combinatorial problem.

Later, Claude Shannon introduced the information entropy H as a measure of the uncertainty in information transfer. For a long sequence of characters of M kinds, H is given as

$$H = -c \sum_{m=1}^{M} p_m \ln p_m, \tag{2.79}$$

where p_m is the probability of finding the mth character in the sequence and c is an arbitrary positive constant to be adjusted for the usage of H. Shannon's entropy has extended the applicability of the concept of entropy to fields other than physics, including economics (Foley, 1994, 1996) and sociology (Montroll, 1978, 1981, 1987).

2.3 Statistical Equilibrium

Given that the number of all possible ways to allocate N different balls to K different boxes is K^N, the probability that a particular allocation $\boldsymbol{n}$ occurs is calculated as

$$P_{\boldsymbol{n}} = \frac{W_N(\boldsymbol{n})}{K^N} = \frac{\exp(S/k_B)}{K^N} \tag{2.80}$$

A fundamental postulate of statistical physics is that the observed state or allocation $\boldsymbol{n}$ which maximizes the probability $P_{\boldsymbol{n}}$ of Eq. (2.80), *under macro-constraints*, must be realized. More precisely speaking, it must be realized in the sense of expected value. In physics, the variance is usually so small relative to the expected value that one practically always observes the expected value. The idea of maximizing $P_{\boldsymbol{n}}$ is analogous to the maximum likelihood in statistics/econometrics. Maximizing $P_{\boldsymbol{n}}$ is equivalent to maximizing $\ln P_{\boldsymbol{n}}$ and hence S.

To maximize entropy S *under two macro-constraints*, Eqs (2.71) and (2.72), we set up the following Lagrangian form L:

$$L = -N \sum_{k=1}^{K} \left(\frac{n_k}{N}\right) \ln \left(\frac{n_k}{N}\right) - \alpha N - \beta E, \tag{2.81}$$

with two Lagrangian multipliers, α and β, corresponding to conservation of the total number and the total energy, respectively. Maximization of this Lagrangian form with respect to n_k leads us to the first-order variational condition:

$$\delta L = \frac{\delta S}{k_B} - \alpha \delta N - \beta \delta E = 0, \tag{2.82}$$

with the following variations,

$$\frac{\delta S}{k_B} = \sum_{k=1}^{K} \delta n_k \ln \left(\frac{n_k}{N}\right), \tag{2.83}$$

$$\delta N = \sum_{k=1}^{K} \delta n_k, \tag{2.84}$$

$$\delta E = \sum_{k=1}^{K} \epsilon_k \delta n_k. \tag{2.85}$$

Equation (2.82) coupled with Eqs (2.83), (2.84), and (2.85) determines n_k as

$$\ln\left(\frac{n_k}{N}\right) = -\alpha - \beta \epsilon_k \quad (k = 1, 2, \cdots, K). \tag{2.86}$$

Because n_k/N sums up to 1, one obtain

$$\frac{n_k}{N} = e^{-\alpha - \beta \epsilon_k} = \frac{e^{-\beta \epsilon_k}}{\sum_{k=1}^{K} e^{-\beta \epsilon_k}}. \tag{2.87}$$

Thus, the number of particles at energy level ϵ_k is exponentially distributed. This distribution is known as the Maxwell–Boltzmann distribution in physics.

The Lagrange parameters, α and β, are determined by comparing the variational condition, Eq. (2.82), with the following thermodynamic relation,

$$dS = \frac{1}{T} dE - \frac{\mu}{T} dN, \tag{2.88}$$

where T and μ are the temperature and chemical potential of a system. The results are given by

$$\beta = \frac{1}{k_{\rm B} T}, \tag{2.89}$$

$$\alpha = -\frac{\mu}{k_{\rm B} T}. \tag{2.90}$$

2.4 Stationarity in Time Series

The first step in the study of financial time series such as stock prices is to draw a graph. Looking at the graph, one will notice at first that the price fluctuates significantly with time. This is often called the property of time variation or **fluctuation**. Besides this property, time series have other important properties such as **non-stationarity** and **long-term memory**.

2.4.1 Stationary and non-stationary time series

First, we explain the important property of non-stationarity in time series (Box and Jenkins, 1970). Figure 2.3a depicts the time fluctuation of the closing price for the Dow Jones industrial average from Oct 1, 1928 to Dec 31, 2007. This figure shows that the price did not change greatly from 1928 to 1980, but it increased rapidly from the late 1980s to 2000. This increase was because of inflation and economic growth. The price dropped in 2004 and increased again until 2007. The drop in 2004 was caused by the burst of the dot-com bubble.

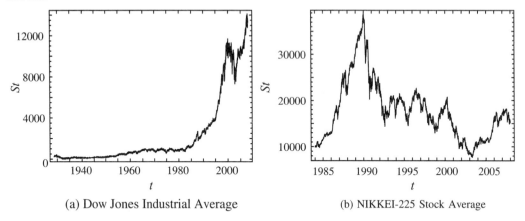

(a) Dow Jones Industrial Average (b) NIKKEI-225 Stock Average

Fig. 2.3 The time fluctuation of the closing price for the Dow Jones industrial average shows that the price did not change significantly from 1928 to 1980, but the price increased rapidly from the late 1980s to 2000. The time fluctuation of the closing price for the NIKKEI-225 shows that the stock price increased significantly from 1984 to 1990, then decreased gradually through a series of small rises and dips.

We take a look at the NIKKEI-225 stock average to capture the price fluctuation in Japan. Figure 2.3b depicts the time fluctuation of the closing price for the NIKKEI-225 from Jan. 4, 1984 to Dec. 28, 2007. This figure shows that the stock price increased significantly from 1984 to 1990, then decreased gradually over a series of small rises and dips. This was because of the bursting of the Japanese asset price bubble. We interpret these observed small rises and dips of price as economic growth and recession rather than inflation and deflation. From 2004, the stock price turned upward, but the Japanese stock market seems to have been left out of the trend of global economic growth from December 2007.

We are able to notice many properties by examining these price fluctuations. This is made possible by finding the price change called trend in stock time series and finding the corresponding relation between the price trend and macroeconomic change. Data which changes over time

$$(x_1, x_2, \ldots, x_T) \tag{2.91}$$

is called time series. Time series with a trend is called a **non-stationary process** in statistics.

The opposite of a non-stationary process is a **stationary process**. When time series x_t satisfies the following three conditions, x_t is called weakly stationary.

- The mean value does not depend on time;

$$\mu_x = \frac{1}{T}\sum_{t=1}^{T} x_t = \text{const.} \tag{2.92}$$

- The variance does not depend on time;

$$\sigma_x^2 = \frac{1}{T-1}\sum_{t=1}^{T}(x_t - \mu_x)^2 = \text{const.} \tag{2.93}$$

- The covariance between different time depends only on lag τ;

$$\text{Cov}(x_t, x_{t+\tau}) = \frac{1}{T-1}\sum_{t=1}^{T}(x_t - \mu_x)(x_{t+\tau} - \mu_x) = \gamma(\tau). \tag{2.94}$$

When the joint distribution function $P(x_1,\ldots,x_T)$ of time series $(x_1,\ldots,x_T)$ and the joint distribution function $P(x_{1+\tau},\ldots,x_{T+\tau})$ of the time series shifted by time τ $(x_{1+\tau},\ldots,x_{T+\tau})$ is identical for all natural numbers t and all integers τ, $(x_1,\ldots,x_T)$ is called **strongly stationary**. A stationary process is often used to mean **weakly stationary** in time series analysis. In this book, we use stationary process in the same sense.

The stationary process is defined by Eq. (2.92) through Eq. (2.94), but these are not very useful in practice. For this reason, a simple statistical test called the **unit root test** to identify a stationary process was developed. We consider the time series generated using

$$x_t = ax_{t-1} + \epsilon_t. \tag{2.95}$$

The characteristic equation of Eq. (2.95) is

$$1 - az = 0. \tag{2.96}$$

The solution of the characteristic equation for z is called a **root**. The condition for stationarity is $|z| > 1$, therefore $|a| < 1$. $z = 1$ is called the **unit root**. If the characteristic equation, Eq. (2.95), has the unit root as a solution, we have $a = 1$ and the time series x_t is non-stationary. We do not have to consider the case $|a| > 1$, because time series will diverge in this case.

We rewrite Eq. (2.95) as follows,

$$x_t - x_{t-1} = (a-1)x_{t-1} + \epsilon_t = bx_{t-1} + \epsilon_t. \tag{2.97}$$

We note that $a < 1$ and $a = 1$ corresponds to $b < 0$ and $b = 0$, respectively. Therefore, we have the unit root when $b = 0$. We postulate a null hypothesis H_0: the unit root is present. t value is defined as the estimated value of parameter b divided by the standard error of estimation. We conduct a one-side test using a probability distribution for the t value shown in Fig. 2.4. Probability p is $P(< t)$ to have t value smaller than the t value obtained for the estimation. The one-side test compares probability p and a specified significance level α. If $p < \alpha$, null hypothesis H_0 is rejected. This means that the unit root is not present. Therefore, the time series is stationary. On the other hand, if $p > \alpha$, null hypothesis H_0 is adopted. This means that we cannot say that the unit root is not present. Therefore, time series is not stationary (in practice).

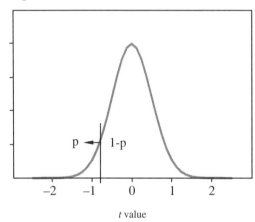

Fig. 2.4 Probability distribution for t value, which is defined as the estimated value of parameter b divided by the standard error of estimation, is shown to conduct one-side test. One-side test compares probability p and a specified significance level α. If $p < \alpha$, the time series is stationary.

2.4.2 Methods to obtain stationary time series

Many economic time series beyond the stock price and the foreign exchange rate are non-stationary processes. First, we need to obtain a stationary time series from the original non-stationary time series because most methods in the time series analysis assume a stationary process. One way to obtain a stationary time series from the original non-stationary time series is to remove the trend from the original time series. We now explain various methods to obtain a stationary time series and discuss their pros and cons.

Moving average

Moving average

$$m_t(k) = \frac{s_{t-k} + \cdots + s_{t-1} + s_t + s_{t+1} + \cdots + s_{t+k}}{2k + 1} \qquad (2.98)$$

is calculated using $2k + 1$ data centering on time t. The residual time series obtained by removing the **moving average** $m_t(k)$ from the original time series s_t

$$x_t = s_t - m_t(k) \qquad (2.99)$$

is a stationary time series.

Basic Concepts in Statistical Physics and Stochastic Models

Finite difference (Price difference)

When media broadcasts market conditions, they report not only the price level but also the price difference from the closing price of the previous day. In this sense, price difference is rather familiar to us. The price difference (**finite difference** of price) Δs_t is defined as follows:

$$\Delta s_t := s_{t+\Delta t} - s_t. \tag{2.100}$$

A stationary time series obtained by taking the finite difference is known as the method of Box–Jenkins. The first order difference removes a linear trend. Similarly, the second order difference

$$\Delta^2 s_t = (s_t - s_{t-1}) - (s_{t-1} - s_{t-2}) = s_t - 2s_{t-1} + s_{t-2} \tag{2.101}$$

removes a quadratic trend. The advantage of a finite difference is that this method does not require both non-linear transformation and probabilistic transformation. On the other hand, the drawback of a finite difference is that this method depends on change of scale. For instance, if we calculate the price difference of the Dow Jones industrial average shown in Fig. 2.3a, we obtain the time series shown in Fig. 2.5a. This figure shows that the time series oscillates around $\Delta s_t = 0$ and does not show the trend observed in Fig. 2.3a. However, we note that the amplitude of oscillation increased after the late 1980s. This was because of the change in scale of inflation. Therefore, it is problematic to analyze Δs_t.

Normalized finite difference

We will overcome the problem of scale change observed in the finite difference time series by normalizing the effect of inflation and deflation. The factor to absorb the scale change in price difference time series, the normalized finite difference, is defined as follows:

$$\Delta s_t^d := \frac{s_{t+\Delta t} - s_t}{d_t} = \frac{\Delta s_t}{d_t}. \tag{2.102}$$

An advantage of finite difference is that this method does not require both non-linear transformation and probabilistic transformation. Another advantage is that the value of money is constant irrespective of the passage of time if d_t is properly determined. The problem is how to determine d_t uniquely. In fact, we have various candidates for d_t; such as the consumer price index (CPI), the gross domestic product (GDP), gold price, etc., but d_t cannot be determined uniquely.

We calculate the price difference Δs_t for the Dow Jones industrial average shown in Fig. 2.3a. Then, we evaluate the normalizing factor every year by assuming GDP in the US in 1970 was equal to 1. Therefore, denoting y as the year after 1970, we normalize Δs_t using the normalized factor $d_y = \text{GDP}_y/\text{GDP}_{1970}$. As a consequence, we obtain Fig. 2.5b. Note that this figure is drawn to the same scale as Fig. 2.5a. A comparison of this figure and Fig. 2.5a shows that the large amplitude after the 1980s observed in Fig. 2.5a disappeared. Thus, the problem of scale change is overcome.

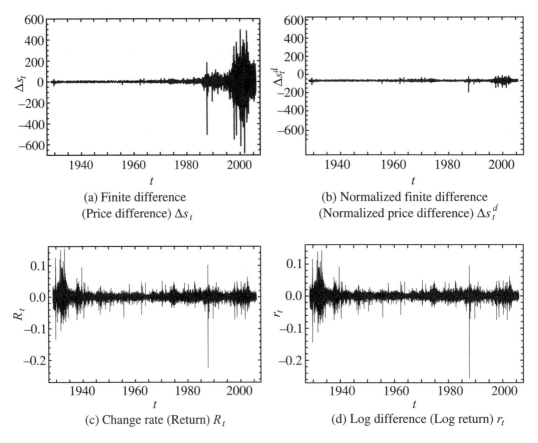

Fig. 2.5 A method to obtain a stationary time series from the original non-stationary time series is to remove the trend from the original time series. We explain various methods (price difference, normalized price difference, return, and log return) to obtain a stationary time series from the Dow Jones industrial average and discuss their pros and cons.

Change rate (Return)

It is important for practitioners to know how much money is earned or lost for the invested money, rather than how the price changes. In fact, it is more meaningful to earn 10,000 yen by investing 100,000 yen than to earn the same amount of money by investing 10 million yen. The measure to scale this meaning is called **return** R_t, and is defined as follows:

$$R_t := \frac{s_{t+\Delta t} - s_t}{s_t} = \frac{\Delta s_t}{s_t}. \tag{2.103}$$

The advantage of return is that this is the direct measure of return or loss during period Δt. The drawback is that it is sensitive to scale change when period Δt is long. If we calculate the price difference Δs_t of the Dow Jones industrial average shown in Fig. 2.3a, we obtain the time series shown in 2.5c. The figure shows that the trend observed in Fig. 2.3a is removed. Note that this figure is drawn to a scale that is different from Fig. 2.5a and Fig. 2.5b. We notice a strong dip at 1988. This is because of the market crash called "**Black Monday**" on Oct. 19, 1987.

Log difference (Log return)

The **log return** is given by

$$r_t := \ln s_{t+\Delta t} - \ln s_t. \tag{2.104}$$

The advantage of using the log return is that it considers the effect of an average scale change without the normalization factor in Eq. (2.102) when the scale change owing to economic growth and recession is constant. In general, the nonlinear transformation of Eq. (2.104) might affect the statistical property of the time series significantly.

When we calculate the log return r_t of the Dow Jones industrial average shown in Fig. 2.3a, we obtain the time series shown in Fig. 2.5d. This figure shows that the trend observed in Fig. 2.3a is removed, just as in the case shown in Fig. 2.5c. Comparison of Fig. 2.5c and Fig. 2.5d shows that these time series look very similar.

Although so far we have discussed Δs_t, Δs_t^d, R_t, and r_t for the Dow Jones industrial average, a similar discussion is possible for the NIKKEI-225 stock average. The results of the analysis are summarized in Fig. 2.6. Figure 2.6a and Fig. 2.6b show Δs_t and Δs_t^d, respectively. These figures are drawn to the same scale. The normalization factor d_t used to draw Fig. 2.6b is $d_t = \text{GDP}_t/\text{GDP}_{1985}$ to make GDP in 1985 equal to 1. A comparison of these figures shows that these time series are very similar. Figure 2.6c and Fig. 2.6d are drawn with the same scale for R_t and r_t, respectively. These again look very similar.

Case of a short time step

Equation (2.104) can be rewritten as

$$r_t = \ln \frac{s_{t+\Delta t}}{s_t} = \ln \left[1 + \frac{\Delta s_t}{s_t}\right]. \tag{2.105}$$

Here, we assume that the time step Δt is very short and s_t is **high-frequency data**. Then, we have $\Delta s_t \ll |s_t|$. Consequently, we obtain

$$r_t = \ln \left[1 + \frac{\Delta s_t}{s_t}\right] \simeq \frac{\Delta s_t}{s_t} = R_t. \tag{2.106}$$

A comparison of Fig. 2.3a and Fig. 2.5a shows that Δs_t is a rapidly changing variable and s_t is a moderately changing variable for short Δt. Thus, we obtain the approximate relation

$$R_t \simeq C \Delta s_t, \tag{2.107}$$

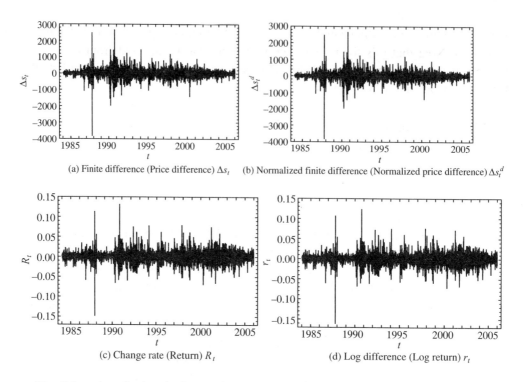

Fig. 2.6 A method to obtain a stationary time series from the original non-stationary time series is to remove the trend from the original time series. We explain various methods (price difference, normalized price difference, return, and log return) to obtain a stationary time series calculated from the NIKKEI-225 stock average and discuss their pros and cons.

with a constant C. Furthermore, when Δt is very small and D_t is assumed to be almost equal to 1, we obtain,

$$\Delta s_t \simeq \Delta s_t^d. \tag{2.108}$$

In summary, we have

$$r_t \simeq R_t \simeq C\Delta s_t \simeq \Delta s_t^d. \tag{2.109}$$

This relation is the reason why Fig. 2.5 and Fig. 2.6 look very similar.

For a fluctuating non-stationary time series, we have a stationary time series by taking the first order difference of log transformation of the original time series. Variance of the stationary time series obtained by the aforementioned procedure is called **volatility**. In other cases, the price difference of the time series is sometimes called volatility. Volatility is an index to express the magnitude of uncertainty in general. It is empirically known that volatility of the log return r_n calculated by taking the first order difference of log transformation of the price time series is not constant in time. In fact, Fig. 2.5d and

Fig. 2.6d show that the variance of log-return r_n do not have a clear trend but fluctuate largely. Once the variance becomes small (large), a period of small (large) variance will ensue. This phenomenon is called **volatility clustering**. Time series with volatility clustering cannot be regarded as a stationary process in a rigorous sense.

2.5 Long-Term Memory in Time Series

Next, we will explain another important statistical property, the **long-term memory**.

2.5.1 Autocorrelation function and average mutual information

An **autocorrelation function** $\rho(\tau)$ is defined as

$$\rho(\tau) = \frac{\gamma(\tau)}{\gamma(0)} = \frac{\mathrm{Cov}(x_t, x_{t+\tau})}{\sigma_x^2} \tag{2.110}$$

for time series x_t. When an autocorrelation function $\rho(\tau)$ has the following property,

$$\sum_{\tau=-\infty}^{\infty} \rho(\tau) = \mathrm{const.} \tag{2.111}$$

the time series x_t is called a **short-term memory** process. On the other hand, when an autocorrelation function $\rho(\tau)$ has the following property,

$$\sum_{\tau=-\infty}^{\infty} \rho(\tau) = \infty, \tag{2.112}$$

the time series x_t is called a long-term memory process. Yearly temporal change of the lowest water level of the Nile River is widely cited as an example of a long-term memory process.

Given its similarity to the autocorrelation function $\rho(\tau)$, the **average mutual information** $I(\tau)$

$$I(\tau) = \sum_{x_t, x_{t+\tau}} P(x_t, x_{t+\tau}) \log_2 \frac{P(x_t, x_{t+\tau})}{P(x_t) P(x_{t+\tau})} \tag{2.113}$$

is often used. This quantity corresponds to the Kullback–Leibler distance between statistically independent distributions. The average mutual information $I(\tau)$ is calculated by counting the number of data $P(x_t, x_{t+\tau})$ that pass through a small region $(\Delta x_t, \Delta x_{t+\tau})$ in plane $(x_t, x_{t+\tau})$. $I(\tau)$ is used for noisy time series. Time series x_n and the definition of lag τ is shown in Fig. 2.7.

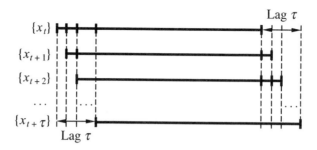

Fig. 2.7 The average mutual information $I(\tau)$ is calculated by counting the number of data $P(x_t, x_{t+\tau})$ that pass through a small region $(\Delta x_t, \Delta x_{t+\tau})$ in plane $(x_t, x_{t+\tau})$. $I(\tau)$ is used for noisy time series. This figure shows time series x_n and the definition of lag τ.

2.5.2 Volatility of return time series

We will now explain volatility of log-return time series r_t. The left panel of Fig. 2.8 shows the daily temporal change of volatility of the return time series calculated for every 10 business days using the closing price of the NIKKEI-225 from Jan. 1984 to Aug. 2007 (5740 business days).

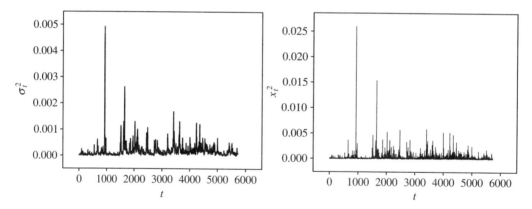

Fig. 2.8 The daily temporal change of volatility and its surrogate variable of return time series was calculated for every 10 business days using the closing price of the NIKKEI-225 from Jan. 1984 to Aug. 2007 (5740 business days). Here, the surrogate variable of volatility is the square of log-return.

The right panel of Fig. 2.8 shows the daily temporal change of the **surrogate variable** of volatility for the same period. Here, the surrogate variable of volatility is the square of log-return,

$$\sigma_t^2 = \left(\frac{r_t - r_{t-1}}{r_{t-1}} \right)^2 = x_t^2. \tag{2.114}$$

Since volatility and the surrogate variable are very similar, this definition of the surrogate variable is considered appropriate.

Then, we calculate the autocorrelation functions $\rho(\tau)$ for log-return of price and its volatility. The calculated autocorrelation functions $\rho(\tau)$ are shown in Fig. 2.9. The horizontal dotted line indicates the significance level. The autocorrelation for log-return will become non-significant after one day. Thus, daily data is regarded as a random walk. On the other hand, autocorrelation for volatility is statistically significant with positive correlation for longer than a couple of ten days. This means that once the volatility of log-return becomes large (small), the large (small) value continues for a while. Therefore, volatility of log-return shows the property of long-term memory.

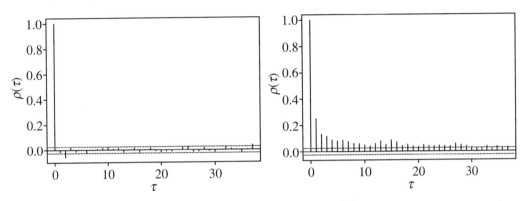

Fig. 2.9 We calculate the autocorrelation functions $\rho(\tau)$ for log-return of price and its volatility. The horizontal dotted line indicates the significance level. The autocorrelation for log-return will become non-significant after one day. On the other hand, autocorrelation for volatility is statistically significant with positive correlation for longer than a couple of ten days.

2.6 Basic Model of Time Series

2.6.1 Autoregression model

The most basic linear model of time series is the **autoregression model** (Box and Jenkins,1970). The pth order autoregression model is given by

$$x_t = \xi_t + \sum_{i=1}^{p} \phi_i x_{t-i}. \tag{2.115}$$

We refer to this model as AR(p). An autoregression model is based on the idea that future forecast is possible using past observed data. The qth order **moving average model** is given by

$$x_t = \xi_t + \sum_{i=1}^{q} \pi_i \xi_{t-i} \tag{2.116}$$

We refer to this model as MA(q). Here, ξ_t is the random shock generated by a Gaussian process with mean 0 and variance σ^2. The moving average model is based on the premise that past **shock** drives future value. We note that the moving average of Eq. (2.116) is not related to Eq. (2.98).

The (p, q)th order **autoregressive moving average model** is given by

$$x_t = \xi_t + \sum_{i=1}^{p} \phi_i x_{t-i} + \sum_{i=1}^{q} \pi_i \xi_{t-i} \tag{2.117}$$

We refer to this model as ARMA(p, q); it is a model that can be applied to stationary time series. For non-stationary time series, ARMA(p, q) can be applied after taking the dth order difference of the original time series or the log-transformation of the original time series. This is called an **autoregressive integrated moving average model** and we refer to this model as ARIMA(p, d, q).

We made maximum likelihood estimation of the parameters of the ARIMA($p, 1, q$) model for stock price s_t and the ARIMA($p, 0, q$) model for log-return r_t. AICs of the ARIMA($p, 1, q$) model for stock price and the ARIMA($p, 0, q$) model for log-return are shown in Table 2.1 and Table 2.2, respectively. See Chapter 3 for the concept of AIC.

Table 2.1 AIC of the ARIMA ($p, 1, q$) Model for Stock Price

p \ q	0	1	2
0	–	79640.	79616.
1	79628.	79625.	79603.
2	79590.	79590.	79592.

Table 2.2 AIC of the ARIMA ($p, 0, q$) Model for Log Return

p \ q	0	1	2
0	–	−32992.	−33009.
1	−32985.	−32989.	−33004.
2	−32995.	−32997.	−32995.

Table 2.2 shows that ARIMA(1,0,2) (ARMA(1,2)) is the best among various order of ARIMA($p, 0, q$) model for log-return. Daily temporal change of of log-return calculated by the ARMA(1,2) model with maximum likelihood estimator of parameters:

$$\begin{aligned}\phi_1 &= -0.432, \\ \pi_1 &= -0.417, \\ \pi_2 &= 0.0640,\end{aligned} \tag{2.118}$$

is shown in Fig. 2.10. A comparison of the model calculation and the actual time series shown in Fig. 2.6d shows that the order of value is not reproduced. Thus, we conclude that a simple linear model dose not reproduce the characteristics of price and log-return time series.

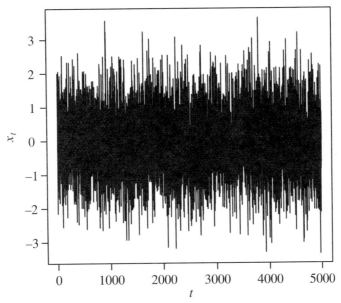

Fig. 2.10 This shows the daily temporal change of of log-return calculated by the ARMA(1,2) model with maximum likelihood estimators. A comparison of the model calculation and the actual time series shown in Fig. 2.6d shows that the order of value is not reproduced. A simple linear model does not reproduce the characteristics of price and log-return time series.

2.6.2 Time dependence of forecast error

The characteristics of time dependence of forecast error are significantly different for stationary time series and non-stationary time series. We will study stationary and non-stationary time series using ARMA(1,0) and ARIMA(1,1,0), respectively.

In ARMA(1,0), x_t at three consecutive time points are written by

$$x_t = \phi_1 x_{t-1} + \xi_t, \tag{2.119}$$

$$x_{t-1} = \phi_1 x_{t-2} + \xi_{t-1}, \tag{2.120}$$

$$x_{t-2} = \phi_1 x_{t-3} + \xi_{t-2}. \tag{2.121}$$

By substituting Eq. (2.120) and Eq. (2.121) into Eq. (2.119), we obtained the following relation,

$$x_t = \phi_1^3 x_{t-3} + \sum_{i=0}^{2} \phi_1^{2-i} \xi_{t-2+i}, \tag{2.122}$$

Similarly, for consecutive h time points we obtain

$$x_t = \phi_1^h x_{t-h} + \sum_{i=0}^{h-1} \phi_1^{h-1-i} \xi_{t-h+1+i}. \tag{2.123}$$

Here we rewrite the error term in Eq. (2.119) using standard deviation σ_1 as follows,

$$\xi_t = \sigma_1 \epsilon_t. \tag{2.124}$$

Using this relation, we obtain the standard deviation of the error term in Eq. (2.123)

$$\sigma_h^2 = \sigma_1^2 \sum_{i=0}^{h} \left(\phi_1^{h-i}\right)^2. \tag{2.125}$$

In Eq. (2.125), by taking $h \to \infty$, we obtain

$$\sigma_h^2 = \frac{\sigma_1^2}{1 - \phi_1^2}. \tag{2.126}$$

Thus, the forecast error converges to a constant value for stationary time series.

On the other hand, in ARIMA(1,1,0), x_t at three consecutive time points are written by

$$x_t = (1 + \phi_1)x_{t-1} - \phi_1 x_{t-2} + \xi_t, \tag{2.127}$$

$$x_{t-1} = (1 + \phi_1)x_{t-2} - \phi_1 x_{t-3} + \xi_{t-1}, \tag{2.128}$$

$$x_{t-2} = (1 + \phi_1)x_{t-3} - \phi_1 x_{t-4} + \xi_{t-2}. \tag{2.129}$$

Thus, we obtain

$$x_t = (1 + \phi_1)(1 + \phi_1^2)x_{t-3} - (\phi_1 + \phi_1^2 + \phi_1^3)x_{t-4}$$
$$+ \xi_t + (1 + \phi_1)\xi_{t-1} + (1 + \phi_1 + \phi_1^2)^2 \xi_{t-2}. \tag{2.130}$$

Standard deviation of the error term in Eq. (2.130) is written as follows,

$$\sigma_h^2 = \sigma_1^2 \sum_{i=0}^{h} \left(\sum_{j=0}^{i} \phi_1^j\right)^2, \tag{2.131}$$

where $h = 2$. Thus, for a non-stationary time series, the forecast error grows over time and does not converge to a constant. We can only say that the future forecast value falls within a certain range.

2.7 Stochastic Model of Non-Stationary Time Series

What causes the long-term memory process? The long-term memory depends directly on linearity or non-linearity. We have a nonlinear model to reproduce the short-term memory process and a linear model to reproduce the long-term memory process. While the standard Brownian motion is a linear model to reproduce the short-memory process (Chandrasekhar, 1943), the fractal Brownian motion is a linear model augmented to reproduce the long-term memory process. The idea of a fractal was first proposed by Mandelbrot et al. (2005).

2.7.1 Standard brownian motion

Change Δz of time series z_t during a short period $[t, t + \Delta t]$ is given by

$$z_{t+1} - z_t = \Delta z_t \tag{2.132}$$

$$\Delta z_t = \epsilon_t \sqrt{\Delta t} \tag{2.133}$$

where ϵ_t is the standard Gaussian random number with mean 0 and variance 1 and satisfies the relation,

$$\langle \epsilon_s, \epsilon_t \rangle = \delta_{st}. \tag{2.134}$$

From Eq. (2.132), change of time series z_t during the finite period $[t, t + \tau]$ is given by

$$z_{t+\tau} - z_t = \sum_{i=t}^{\tau} \epsilon_i \sqrt{\Delta t}, \tag{2.135}$$

where $\tau = n\Delta t$. Time series z_t possesses the following properties:

$$\langle z_{t+\tau} - z_t \rangle = 0, \tag{2.136}$$

$$\langle (z_{t+\tau} - z_t)^2 \rangle \propto \tau. \tag{2.137}$$

Here $\langle \cdot \rangle$ means time average. A time series satisfying Eq. (2.136) and Eq. (2.137) is called the **standard Brownian motion** or **Wiener process**.

Next, a new time series x_t is generated based on the standard Brownian motion z_t as follows:

$$\Delta x = a(x,t)\Delta t + b(x,t)\Delta z. \tag{2.138}$$

By taking the limit $\Delta t \to 0$, the relation in continuous time

$$dx = a(x,t)dt + b(x,t)dz, \tag{2.139}$$

is obtained. Here $a(x,t)$ and $b(x,t)$ are parameters corresponding to **drift** and variance, respectively. Both parameters depend on x and t. A time series that satisfies Eq. (2.139) is called an **Ito process**.

Specifically, parameters are constant: $a(x,t) = \mu$, $b(x,t) = \sigma$, giving us

$$dx = \mu dt + \sigma dz. \tag{2.140}$$

The time series described in Eq. (2.140) is a standard Brownian motion with drift, or simply a standard Brownian motion. Here we have $\mu > 0$ and $\sigma > 0$.

Furthermore, when $a(x,t) = \mu x$ and $b(x,t) = \sigma x$, we obtain

$$dx = \mu x dt + \sigma x dz. \tag{2.141}$$

The time series described in Eq. (2.141) is a **geometrical Brownian motion**.

Next, we consider function $F(x,t)$ for which variable x_t is an Ito process satisfying Eq. (2.139). It is inappropriate to write the total differentiation of function $F(x,t)$ as

$$dF = \frac{\partial F}{\partial x} dx + \frac{\partial F}{\partial t} dt, \tag{2.142}$$

because dx in the first term of the right-hand side includes $dz \propto \sqrt{dt}$. Then we rewrite the total differentiation by including terms of the higher order for dx as follows;

$$dF = \frac{\partial F}{\partial x} dx + \frac{\partial F}{\partial t} dt + \frac{1}{2} \frac{\partial^2 F}{\partial x^2} (dx)^2 + \frac{1}{6} \frac{\partial^3 F}{\partial x^3} (dx)^3 + \cdots. \tag{2.143}$$

Here we take only the first order for dt and ignore the higher order terms in the following relations:

$$(dx)^2 = a^2 (dt)^2 + 2ab(dt)^{3/2} + b^2 dt, \tag{2.144}$$

$$(dx)^3 = a^3 (dt)^3 + 3a^2 b (dt)^{5/2} + 3ab^2 (dt)^2 + b^3 (dt)^{3/2}. \tag{2.145}$$

Consequently, we obtain

$$dF = \frac{\partial F}{\partial x} dx + \frac{\partial F}{\partial t} dt + \frac{1}{2} b^2 \frac{\partial^2 F}{\partial x^2} dt. \tag{2.146}$$

Equation (2.146) is called **Ito's lemma**.

2.7.2 Fractal brownian motion

The variance of time series z_t in a finite period $[t, t+\tau]$ behaves like

$$\langle (z_{t+\tau} - z_t)^2 \rangle \propto \tau^{2H}, \tag{2.147}$$

where H is called the **Hurst index**. When H is equal to $1/2$, Eq. (2.147) is identical to Eq. (2.137). Therefore, the time series z_t in Eq. (2.147) becomes a standard Brownian motion. When power index $2H$ of τ is other than 1, the time series z_t is called a **fractional Brownian motion** (Mandelbrot, 1983). If we multiply the period by b times in Eq. (2.147):

$$\langle (z_{t+b\tau} - z_t)^2 \rangle \propto b^{2H} \tau^{2H}, \tag{2.148}$$

we obtain

$$z_{t+b\tau} - z_t = b^H (z_{t+\tau} - z_t). \tag{2.149}$$

More specifically, if we multiply time by b, z will increase by b^H. **Self-similarity** implies that the lengths L_x and L_y in each dimension are written as

$$L_x \propto S^{\mu_x}, \tag{2.150}$$

$$L_y \propto S^{\mu_y}, \tag{2.151}$$

using scale of length S. If the pattern looks the same as the original with a different scale S, the pattern is called a **fractal**. Specifically, a pattern with $\mu_x = \mu_y$ is called a fractal, narrowly defined, and a pattern with $\mu_x \neq \mu_y$ is called **self-affine**. A fractal Brownian motion corresponds to $\mu_x = 1$ and $\mu_y = H$.

We calculate the autocorrelation function $\rho(\tau)$ for fractal Brownian motion. By taking into account $\gamma(0) = \langle z_t^2 \rangle = \langle z_{t+\tau}^2 \rangle$, we obtain the relation;

$$2\gamma(0) - 2\gamma(\tau) = A\tau^{2H} \tag{2.152}$$

from Eq. (2.147). Thus, the autocorrelation function $\rho(\tau)$ is given by

$$\rho(\tau) = \frac{\gamma(\tau)}{\gamma(0)} = 1 - \frac{A\tau^{2H}}{2\gamma(0)}. \tag{2.153}$$

From Eq. (2.153), the autocorrelation function shows power-law time dependence for $H > 1/2$. Therefore, a fractal Brownian motion is a linear model that reproduces long-term memory.

The Hurst index characterizing fractal Brownian motion is one of fractal dimension. Dimension is an integer in **Euclidean geometry**. For instance, a point is 0 dimension, a line is 1 dimension, a plane is 2 dimension, and a cuboid is 3 dimension. Fractal dimension is a generic term for non-integer dimension. Various fractal dimensions are defined depending on the subjects of research and those estimation methods vary based on definitions.

2.7.3 Fractal dimension

The well-known **fractal dimension** is explained as follows:

Self-Similarity dimension

The self-similarity dimension D_s is used for contracting a pattern by a constant rate r. D_s is defined by

$$a = \frac{1}{r^{D_s}}, \tag{2.154}$$

or

$$D_s = \lim_{r \to 0} \frac{\ln a}{\ln 1/r}, \tag{2.155}$$

where a is the number of contracted patterns that are geometrically similar to the original pattern. For instance, if $a = 2$ for $r = 1/2$, we have $D_s = 1$. In the same manner, if $a = 4$ for $r = 1/2$, we have $D_s = 2$. Furthermore, a **Cantor set** is obtained by removing the central line segment among three line segments divided equally as shown in Fig. 2.11. For a Cantor set, we have $a = 2$ for $r = 1/3$, and thus, we have $D_s = 0.631$.

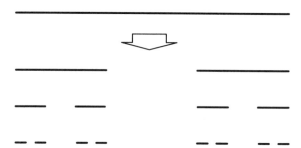

Fig. 2.11 A Cantor set is obtained by removing the central line segment among three line segments divided equally. For a Cantor set, we have $a = 2$ for $r = 1/3$, and thus, the self-similarity dimension $D_s = 0.631$.

Box-Counting dimension

A contracted pattern with self-similarity observed in nature does not completely match a part of the original pattern. For instance, imagine the pattern of a coastal line, and consider a curve on a two-dimensional plane. Divide the whole plane by a square lattice of width ϵ (box) and count the number of boxes the curve passes through $N(\epsilon)$. Here, the box-counting dimension D_b is defined as follows,

$$N(\epsilon) = \frac{1}{\epsilon^{D_b}} \tag{2.156}$$

or

$$D_b = \lim_{\epsilon \to 0} \frac{\ln N(\epsilon)}{\ln(1/\epsilon)}. \tag{2.157}$$

The **box-counting dimension** D_b is equivalent to the **capacity dimension** D_c.

Note that it is inappropriate to apply the box-counting dimension D_b to self-affine data such as fractal Brownian motion, since the definition of box-counting dimension D_b assumes the square lattice where the scale of each dimension is the same. For fractal Brownian motion, the scale of the horizontal axis is $\mu_x = 1$ and the scale of the vertical axis is Hurst index $\mu_y = H$. Therefore, self-similarity of a pattern is not guaranteed for the square lattice with the same scale for each dimension.

Hurst index

The fractal dimension appropriate for self-affine data is the Hurst index (Hurst, 1951). Suppose $x_j(i)$ is the ith data starting from point j ($x_j(i) = x_{j+i-1}$). The Hurst index H is defined as

$$S_j(\tau) = \max\left\{\sum_{i=1}^{k}(x_j(i) - \langle x_j\rangle_\tau); k = 1, \cdots, \tau\right\}$$

$$- \min\left\{\sum_{i=1}^{k}(x_j(i) - \langle x_j\rangle_\tau); k = 1, \cdots, \tau\right\}, \quad (2.158)$$

$$V_j(\tau) = \left(\frac{1}{\tau}\sum_{i=1}^{\tau}(x_j(i) - \langle x_j\rangle_\tau)^2\right)^{1/2} \quad (2.159)$$

$$H = \lim_{\tau \to \infty} \frac{\ln\langle S_j(\tau)/V_j(\tau)\rangle}{\ln(\tau)}. \quad (2.160)$$

We estimated the Hurst index H for daily return time series x_n calculated for the TOPIX closing price from 1975 to 2006 (8282 business days). The results are shown for periods $N = 4, 8, 16, 32, 64, 128, 256, 512$, and 1024 in Fig. 2.12 and Table 2.3. The normalized range varies smoothly for the whole period and we obtain the Hurst index $H = 0.5594$. The estimated index is close to $1/2$. Therefore, standard Brownian motion is a good approximation for daily return time series x_n for the TOPIX closing price.

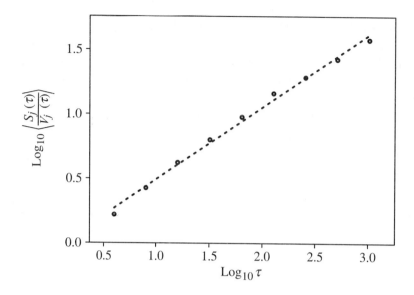

Fig. 2.12 We estimated the Hurst index H for daily return time series x_n calculated for the TOPIX closing price from 1975 to 2006 (8282 business days). The normalized range varies smoothly in the whole period and we obtain the Hurst index $H = 0.5594$. The estimated index is close to $1/2$. Therefore, standard Brownian motion is a good approximation for daily return time series x_n for the TOPIX closing price.

Table 2.3 Hurst Index H

Item	Estimation	Error	t Value	p Value
intercept	−0.0658	0.0311	−2.11	0.0723
H_H	0.559	0.0158	35.3	0.00
R^2	0.994			

2.8 Advanced Model of Time Series

2.8.1 Autoregressive fractionally integrated moving average model

An **autoregressive fractionally integrated moving average (ARFIMA) model** is obtained by augmenting the ARIMA model to make possible the use of the fractional difference d (Granger and Joyeaux, 1980). ARFIMA is a linear model that reproduces long-term memory. The dth order difference of time series x_t is written as

$$\Delta^d x_t = (1 - B)^d x_t, \tag{2.161}$$

where B is the **lag operator**. The following two operators are defined using lag operator B:

$$\phi(B) = 1 - \sum_{i=1}^{p} \phi_i B^i \qquad (2.162)$$

$$\pi(B) = 1 + \sum_{j=1}^{q} \pi_j B^j. \qquad (2.163)$$

Using these operators, the ARIMA(p,d,q) model is written as

$$\phi(B)(1-B)^d x_t = \pi(B)\xi_t. \qquad (2.164)$$

Then operator $(1-B)^d$ is rewritten using **binomial expansion**;

$$(1-B)^d = \sum_{k=0}^{d} \binom{d}{k} (-B)^k, \qquad (2.165)$$

$$\binom{d}{k} = \frac{d!}{k!(d-k)!} = \frac{\Gamma(d+1)}{\Gamma(k+1)\Gamma(d-k+1)}. \qquad (2.166)$$

The ARFIMA model extends the order of difference d to fractional from integer in Eq. (2.164). Operator $(1-B)^d$ is given by Eq. (2.165) and Eq. (2.166). Note that the fractional number is allowed for the variable in the **Gamma function** in Eq. (2.166).

The autocorrelation function $\rho(\tau)$ is given by

$$\rho(\tau) = \frac{\Gamma(\tau+d)\Gamma(1-d)}{\Gamma(\tau+1-d)\Gamma(d)}. \qquad (2.167)$$

For large τ, we have

$$\rho(\tau) = \tau^{2d-1} \frac{\Gamma(1-d)}{\Gamma(d)}. \qquad (2.168)$$

From Eq. (2.168), for $0 < d < 1/2$, the autocorrelation function $\rho(\tau)$ shows the power-law time dependence and is positive for all τ. Therefore, ARFIMA is a linear model that reproduces the long-term memory process.

We made maximal likelihood estimation of parameters in the ARFIMA$(1,2)$ model for daily return time series r_n of the NIKKEI-225 from January 1984 to August 2007 (5740 business days). Figure 2.13 shows the daily return calculated using ARFIMA$(1,2)$ with the estimated parameters:

$$\begin{aligned} \phi_1 &= -0.446, \\ \pi_1 &= -0.429, \\ \pi_2 &= 0.0649, \\ d &= 0.00150. \end{aligned} \qquad (2.169)$$

Comparison with the model calculation and the actual time series shown in Fig. 2.6d shows that volatility clustering is not well reproduced in the model calculation.

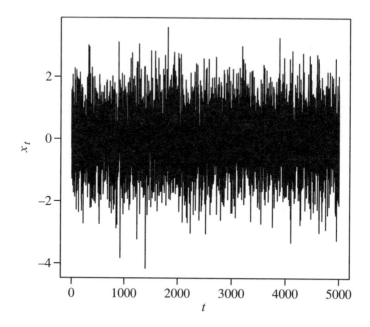

Fig. 2.13 We made maximal likelihood estimation of parameters in the ARFIMA$(1,2)$ model for daily return time series r_n of the NIKKEI-225 from January 1984 to August 2007 (5740 business days). We show the daily return calculated using ARFIMA$(1,2)$. Comparison of the model calculation and the actual time series shown in Fig. 2.6d shows that volatility clustering is not well reproduced in the model calculation.

2.8.2 ARCH model and GARCH model

The **autoregressive conditional heteroscedasticity (ARCH) model** and the **generalized autoregressive conditional heteroscedasticity (GARCH) model** are nonlinear models that reproduce volatility clustering (Bollerslev, 1986; Engle, 1982). These models were first introduced into econophysics by Mantegna and Stanley (1999b).

Return of stock x_t at time t is written as a sum of two terms;

$$x_t = E(x_t|\boldsymbol{I}_{t-1}) + \epsilon_t \tag{2.170}$$

where $E(x_t|\boldsymbol{I}_{t-1})$ is the forecast value at time $t-1$ and unpredictable fluctuation ϵ_t. $\boldsymbol{I}_{t-1}$ means the information set usable at time $t-1$. Unpredictable fluctuation ϵ_t is written using standard Brownian motion z_t as follows;

$$\epsilon_t = \sigma_t z_t. \tag{2.171}$$

In the ARCH(q) model, volatility σ_t^2 at time t is formulated as a function of variables whose values are known at time $t-1$;

$$\sigma_t^2 = \omega + \sum_{j=1}^{q} \alpha_j x_{t-j}^2, \tag{2.172}$$

where x_{t-j}^2 is a surrogate variable of volatility. The first term of the right-hand side in Eq. (2.172) is long-term volatility and the second term is fluctuation around the long-term volatility. Parameters ω, $\alpha_j (j = 1, \ldots, q)$ are all non-negative.

In the GARCH(p, q) model, volatility σ_t^2 at time t is formulated by adding volatility σ_{t-i}^2 estimated in the past period to the right-hand side of Eq. (2.172) as follows;

$$\sigma_t^2 = \omega + \sum_{j=1}^{q} \alpha_j x_{t-j}^2 + \sum_{i=1}^{p} \beta_i \sigma_{t-i}^2. \tag{2.173}$$

Parameters ω, $\alpha_j (j = 1, \ldots, q)$, $\beta_i (i = 1, \ldots, p)$ are all non-negative. Note that the ARCH(p, q) model and the GARCH(p, q) model are non-linear in terms of return x_t, but are linear for volatility x_t^2.

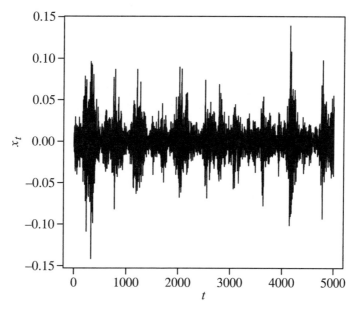

Fig. 2.14 We made a maximal likelihood estimation of parameters in the GARCH($1, 2$) model for daily return time series r_n of the NIKKEI-225 from January 1984 to August 2007 (5740 business days). Daily return calculated using the GARCH($1, 2$) model with the estimated parameters is shown. Comparison of the model calculation and the actual time series shown in Fig. 2.6d shows that the GARCH($1, 2$) model reproduces volatility clustering successfully.

We made a maximal likelihood estimation of parameters in the GARCH(1, 2) model for daily return time series r_n of the NIKKEI-225 from January 1984 to August 2007 (5740 business days). Figure 2.14 shows the daily return calculated using GARCH(1, 2) with the estimated parameters:

$$\mu = 7.439 \times 10^{-4},$$
$$\omega = 3.22 \times 10^{-6},$$
$$\alpha_1 = 1.69 \times 10^{-1}, \qquad (2.174)$$
$$\beta_1 = 4.60 \times 10^{-1},$$
$$\beta_2 = 3.67 \times 10^{-1}.$$

Comparison of the model calculation and the actual time series shown in Fig. 2.6d shows that the GARCH(1, 2) model reproduces volatility clustering successfully.

Table 2.4 Comparrison of Hurst Index H

Item	Return	Volatility
NIKKEI-225	0.559	0.694
GARCH(1,2)	0.573	0.824

Finally, we study the statistical property of long-term memory quantitatively and qualitatively. Hurst index H calculated for return and its volatility of real time series and calculated time series using the GARCH(1, 2) model are summarized in Table 2.4. This table shows that Hurst index H calculated for return of real time series and calculated time series using the GARCH(1, 2) model are close to $1/2$. Therefore, these time series are regarded as standard Brownian motion. Furthermore, the Hurst index H calculated for volatility of real time series and calculated time series using the GARCH(1, 2) model are 0.7 to 0.8. This means that volatility shows positive autocorrelation. However, the autocorrelation function of volatility shows qualitative difference between real time series and calculated time series using the GARCH(1, 2) model. Figure 2.15 shows the autocorrelation function of volatility for real time series and calculated time series using the GARCH(1, 2) model. The autocorrelation function of volatility shows the power-law time dependence for real data, and the exponential time dependence for calculated time series using the GARCH(1, 2) model. For this reason, we cannot say that the GARCH(1, 2) reproduces the long-term memory of volatility very well. The GARCH(1, 2) model reproduces volatility clustering but does not reproduce the long-term memory of volatility. Therefore, calculated time series using the GARCH(1, 2) model shows qualitative difference from real time series and does not capture the properties of real data sufficiently.

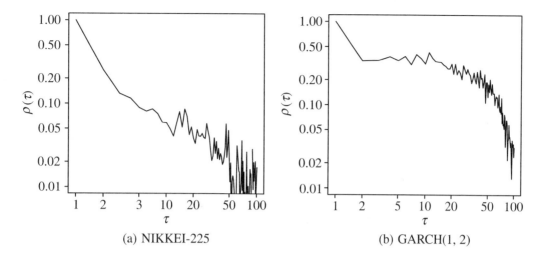

Fig. 2.15 Autocorrelation functions of volatility for real time series and calculated time series using GARCH$(1, 2)$ model are shown. The autocorrelation function of volatility shows the power-law time dependence for real data and the exponential time dependence for calculated time series using the GARCH$(1, 2)$ model. The GARCH$(1, 2)$ model reproduces volatility clustering but does not reproduce long-term memory of volatility.

2.9 Exercise

2.1 The modified Yule model with the processes of **Y1** and **Y2'** was shown to be equivalent to Simon's model. Consider the case of the original Yule model with the processes of **Y1** and **Y2**.

2.2 For the extension of Simon's model with the probability given by Eq. (2.28), write master equations, verify Eq. (2.29) and obtain the stationary distribution in Eq. (2.30).

2.3 Suppose that an ideal monatomic gas in thermal equilibrium with temperature T is drifting at a constant velocity u. According to Maxwell, the velocity distribution function $f(v)$ of the gas molecules with mass m in the drift direction is given by

$$f(v) = C \exp\left[-\frac{m(v-u)^2}{2k_\mathrm{B} T}\right], \tag{2.175}$$

where C is the normalization constant of $f(v)$. Work out a formulation in terms of entropy to derive the Maxwell distribution.

2.4 When the return of stock price dx is in accordance with the standard Brownian process, Eq. (2.140), derive the stochastic differential equation to describe the stock price $s = e^x$ using Ito's lemma, Eq. (2.146).

2.5 Two pieces ($a = 2$) of line segments are obtained by dividing a line segment into four ($r = \frac{1}{4}$) and removing every other line segment of $\frac{1}{4}$. Estimate the fractal dimension D_s using Eq. (2.155) for the figure obtained by repeating this procedure.

2.6 Rewrite the ARIMA(p, d, q) model Eq. (2.164) for x_t.

2.7 Derive the asymptotic relation, Eq. (2.168), when τ is large, from the autocorrelation function Eq. (2.167) for the ARFIMA time series.

3
Income and Firm-size Distributions

> Society is not homogeneous, and those who do not deliberately close their eyes have to recognize that men differ greatly from one another from the physical, moral, and intellectual viewpoints.
>
> Vilfredo Pareto

Distribution of income is a vital issue in every country. The stability of society crucially depends on it. Piketty (2014) in his worldwide bestseller *Capital in the Twenty-First Century* once again reminded us of the fact that income and wealth are very unevenly distributed in most advanced countries. In the late 19th century, Pareto (1897) already found that income distribution is so skewed that it is actually what we now call the **Pareto distribution**. Exploration into the mechanisms generating the observed empirical income distribution was started by Gibrat (1931) and Champernowne (1952, 1973). Following their lead, this chapter presents modern treatments and our own results.

What is said about wealth can also be said about firms: The size of firms, measured by the profit they make, the number of employees or assets, differ from firm to firm, by much more difference than people's income or wealth. For these reasons, we first present many distribution functions which were proposed to explain the distribution of income, wealth, and firm size (for details, see Kleiber and Kotz, 2003). And then we will study the properties of large values that result from power-law distributions in detail, as power-law distributions are the most common class of distributions that we observe in economic systems, both people's income and firm sizes.

One caution: in summarizing various distribution functions, the reader may be tempted to play the fitting game. One may use all the possible functions in the following section or that he/she can think of to find the best-fit to a given distribution. It is a fruitless effort beyond a certain point. Imagine that a variable x takes only seven values. Then we have seven data, the number of occurrences at those seven values of x. Anyone clever enough can think of a function with seven parameters, that fits the data *perfectly*. But it is evidently fruitless. Nothing was revealed by such a fit. Only when it is fitted with a small number of parameters, does it lead to some insight to the deep nature of the data (Aoyama and Constable, 1999). This is the issue of **information criteria**, which deals with the "quality" of the fit. **Akaike information criteria** (AIC) (Akaike, 1973, 1974a), part of this information criteria, is defined by the following:

$$\text{AIC} = -2\ell + 2k, \tag{3.1}$$

where ℓ is the maximized log-likelihood and k is the number of parameters in a statistical model. The larger the AIC, the better is the result of the fit, as it means a better fit was obtained with a smaller number of parameters. Although we shall not devote space to this matter any further, this should be kept in mind in any effort to find the best model that fits the data.

Distributions of income and firm size are known to have "fat-tails." Thus, we begin by explaining such distributions.

3.1 Distributions with Fat-Tails

We denote a **stochastic variable** as x, and a **probability density function** (PDF) as $f(x)$. A **cumulative distribution function** (CDF) $F(x)$ is defined by

$$F(x) := \int_{-\infty}^{x} f(x')\,dx', \tag{3.2}$$

and a **complementary cumulative distribution function** (CCDF) $\bar{F}(x)$ is defined by

$$\bar{F}(x) := 1 - F(x) \tag{3.3}$$

$$= \int_{x}^{+\infty} f(x')\,dx'. \tag{3.4}$$

Thus, PDF and CCDF have the relation:

$$f(x) = -\frac{d\bar{F}(x)}{dx}. \tag{3.5}$$

A **raw moment** of order n denoted by μ'_n is defined about $x = 0$ as

$$\mu'_n := \int_{-\infty}^{+\infty} x^n f(x)\,dx. \tag{3.6}$$

A **mean** $\langle x \rangle$ is defined by the first moment of Eq. (3.6):

$$\langle x \rangle := \mu'_1 = \int_{-\infty}^{+\infty} x f(x)\,dx. \tag{3.7}$$

Moments are sometimes calculated about the mean, and a **central moment** of order n denoted by μ_n is defined about $x = \langle x \rangle$ as

$$\mu_n := \int_{-\infty}^{+\infty} (x - \langle x \rangle)^n f(x) dx. \tag{3.8}$$

A **variance** σ^2 is defined by the second moment of Eq. (3.8):

$$\sigma^2 := \mu_2 = \int_{-\infty}^{+\infty} (x - \langle x \rangle)^2 f(x) dx, \tag{3.9}$$

and a **standard deviation** is the square root of the variance, i.e., σ. A **skewness** γ_1 is defined by

$$\gamma_1 := \frac{\mu_3}{\sigma^3} = \frac{\mu_3}{\mu_2^{3/2}}, \tag{3.10}$$

and a **kurtosis** γ_2 is defined by

$$\gamma_2 := \frac{\mu_4}{\sigma^4} - 3 = \frac{\mu_4}{\mu_2^2} - 3, \tag{3.11}$$

when we assume that the kurtosis of the normal distribution equals 0.

3.1.1 Pearson system

PDFs of **Pearson distributions** (Pearson, 1895) are derived from a differential equation given by

$$\frac{df(x)}{dx} = \frac{(x-a)f(x)}{c_0 + c_1 x + c_2 x^2}, \tag{3.12}$$

where a, c_0, c_1, and c_2 are constants. If we assume $1 \ll x$ and $p(x)$ remains positive and finite, we can write Eq. (3.12) as

$$\frac{d \ln f(x)}{dx} \approx \frac{1}{c_2 x}. \tag{3.13}$$

The solution of Eq. (3.13) is

$$f(x) \propto x^{1/c_2}, \tag{3.14}$$

and this is one of the **Pareto distribution** with exponent $1/c_2$. If we assume $c_2 < -1$, the corresponding CCDF is given by

$$\bar{F}(x) \propto x^{1+1/c_2}. \tag{3.15}$$

If we denote $1/c_2$ as $-\mu - 1$, then we obtain the general form of Pareto distributions:

$$f(x) \propto x^{-\mu-1}, \quad \bar{F}(x) \propto x^{-\mu}, \quad \text{for } x \to \infty. \tag{3.16}$$

In a **Pearson system**, a, c_0, c_1, and c_2 are determined up to the fourth moment of $f(x)$, as shown here. If we multiply x^n to Eq. (3.12) and integrate it with respect to x, then we obtain

$$\{c_2(n+2) + 1\} \int x^{n+1} f(x) dx + \{c_1(n+1) - a\} \int x^n f(x) dx$$

$$+ c_0 n \int x^{n-1} f(x) dx = 0, \tag{3.17}$$

when the boundary condition satisfies

$$\left(x^n c_0 + c_1 x^{n+1} + c_2 x^{n+2} \right) f(x) \Big|_{-\infty}^{+\infty} = 0. \tag{3.18}$$

Thus, Eq. (3.17) is written in terms of μ'_n:

$$\{c_2(n+2) + 1\} \mu'_{n+1} + \{c_1(n+1) - a\} \mu'_n + c_0 n \mu'_{n-1} = 0. \tag{3.19}$$

For $n = 0, \ldots, 3$, we obtain

$$(2c_2 + 1)\mu'_1 + c_1 - a = 0, \tag{3.20}$$

$$(3c_2 + 1)\mu'_2 + (2c_1 - a)\mu'_1 + c_0 = 0, \tag{3.21}$$

$$(4c_2 + 1)\mu'_3 + (3c_1 - a)\mu'_2 + 2c_0\mu'_1 = 0, \tag{3.22}$$

$$(5c_2 + 1)\mu'_4 + (4c_1 - a)\mu'_3 + 3c_0\mu'_2 = 0, \tag{3.23}$$

where we defined $\mu'_0 = 1$ and $\mu'_{-1} = 0$. These equations suggest that constants a, c_0, c_1, and c_2 are determined up to the fourth moment, and are therefore, determined by $\langle x \rangle, \sigma^2, \gamma_1$, and γ_2. For different values of a, c_0, c_1, and c_2, we obtain the Pearson distributions listed in Table 3.1.

Table 3.1 Pearson distributions (for details, see Pearson, 1895 and Kleiber and Kotz, 2003). Here we abbreviate the normalization constant of PDF.

Type	PDF	Support
Normal	$\exp\left(-\frac{1}{2}x^2\right)$	$-\infty < x < \infty$
I	$(1+x)^{m_1}(1-x)^{m_2}$	$-1 \leq x \leq 1$
II	$(1-x^2)^m$	$-1 \leq x \leq 1$
III	$x^m \exp(-x)$	$0 \leq x < \infty$
IV	$(1+x^2)^{-m}\exp[-v\tan^{-1}(x)]$	$-\infty < x < \infty$
V	$x^{-m}\exp(-1/x)$	$0 \leq x < \infty$
VI	$x^{m_2}(1+x)^{-m_1}$	$0 \leq x \leq \infty$
VII	$(1+x^2)^{-m}$	$-\infty < x < \infty$
VIII	$(1+x)^{-m}$	$0 \leq x \leq 1$
IX	$(1+x)^m$	$0 \leq x \leq 1$
X	$\exp(-x)$	$0 \leq x < \infty$
XI	x^{-m}	$1 \leq x < \infty$
XII	$[(g+x)(g-x)]^h$	$-g \leq x \leq g$

3.1.2 Burr and dagum system

CDFs $F(x)$ of **Burr distributions** (Burr, 1942) are derived from a differential equation given by

$$\frac{dy}{dx} = y(1-y)g(x,y), \qquad y = F(x). \tag{3.24}$$

Here $g(x,y)$ will be positive for $0 \leq y \leq 1$ and x in the range over which the solution will be used. If we take $y = p(x)$ and

$$g(x,y) = -\frac{(x-a)}{(1-y)(c_0 + c_1 x + c_2 x^2)}, \tag{3.25}$$

then Eq. (3.24) equals the Pearson system defined by Eq. (3.12). Burr (1942) derived 12 types of CDFs and they are listed in Table 3.2.

Table 3.2 Burr distributions (for details, see Burr, 1942 and Kleiber and Kotz, 2003).

Type	CDF	Support
I	x	$0 < x < 1$
II	$(1+e^{-x})^{-p}$	$-\infty < x < \infty$
III	$(1+x^{-a})^{-p}$	$0 < x < \infty$
IV	$\left[1+\left(\frac{c-x}{x}\right)^{1/c}\right]^{-q}$	$0 < x < c$
V	$[1+c\exp\{-\tan(x)\}]^{-q}$	$-\pi/2 < x < \pi/2$
VI	$[1+\exp\{-c\sinh(x)\}]^{-q}$	$-\infty < x < \infty$
VII	$2^{-q}\{1+\tanh(x)\}^{q}$	$-\infty < x < \infty$
VIII	$\left[\frac{2}{\pi}\tan^{-1}(e^{x})\right]^{q}$	$-\infty < x < \infty$
IX	$1 - \frac{2}{2+c[(1+e^{x})^{q}-1]}$	$-\infty < x < \infty$
X	$[1-\exp(-x^{2})]^{a}$	$0 \leq x < \infty$
XI	$\left[x-\frac{1}{2\pi}\sin(2\pi x)\right]^{q}$	$0 < x < 1$
XII	$1-(1+x^{a})^{-q}$	$0 \leq x < \infty$

The **Burr system** is divided into three categories. The differential equation of the first category is defined by

$$\frac{dy}{dx} = 1, \quad (3.26)$$

and, as easily shown, the type I Burr distribution, i.e., the uniform distribution, is derived from Eq. (3.26).

The differential equation of the second category is defined by

$$\frac{dy}{dx} = yg(x), \quad (3.27)$$

and the Burr distributions of type II, III, IV, V, VI, VII, VIII, X, and XI are derived from this equation. The general solution of Eq. (3.27) is given by

$$y = K\exp\left[\int g(x)dx\right], \quad (3.28)$$

where K is determined by the boundary conditions:

$$\lim_{x \to 0} y = 0 \quad \text{and} \quad \lim_{x \to \infty} y = 1, \qquad (3.29)$$

if y is defined in the range $0 \leq x < \infty$. For example, if $g(x)$ is given by

$$g(x) = \frac{\mu e^{-x}}{1 + e^{-x}}, \qquad (3.30)$$

then the type II Burr distribution is derived.

The differential equation of the third category is defined by

$$\frac{dy}{dx} = (1 - y)g(x), \qquad (3.31)$$

and the type IX and XII Burr distributions are derived from Eq. (3.31). The general solution of Eq. (3.31) is given by

$$y = 1 - K \exp\left[-\int g(x)dx\right], \qquad (3.32)$$

where K is determined by the boundary conditions:

$$\lim_{x \to -\infty} y = 0 \quad \text{and} \quad \lim_{x \to \infty} y = 1, \qquad (3.33)$$

if y is defined in the range $-\infty < x < \infty$. For example, if $g(x)$ is given by

$$g(x) = \frac{cq(1 + e^x)^{k-1} e^x}{c[(1 + e^x)^k - 1] + 2}, \qquad (3.34)$$

then

$$y = 1 - \frac{K}{2 + c[(1 + e^x)^q - 1]}. \qquad (3.35)$$

From the normality condition given by Eq. (3.33), we obtain $K = 2$. Thus, the type IX Burr distribution is derived.

In general, an income elasticity $\epsilon(F(x), x)$ of CDF is a decreasing and concave function of $F(x)$. Based on this observation, Dagum (1977, 1980, 1990, 2008) proposed the **Dagum system** defined by

$$\epsilon(y - \alpha, x) = \frac{d \ln(y - \alpha)}{d \ln(x)} = \psi(x)\phi(y) \leq k, \qquad y = F(x) \qquad (3.36)$$

where parameters satisfy

$$0 < k, \quad 0 \leq x_0 < x < \infty, \quad \alpha < 1, \quad 0 < \psi(x), \quad 0 < \phi(y),$$

$$\frac{d\psi(x)\phi(y)}{dy} < 0, \quad \text{hence} \quad \frac{d\psi(x)\phi(y)}{dx} < 0. \qquad (3.37)$$

Dagum (1980) found Eq. (3.36) for some cases, and obtained the CDFs listed in Table 3.3.

Table 3.3 Dagum distributions (for details, see Dagum (1977, 1980, 1990)).

Income Distribution	$\psi(x)$	$\phi(y)$	CDF	Support
Pareto Type I (1895)	δ	$(1-y)/y$	$1-\left(\frac{x}{x_0}\right)^{-\delta}$	$0 < x_0 \leq x < \infty, 1 < \delta$
Pareto Type II (1896)	$\frac{\delta x}{x-c}$	$(1-y)/y$	$1-\left(\frac{x-c}{x_0-c}\right)^{-\delta}$	$c < x_0 \leq x < \infty, 1 < \delta$
Pareto Type III (1896)	$\beta x + \frac{\delta x}{x-c}$	$(1-y)/y$	$1-\left(\frac{x-c}{x_0-c}\right)^{-\delta}\exp(-\beta x)$	$c < x_0 \leq x < \infty, 0 < (\beta, \delta)$
Benini (1906)	$2\delta \ln(x)$	$(1-y)/y$	$1-\left(\frac{x}{x_0}\right)^{-\delta \ln(x/x_0)}$	$0 < x_0 \leq x < \infty, 0 < \delta$
Weibull (1939)	$\beta x(x-c)^{\delta-1}$	$(1-y)/y$	$1-\exp\left[-\frac{(x-x_0)^\delta}{x_0}\right]$	$0 < x_0 \leq x < \infty, 0 < \delta$
Loglogistic (Fisk, 1961)	δ	$1-y$	$(1+\lambda x^{-\delta})^{-1}$	$0 \leq x < \infty, 1 < \delta, 0 < \lambda$
Singh-Maddala (1976)	δ	$\frac{1-(1-y)^\beta}{y(1-y)^{-1}}$	$1-(1+\lambda x^\delta)^{-\beta}$	$0 \leq x < \infty, 0 < (\beta, \lambda, \delta), 1 < \beta\delta$
Log-Gompertz (Dagum, 1980)	$-\ln(\delta)$	$-\ln(y)$	$\exp(-\lambda x^{-\delta})$	$0 \leq x < \infty, 1 < \delta, 0 < \lambda$
Dagum Type I (1977)	δ	$1-y^{1/\beta}$	$(1+\lambda x^{-\delta})^{-\beta}$	$0 \leq x < \infty, 0 < (\beta, \lambda), 1 < \delta$
Dagum Type II (1977)	δ	$1-\left(\frac{y-\alpha}{1-\alpha}\right)^{1/\beta}$	$\alpha+(1-\alpha)(1+\lambda x^{-\delta})^{-\beta}$	$0 \leq x < \infty, 0 < (\beta, \lambda), 1 < \delta, 0 < \alpha < 1$
Dagum Type III (1980)	δ	$1-\left(\frac{y-\alpha}{1-\alpha}\right)^{1/\beta}$	$\alpha+(1-\alpha)(1+\lambda x^{-\delta})^{-\beta}$	$0 < x_0 \leq x < \infty, 0 < (\beta, \lambda), 1 < \delta, \alpha < 0$

The Dagum system defined by Eq. (3.36) is reduced to

$$\frac{dy}{dx} = \frac{\psi(x)}{x}(y-\alpha)\phi(y), \tag{3.38}$$

so, if we take

$$g(x,y) = \frac{xy(1-y)}{(y-\alpha)}\psi(x)\phi(y), \tag{3.39}$$

then Eq. (3.24) equals Eq. (3.38). Hence, the Burr system includes the Dagum system.

3.1.3 The Beta-type distributions

Many distributions have been proposed to explain the distribution of income, wealth, and firm sizes. However, almost all of them belong to the **beta-type distributions**. Here, we summarize the hierarchical structure of the beta-type distributions. The beta-type distribution tree is depicted in Fig. 3.1 (McDonald and Xu, 1995). Now, we will explain some distribution functions regarding income, wealth, and firm sizes.

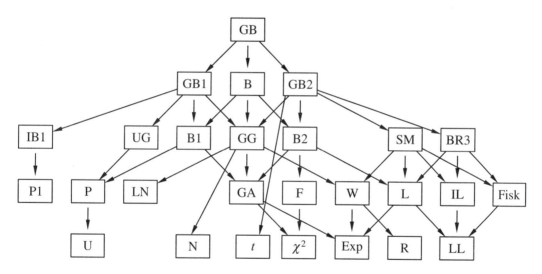

Fig. 3.1 Beta-type distribution tree. The abbreviations stand for generalized beta (GB), generalized beta of the first kind (GB1), beta (B), generalized beta of the second kind (GB2), inverse beta of the first kind (IB1), unit gamma (UG), beta of the first kind (B1), generalized gamma (GG), beta of the first kind (B2), Singh–Maddala (SM), Burr III (BR3), type I Pareto (P1), Power (P), log-normal (LN), gamma (GA), F statistics (F), Weibull (W), the first Lomax (L), inverse Lomax (IL), Fisk, uniform (U), Half normal (N), Student's t (t), χ^2, exponential (Exp), Rayleigh (R), log-logistic (LL).

Five-parameter distribution

The PDF of a generalized beta distribution (GB) is defined by

$$f_{\text{GB}}(x;a,b,c,\mu,\nu) = \frac{|a|x^{a\mu-1}}{b^{a\mu}B(\mu,\nu)}\left[1-(1-c)\left(\frac{x}{b}\right)^a\right]^{\nu-1}\left[1+c\left(\frac{x}{b}\right)^a\right]^{-(\mu+\nu)}, \quad (3.40)$$

for $0 < x^a < b^a/(1-c)$ and 0 otherwise with $0 \leq c \leq 1$, and b, μ and ν positive. Here, $B(\mu,\nu)$ is the beta function defined by Eq. (2.6). The parameter a affects the peakedness of the PDF, b is a scale parameter, and μ and ν control the shape and skewness. As explained here, the parameter c determines the mixing ratio of GB1 ($c = 0$) and GB2 ($c = 1$). GB was introduced by McDonald and Xu (1995) to give a unified framework to beta-type distributions. McDonald and Xu (1995) applied GB to 1985 nominal family income obtained from the Census Population Report. This data contains 63,558 sample families combined into 21 income groups. McDonald and Xu (1995) suggested that the data was well explained by GB and GB2. Bordley et al. (1997) applied 15 distributions to five sets of US family income data for 1970, 1975, 1980, 1985, and 1990 and suggested that the best fit by GB has $c \approx 1$, i.e., GB2, except for 1970. This means that GB chose GB2 in the data investigated, except for one data point. Bandourian et al. (2002) applied GB to income data for 23 countries and various years, and suggested that the Weibull, Gabum, and generalized beta distribution of the second kind were the best fitting of the two-, three-, and four-parameter distributions respectively.

Four-parameter distributions

The PDF of the generalized beta distribution of the first kind (GB1) is given by Eq. (3.40) with $c = 0$:

$$f_{\text{GB1}}(x;a,b,\mu,\nu) = f_{\text{GB}}(x;a,b,c=0,\mu,\nu) \quad (3.41)$$

$$= \frac{|a|x^{a\mu-1}}{b^{a\mu}B(\mu,\nu)}\left[1-\left(\frac{x}{b}\right)^a\right]^{\nu-1}, \quad (3.42)$$

for $0 < x^a < b^a$ and zero otherwise. Here, the parameters b, μ, and ν are positive. McDonald (1984) was the first to introduce GB1. It was also applied to the distribution of income by McDonald and Xu (1995); Bordley et al. (1997); and Bandourian et al. (2002). Its explanatory power was lower than that of any other four-parameter distributions.

The PDF of *beta* distribution (B) is given by Eq. (3.40) with $a = 1$:

$$f_{\text{B}}(x;b,c,\mu,\nu) = f_{\text{GB}}(x;a=1,b,c,\mu,\nu) \quad (3.43)$$

$$= \frac{x^{\mu-1}}{b^{\mu}B(\mu,\nu)}\left[1-(1-c)\left(\frac{x}{b}\right)\right]^{\nu-1}\left[1+c\left(\frac{x}{b}\right)\right]^{-(\mu+\nu)}, \quad (3.44)$$

for $0 < x < b/(1-c)$ and zero otherwise with $0 \le c \le 1$, and b, μ and ν positive. If we take $b = 1$ in Eq. (3.44), then we obtain

$$f(x; c, \mu, \nu) = \frac{x^{\mu-1}}{B(\mu, \nu)} \left[1 - (1-c)x\right]^{\nu-1} (1+cx)^{-(\mu+\nu)}, \tag{3.45}$$

referred to as the *standard form* of the beta distribution. B was introduced by McDonald and Xu (1995). As explained earlier, Bordley et al. (1997) applied 15 distributions to US family income data for each of the five years from 1970 to 1990, and suggested that the best fit by B has $c \approx 1$, i.e., GB1, for each of five years. This means that B chose GB1 for the data investigated.

The PDF of generalized beta distributions of the second kind (GB2) is given by Eq. (3.40) with $c = 1$:

$$f_{\text{GB2}}(x; a, b, \mu, \nu) = f_{\text{GB}}(x; a, b, c = 1, \mu, \nu) \tag{3.46}$$

$$= \frac{ax^{a\mu-1}}{b^{a\mu}B(\mu, \nu)} \left[1 + \left(\frac{x}{b}\right)^a\right]^{-(\mu+\nu)}, \tag{3.47}$$

for $0 < x < \infty$ and zero otherwise with a, b, μ and ν positive. McDonald (1984) was the first to introduce GB2. It was also applied to the distribution of income by Butler and McDonald (1989); Majumder and Chakravarty (1990); McDonald and Mantrala (1995); Bordley et al. (1997); Parker (1997); and and Bandourian et al. (2002). It seems GB2 was the best fitting among four-parameter distributions.

Three-parameter distributions

The PDF of an inverse beta distribution of the first kind (IB1) is given by Eq. (3.42) with $a = -1$:

$$f_{\text{IB1}}(x; b, \mu, \nu) = f_{\text{GB1}}(x; a = -1, b, \mu, \nu) \tag{3.48}$$

$$= \frac{x^{-\mu-1}}{b^{-\mu}B(\mu, \nu)} \left[1 - \left(\frac{x}{b}\right)^{-1}\right]^{\nu-1}, \tag{3.49}$$

for $0 < b < x$ and zero otherwise. Here, the parameters $b, \mu,$ and ν are positive. Although IB1 appears in McDonald and Mantrala (1995), there seems to have been no application to the distribution of income, wealth, and firm sizes.

The PDF of a unit gamma distribution (UG) is given by

$$f_{\text{UG}}(x; b, \delta, q) = \lim_{a \to 0} f_{\text{GB1}}(x; a, b, \mu = \delta/a, \nu) \tag{3.50}$$

$$= \frac{\delta^\nu x^{\delta-1}}{b^\delta \Gamma(\nu)} \left[\log(b/x)\right]^{\nu-1}, \tag{3.51}$$

for $0 < x < b$ and zero otherwise. Here the parameters b, μ, and ν are positive. Again, UG appears in McDonald and Mantrala (1995), but there seems to have been no application to the distribution of income, wealth, and firm sizes.

The PDF of a beta distribution of the first kind (B1) is given by Eq. (3.42) with $a = 1$, or Eq. (3.44) with $c = 0$:

$$f_{B1}(x; b, \mu, \nu) = f_{GB1}(x; a = 1, b, \mu, \nu) \tag{3.52}$$

$$= f_B(x; b, c = 0, \mu, \nu) \tag{3.53}$$

$$= \frac{x^{\mu-1}}{b^\mu B(\mu, \nu)} \left[1 - \left(\frac{x}{b}\right)\right]^{\nu-1}, \tag{3.54}$$

for $0 < x < b$ and zero otherwise with b, μ and ν positive. B1 is the same as the type I Pearson distribution listed in Table 3.1. B1 is applied to the distribution of income by McDonald (1984); Butler and McDonald (1989); Bordley et al. (1997); Parker (1997); and Bandourian et al. (2002). Its explanatory power is lower than that of any other three-parameter distributions.

If we take $b = \nu^{1/a}\beta$ and take the limit $\nu \to \infty$ of Eq. (3.42), Eq. (3.47), and Eq. (3.44), we obtain the PDF of a generalized gamma distribution (GG):

$$f_{GG}(x; a, \beta, \mu) = \lim_{\nu \to \infty} f_{BG1}(x; a, b = \nu^{1/a}\beta, \mu, \nu) \tag{3.55}$$

$$= \lim_{\nu \to \infty} f_{BG2}(x; a, b = \nu^{1/a}\beta, \mu, \nu) \tag{3.56}$$

$$= \lim_{\nu \to \infty} f_B(x; b = \nu^{1/a}\beta, \forall c, \mu, \nu) \tag{3.57}$$

$$= \frac{ax^{a\mu-1}}{\beta^{a\mu}\Gamma(\mu)} \exp\left[-\left(\frac{x}{\beta}\right)^a\right], \tag{3.58}$$

for $0 \leq x$ and zero otherwise with a, β and μ positive. GG was first introduced by Amoroso (1925) and applied to income distribution in 1912 Prussia. GG was also applied to the distribution of income by McDonald (1984); Atoda et al. (1988); Butler and McDonald (1989); McDonald and Mantrala (1995); Bordley et al. (1997); Parker (1997); and Bandourian et al. (2002). Its explanatory power is lower than that of any other three-parameter distribution.

The PDF of beta distributions of the second kind (B2) is given by Eq. (3.47) with $a = 1$ or Eq. (3.44) with $c = 1$:

$$f_{B2}(x; b, \mu, \nu) = f_{GB2}(x; a = 1, b, \mu, \nu) \tag{3.59}$$

$$= f_\text{B}(x; b, c = 1, \mu, \nu) \tag{3.60}$$

$$= \frac{x^{\mu-1}}{b^\mu B(\mu,\nu)} \left[1 + \left(\frac{x}{b}\right)\right]^{-(\mu+\nu)}, \tag{3.61}$$

for $0 \leq x$ and zero otherwise with b, μ, and ν positive. B2 is the same as the type VI Pearson distribution listed in Table 3.1. Again, B2 is applied to the distribution of income by McDonald (1984); Butler and McDonald (1989); Bordley et al. (1997); Parker (1997); and Bandourian et al. (2002). Its explanatory power is lower than that of any other three-parameter distribution.

The PDF of the Singh–Maddala distribution (SM) is given by Eq. (3.47) with $\mu = 1$:

$$f_\text{SM}(x; a, b, \nu) = f_{\text{B}2}(x; a, b, \mu = 1, \nu) \tag{3.62}$$

$$= \frac{a\nu x^{a-1}}{b^a} \left[1 + \left(\frac{x}{b}\right)^a\right]^{-(1+\nu)}, \tag{3.63}$$

for $0 < x < \infty$ and zero otherwise with a, b, and ν positive. Although SM was introduced by Singh and Maddala (1976) to consider the hazard rate of income, it was repeatedly rediscovered. Burr (1942) was the first to discover the same distribution function as SM; however, it was known as the Burr XII distribution and simply called the Burr distribution as listed in Table 3.2. SM is also known as the type IV Pareto distribution (Arnold, 2015) or the generalized log-logistic distribution. It was applied to the distribution of income by Suruga (1982); McDonald (1984); Butler and McDonald (1989); McDonald and Mantrala (1995); McDonald and Xu (1995); Bordley et al. (1997); Parker (1997); Tachibanaki et al. (1997); and Bandourian et al. (2002). Its explanatory power is higher than that of any other three-parameter distributions, and is comparable to the Burr III (DR3) distribution.

The PDF of the Burr III distribution (BR3) is given by Eq. (3.47) with $\nu = 1$:

$$f_\text{BR3}(x; a, b, \mu) = f_\text{GB2}(x; a, b, \mu, \nu = 1) \tag{3.64}$$

$$= \frac{a\mu x^{a\mu-1}}{b^{a\mu}} \left[1 + \left(\frac{x}{b}\right)^a\right]^{-(\mu+1)}, \tag{3.65}$$

for $0 < x < \infty$ and zero otherwise with a, b, and μ positive. BR3 was introduced by Burr (1942), and is also known as the Dagum distribution (Dagum, 1977) or the generalized log-logistic distribution. BR3 was applied to the distribution of income by McDonald and Mantrala (1995); McDonald and Xu (1995); Bordley et al. (1997); and Bandourian et al. (2002). Its explanatory power is higher than that of any other three-parameter distribution, and is comparable to the Singh–Maddala distribution.

Two-parameter distributions

The PDF of the type I Pareto distribution (P1) is given by Eq. (3.49) with $q = 1$:

$$f_{\text{P1}}(x; b, \mu) = f_{\text{IB1}}(x; b, \mu, \nu = 1) \tag{3.66}$$

$$= \frac{\mu b^\mu}{x^{\mu+1}}, \tag{3.67}$$

for $b < x$ and zero otherwise with b and μ positive. Here, b is called the Pareto scale, and $\mu > 0$, the Pareto index. This distribution was first introduced by Pareto (1895). A short history of the Pareto distribution is given in Section 3.3.1. Pareto distributions were applied to income distributions by Pareto (1895), Pareto (1897), Macgregor (1936), Johnson (1937), Lydall (1959), Harrison (1981), Ransom and Cramer (1983), Creedy (1977), Aoyama et al. (2000), Souma (2001), Souma (2002), Fujiwara et al. (2003), etc. It is considered best applicable to the tail of the income distribution, i.e., the high-income range.

The PDF of a power distribution (P) is given by Eq. (3.51) with $q = 1$:

$$f_{\text{P}}(x; b, \delta) = f_{\text{UG}}(x; b, \delta, q = 1) \tag{3.68}$$

$$= \frac{\delta x^{\delta-1}}{b^\delta}, \tag{3.69}$$

for $0 < x < b$ and zero otherwise with b and δ positive. Almost all the income distributions have a fat-tail in the range of large x, and a power-law exponent greater than one. Thus, it is impossible to apply Eq. (3.69) to personal income distribution. Therefore, it appears there is no application of the power distribution to income distribution.

If we take $\beta^a = \sigma^2 a^2$ and $\mu = (a\langle x \rangle + 1)/\beta^a$, and take the limit $a \to 0$ for Eq. (3.58), we obtain the PDF of the log-normal distribution (LN):

$$f_{\text{LN}}(x; \langle x \rangle, \sigma) = \lim_{a \to 0} f_{\text{GG}}(x; a, \beta^a = \sigma^2 a^2, \mu = (a\langle x \rangle + 1)/\beta^a) \tag{3.70}$$

$$= \frac{1}{\sqrt{2\pi}\sigma x} \exp\left[-\frac{(\ln x - \langle x \rangle)^2}{2\sigma^2}\right], \tag{3.71}$$

for $0 < x < \infty$ and zero otherwise with $\langle x \rangle$ and σ positive. As we will explain in Section 3.3.1, LN was first suggested by Galton (1879) and mathematically formulated by McAlister (1879) (for details, see Aitchison and Brown (1957) and Johnson et al. (1995)). However, Gibrat (1931) is well known. The PDF is given by Eq. (3.71) and is correctly called the two-parameter log-normal distribution. There are three-parameter log-normal distributions (LN3) defined by

$$f_{\text{LN3}} = \frac{1}{\sqrt{2\pi}\sigma(x-b)} \exp\left\{-\frac{1}{2\sigma^2}[\ln(x-b) - \langle x \rangle]^2\right\}, \tag{3.72}$$

for $0 < b < x$ and zero otherwise with b, $\langle x \rangle$, and μ positive. LN3 suggests that $z = \ln(x - b)$ follows a normal distribution. For example, LN was applied to the distribution of income by Gibrat (1931); Kalecki (1945); Champernowne (1952); Steyn (1959, 1966); Lydall (1968); Gastwirth (1972); Kloek and van Dijk (1978); Harrison (1979); McDonald and Ransom (1979); Suruga (1982); Ransom and Cramer (1983); Kmietowicz (1984); McDonald (1984); Atoda et al. (1988); Kmietowicz and Ding (1993); McDonald and Mantrala (1995); McDonald and Xu (1995); Bordley et al. (1997); Tachibanaki et al. (1997); Souma (2001); Bandourian et al. (2002), etc. LN is considered applicable to the middle part of the income distribution.

The PDF of a gamma distribution (GA) is given by Eq. (3.58) with $a = 1$:

$$f_{\text{GA}}(x; \beta, \mu) = f_{\text{GG}}(x; a = 1, \beta, \mu) \tag{3.73}$$

$$= \frac{x^{\mu-1}}{\beta^\mu \Gamma(\mu)} \exp\left[-\left(\frac{x}{\beta}\right)\right], \tag{3.74}$$

for $0 < x$ and zero otherwise with β and μ positive. This is also known as the type III Pearson distribution listed in Table 3.1. GA was applied to the distribution of income and wealth, for example, by March (1898); Salem and Mount (1974); Kloek and van Dijk (1978); Suruga (1982); Ransom and Cramer (1983); McDonald (1984); Atoda et al. (1988); Bordley and McDonald (1993); Victoria-Feser and Ronchetti (1994); Bordley et al. (1997); Tachibanaki et al. (1997); Bandourian et al. (2002), etc. Its explanatory power is relatively better than that of any other two-parameter distributions.

The PDF of a Weibull distribution (W) is given by Eq. (3.58) with $p = 1$:

$$f_{\text{W}}(x; a, \beta) = f_{\text{GG}}(x; a, \beta, \mu = 1) \tag{3.75}$$

$$= \frac{a}{\beta}\left(\frac{x}{\beta}\right)^{a-1} \exp\left[-\left(\frac{x}{\beta}\right)^a\right], \tag{3.76}$$

for $0 \leq x < \infty$ and zero otherwise with a and β positive. This distribution was suggested by Weibull (1939a,b), and applied to the distribution of income by Bartels (1977); McDonald (1984); Atoda et al. (1988); Bordley et al. (1997); Tachibanaki et al. (1997); Bandourian et al. (2002), etc. Its explanatory power is relatively better than that of any other two-parameter distributions.

The PDF of the first Lomax distribution (L) is given by Eq. (3.61) with $\mu = 1$, Eq. (3.63) with $a = 1$, or Eq. (3.65) with $a = -1$:

$$f_{\text{L}}(x; b, \nu) = f_{\text{B2}}(x; b, \mu = 1, \nu) \tag{3.77}$$

$$= f_{\text{SM}}(x; a = 1, b, \nu) \tag{3.78}$$

$$= f_{\text{BR3}}(x; a = -1, b, \mu \to \nu) \tag{3.79}$$

$$= \frac{\nu}{b}\left(1+\frac{x}{b}\right)^{-(1+\nu)}, \tag{3.80}$$

for $0 \leq x < \infty$ and zero otherwise with b and ν positive. This distribution was proposed by Lomax (1954) and is also known as the type II Pareto distribution.

The PDF of a Fisk distribution is given by Eq. (3.63) with $\nu = 1$ or Eq. (3.65) with $\mu = 1$:

$$f_{\text{Fisk}}(x; a, b) = f_{\text{SM}}(x; a, b, \nu = 1) \tag{3.81}$$

$$= f_{\text{BR3}}(x; a, b, \mu = 1) \tag{3.82}$$

$$= \frac{ax^{a-1}}{b^a}\left[1 + \left(\frac{x}{b}\right)^a\right]^{-2}, \tag{3.83}$$

for $0 \leq x < \infty$ and zero otherwise with a and b positive. Fisk (1961a,b) proposed this distribution to investigate weekly earnings in agriculture in 1955–1956 England and Wales and 1954 U.S. income. It was applied to the distribution of personal income by Gastwirth and Smith (1972); Harrison (1979); Suruga (1982); McDonald (1984); Atoda et al. (1988); Tachibanaki et al. (1997), etc.

One-parameter distributions

Although seven distribution functions with one parameter are shown in Fig. 3.1, we only explain the exponential distribution (Exp). This is because, in one-parameter distributions, only the exponential distribution has been applied to the distribution of income, wealth, and firm sizes. The PDF of Exp is given by Eq. (3.74) with $\mu = 1$ or Eq. (3.76) with $a = 1$:

$$f_{\text{Exp}}(x; \beta) = f_{\text{GA}}(x; \beta, \mu = 1) \tag{3.84}$$

$$= f_{\text{W}}(x; a = 1, \beta) \tag{3.85}$$

$$= \frac{1}{\beta}\exp\left[-\left(\frac{x}{\beta}\right)\right], \tag{3.86}$$

for $0 \leq x < \infty$ and zero otherwise. The exponential distribution is also called the exponential Boltzmann–Gibbs distribution in physics. By using tax and census data, Drăgulescu and Yakovenko (2001a) showed that the distribution of individual income in the U.S. is exponential. By investigating the data on wealth and income in the United Kingdom and the data on income in the individual states of the U.S., Drăgulescu and Yakovenko (2001b) showed that the great majority of a population, i.e. those in the low and middle ranges of income, is described by an exponential distribution, whereas the high-end tail follows the power-law.

Income and Firm-size Distributions

3.2 Large Values and their Shares

In this section, we will examine the properties of the largest values among N values that obey the power-law distribution for a stochastic variable $x \in [0, \infty]$. This follows the PDF $f(x)$ and the complementary CDF $\bar{F}(x)$ that has the property Eq. (3.16). We shall find that when $N \gg 1$, their distribution and share depend on the asymptotic behavior of the Pareto distribution, namely, the power exponent.

This topic is related to the discussion of whether events of great magnitude such as economic crisis, or the population of a capital, belong to a power distribution or are an isolated outsider.[1] The point of discussion here is more on the view of inequality resulting from the power law distribution, as it will become clear toward the end of this section.

3.2.1 Maximum value and its share

The probability density $f_N^{(\max)}(x)$ of the maximum value $X_N^{(\max)}$ is given by the following.

$$f_N^{(\max)}(x) = N f(x)(1 - \bar{F}(x))^{N-1}. \tag{3.87}$$

This is derived by multiplying the number of cases of choosing 1 among N, which is N, and the probability $f(x)dx$ that it is between x and $x + dx$, and the probability that all the $N - 1$ other values are less than x, which is $(1 - \bar{F}(x))^{N-1}$. The following is satisfied by Eq. (3.87).

$$f_N^{(\max)}(0) = 0, \tag{3.88}$$

which should hold from the definition of $f_N^{(\max)}(x)$.

The CDF $\bar{F}_N^{(\max)}(x)$ of $X_N^{(\max)}$ is given by the following:

$$\bar{F}_N^{(\max)}(x) = 1 - (1 - \bar{F}(x))^N. \tag{3.89}$$

From this, we find that $\bar{F}_N^{(\max)}(0) = 1$, confirming that the pdf, Eq. (3.87) is normalized. Also, the following holds:

$$f_1^{(\max)}(x) = p(x). \tag{3.90}$$

Equation (3.87) is also obtained by first ordering N values in decreasing order, $x, x_2, x_3, \ldots, x_N$ and integrating over x_2 and others as follows:

$$N! f(x)dx \int_0^x f(x_2)dx_2 \int_0^{x_2} f(x_3)dx_3 \cdots \int_0^{x_{N-1}} f(x_N)dx_N$$

$$= N! f(x)dx \frac{1}{(N-1)!} \left(\int_0^x f(x')dx'\right)^{N-1} = f_N^{(\max)}(x)\,dx. \tag{3.91}$$

---

[1] We refer readers interested in this subject to Pisarenko and Sornette (2012) and other articles contained in this volume.

where we used the following relation.

$$G_N(x) := \int_0^x f(x_2)dx_2 \int_0^{x_2} f(x_3)dx_3 \cdots \int_0^{x_{N-1}} f(x_N)dx_N$$

$$= \frac{1}{(N-1)!}\left(\int_0^x f(x')dx'\right)^{N-1}. \tag{3.92}$$

This holds because the number of configurations of $N-1$ integration variables is $(N-1)!$. The proof of this by induction is part of the exercise.

The expectation value of the largest value is,

$$\langle X_N^{(\max)}\rangle = \int_0^\infty x f_N^{(\max)}(x)\,dx = N\int_0^\infty xf(x)(1-\bar{F}(x))^{N-1}dx. \tag{3.93}$$

The followings are some of points that are easily derived.

1. Equation (3.93) satisfies

$$\langle X_1^{(\max)}\rangle = \langle X\rangle \tag{3.94}$$

   as it should, and it diverges for $\mu \leq 1$ for any N much like $\langle X\rangle$.

2. The expectation value of the sum of all of them is,

$$\langle \sum_{i=1}^N X_i\rangle = N\langle X\rangle. \tag{3.95}$$

   Comparing Eqs. (3.93) and (3.95), we find that the latter is smaller than the former by a factor $(1-\bar{F}(x))^{N-1}$ in the integration.

$$\langle X_N^{(\max)}\rangle \leq \langle \sum_{i=1}^N X_i\rangle. \tag{3.96}$$

   The equality holds if and only if $\bar{F}(x)$ is zero everywhere, which means that X takes only the value 0. This, too, is as it should be.

Let us now examine how the largest value behaves when N is very large. The CDF of the largest value, Eq. (3.89) has the factor $(1-\bar{F}(x))$, which is zero for $x=0$ and approaches 1 for $x \to \infty$. Therefore, the factor $(1-\bar{F}(x))^{N-1}$ is finite only for $x \gg 1$ when $N \gg 1$. Because of this, the asymptotic behavior of the integration in Eq. (3.93) for $N \gg 1$ is determined only by the behavior Eq. (3.16) of $f(x)$ for $x \gg 1$. This is the key point of the discussion, because this property guarantees that the dominant behavior of the distribution of the largest value for $N \gg 1$ is identical *for all power-law distributions*. In fact, the following holds asymptotically:

$$\bar{F}_N^{(\max)}(x) = (1-\bar{F}(x))^N \simeq (1-x^{-\mu})^N \simeq e^{-Nx^{-\mu}}. \tag{3.97}$$

Therefore, the pdf behaves as

$$f^{(\max)}(x) = -\frac{d}{dx} \bar{F}_N^{(\max)}(x) \simeq N\mu\, x^{\mu-1}\, e^{-Nx^{-\mu}}. \qquad (3.98)$$

This is called the Fréchet distribution.

Changing the integration variable from x to a scaled variable $\tilde{x} = N^{-1/\mu}x$, we obtain the CDF and the PDF as follows:

$$\bar{F}_N^{(\max)}(\tilde{x}) \simeq e^{-\tilde{x}^{-\mu}}, \qquad (3.99)$$

$$f^{(\max)}(\tilde{x}) \simeq \mu\,\tilde{x}^{-\mu-1}\, e^{-\tilde{x}^{-\mu}}. \qquad (3.100)$$

By using this variable $\tilde{x}$, N disappears and the main contribution to the pdf comes from $\tilde{x} \simeq O(1)$. This means that the scale transformation from x to $\tilde{x}$ correctly picks up the main contribution for $N \to \infty$: Figure 3.2 shows the plot of $f_N^{(\max)}(\tilde{x})$.

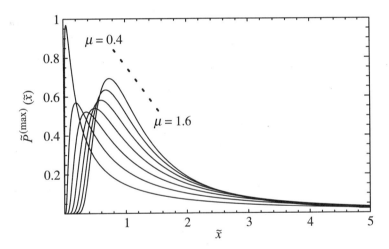

Fig. 3.2 The behavior of $f_N^{(\max)}(\tilde{x})$, where the curves are for $\mu = 0.4, 0.6, 0.8, 1.0, 1.2, 1.4, 1.6$ from left to right.

$\langle X_N^{(\max)} \rangle$ is calculated for $N \gg 1$ as follows:

$$\langle X_N^{(\max)} \rangle = \int_0^\infty x\, f_N^{(\max)}(x)\, dx = \int_0^\infty N^{1/\mu}\tilde{x},\, f_N^{(\max)}(\tilde{x})\, d\tilde{x}$$

$$\simeq N^{1/\mu} \int_0^\infty z^{-1/\mu}\, e^{-z}\, dz = N^{1/\mu}\, \Gamma\left(1 - \frac{1}{\mu}\right), \qquad (3.101)$$

where the integration variable z is defined by

$$z = \tilde{x}^{-\mu}. \qquad (3.102)$$

Equation (3.101) diverges for $\mu \to 1^+$, in agreement with the divergence of the expression (3.93) of $\langle X_N^{(\mathrm{max})} \rangle$.

Some power-law distributions yield analytical results. For example, the Pareto type I distribution in Table 3.3 yields,

$$\langle X_N^{(\mathrm{max})} \rangle = N B\left(N, 1 - \frac{1}{\mu}\right). \tag{3.103}$$

Equation (3.101) agrees with the large N behavior (the dominant term of the expansion around $n = \infty$) of Eq. (3.103). This is a useful observation, because $\langle X_N^{(\mathrm{max})} \rangle$ is determined only by the asymptotic behavior of $\bar{F}(x)$ for $x \to \infty$, and for any example where the exact result is obtained, its asymptotic behavior applies to any other distribution.

Next, let us discuss the share of the largest value;

$$R_N^{(\mathrm{max})} := \frac{X_N^{(\mathrm{max})}}{\sum_{i=1}^N X_i}. \tag{3.104}$$

We first express $\langle R_N^{(\mathrm{max})} \rangle$ as follows, like Eq. (3.91):

$$\langle R_N^{(\mathrm{max})} \rangle = N \int_0^\infty f(x) dx \prod_{i=2}^N \int_0^x f(x_i) dx_i \, \frac{x}{x + \sum_{i=2}^N x_i}$$

$$= N \int_0^\infty f(x) dx \prod_{i=2}^N \int_0^x f(x_i) dx_i \int_0^\infty x\, e^{-(x+\sum_{i=2}^N x_i)t} dt$$

$$= N \int_0^\infty dt \int_0^\infty x\, e^{-xt} f(x) dx \left(\int_0^x e^{-x't} f(x') dx'\right)^{N-1}. \tag{3.105}$$

Evidently,

$$\langle R_1^{(\mathrm{max})} \rangle = 1 \tag{3.106}$$

holds.

We express the integrand of the t integral of Eq. (3.105) as $H_N(t)$;

$$\langle R_N^{(\mathrm{max})} \rangle = \int_0^\infty H_N(t) dt, \tag{3.107}$$

$$H_N(t) := N \int_0^\infty x\, e^{-xt} f(x) dx \left(\int_0^x e^{-x't} f(x') dx'\right)^{N-1}. \tag{3.108}$$

As is easily seen, $H_N(t)$ is a positive-definite monotonically decreasing function and has the following property:

$$H_N(0) = \langle X_N^{(\mathrm{max})} \rangle, \quad H_N(\infty) = 0. \tag{3.109}$$

$\langle X_N^{(\max)} \rangle$ is finite for $\mu > 1$ and diverges for $\mu \leq 1$. Figure 3.3 is the plot of $H_N(t)$ at $\mu = 0.8$ (left) and $\mu = 1.2$ (right) for $N = 1, \ldots, 10$.

As seen here, the t integration of $\langle R_N^{(\max)} \rangle$ receives a dominant contribution from $t \ll 1$ and the behavior of $H_N(t)$ for $t \ll 1$ receives a contribution from $x \gg 1$, as seen from Eq. (3.108). Therefore, we need to access the integrand for $t \ll 1$, $x \gg 1$ correctly.

As can be deduced from the behavior of $\langle X_N^{(\max)} \rangle$, $\langle R_N^{(\max)} \rangle$ behaves very differently for the three cases of $\mu > 1$, $\mu = 1$, and $\mu < 1$. We shall discuss these three cases separately.

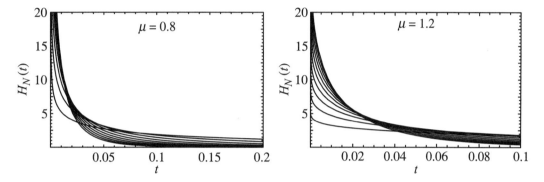

Fig. 3.3 The behavior of $H_N(t)$ at $\mu = 0.8$ (left) and $\mu = 1.2$ (right). On each plot, ten curves show $N = 1, 2, \ldots, 10$ from top to bottom on the left edge.

$\mu > 1$

First, we shall approximate the function to be powered by $(N-1)$ as in the case of $\langle X_N^{(\max)} \rangle$:

$$\int_0^x e^{-x't} f(x') dx' = \int_0^\infty e^{-x't} f(x') dx' - \int_x^\infty e^{-x't} f(x') dx' \qquad (3.110)$$

$$= 1 - G_1(t) - G_2(x, t), \qquad (3.111)$$

$$G_1(t) := \int_0^\infty (1 - e^{-x't}) f(x') dx', \qquad (3.112)$$

$$G_2(x, t) := \int_x^\infty e^{-x't} f(x') dx'. \qquad (3.113)$$

Since the function $G_1(t)$ is

$$\left. \frac{d}{dt} G_1(t) \right|_{t=0} = \langle X \rangle, \qquad (3.114)$$

we evaluate the following for $t \ll 1$,

$$G_1(t) = \langle X \rangle t + O(t^2). \qquad (3.115)$$

The function $G_2(x, t)$ is

$$G_2(x, t) = \bar{F}(x) + O(t) \tag{3.116}$$

Substituting these into Eq. (3.105), we obtain

$$\begin{aligned}
\langle R_N^{(\max)} \rangle &\simeq N \int_0^\infty dt \int_0^\infty dx \, x \, e^{-xt} f(x)(1 - \langle X \rangle t - \bar{F}(x))^{N-1} \\
&\simeq N \int_0^\infty dt \int_0^\infty dx \, x \, e^{-xt} f(x) \, e^{-N\langle X \rangle t - N\bar{F}(x)} \\
&= N\mu \int_0^\infty dx \, x \, f(x) \, e^{-N\bar{F}(x)} \frac{1}{x + N\langle X \rangle} \\
&\simeq \frac{\langle X_N^{(\max)} \rangle}{N\langle X \rangle} \simeq \frac{\Gamma(1 - 1/\mu)}{N^{1-1/\mu} \langle X \rangle}.
\end{aligned} \tag{3.117}$$

Here in the 3rd and 4th line of the approximation, we have neglected the variable x in the denominator of $1/(x + N\langle X \rangle)$, which is valid as $x = O(N^{1/\mu})$ is the dominant part of the contribution in calculating $\langle X_N^{(\max)} \rangle$. Also, since $t = O(1/N)$ yields the dominant contribution in t integration, we have

$$xt = O(N^{-1+1/\mu}). \tag{3.118}$$

These justify the approximation $e^{-x't} \simeq 1$ in Eq. (3.116).

Equation (3.117) is the same result obtained by evaluating the numerator and the denominator of Eq. (3.104) separately. Since $\langle X \rangle$ is finite when $N \to \infty$ owing to $\mu > 1$, we have $\langle R_\infty^{(\max)} \rangle = 0$.

Figure 3.4 shows the $\langle R_N^{(\max)} \rangle$ obtained by Monte Carlo calculation and the behavior of Eq. (3.117). We readily observe that the approximation is good for large μ. This is because as μ approaches 1, $xt = O(1)$ and the uncalculated correction terms become large. In fact, as will be seen here, for $\mu = 1$, the behavior or the dominant term is completely different from $\mu > 1$.

$\mu = 1$

In this case, the main contribution comes from $xt = O(1)$ and the earlier calculation does not hold. Let us now carefully examine the behavior of each term.

First, owing to $\langle X \rangle = \infty$, Eq. (3.115) is useless. Instead, the following holds:

$$G_1(t) = -t \log t + C_1 t + O(t^2). \tag{3.119}$$

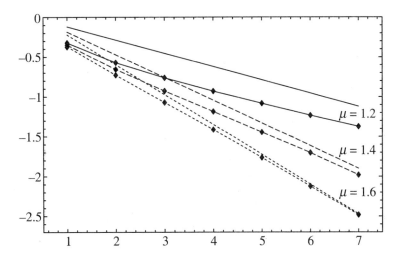

Fig. 3.4 The behavior of $\langle R_N^{(\max)} \rangle$ for $N \to \infty$ in the case of $\mu > 1$. The abscissa is $\log_{10} N$ and the ordinate is $\log_{10} \langle R_N^{(\max)} \rangle$. The top solid line shows the theoretical results Eq. (3.117) for $\mu = 1.2$. The solid lines below connect the results of the Monte Carlo calculation shown by rhombuses. As N becomes larger, they approach each other. Similarly, the broken lines show $\mu = 1.4$ and the dotted lines show $\mu = 1.6$. As μ becomes larger, the theoretical results and the results of the Monte Carlo calculation come closer.

The first term on the left-hand side of this equation comes from the log divergence of $G'(0)$ and is common to all power-law distributions. The second term C_1, however, depends on the behavior for finite x.

Next comes $G_2(x,t)$. Since $x \gg 1$, substituting its asymptotic form x^{-2} into $f(x')$, we obtain

$$G_2(x,t) \simeq \int_x^\infty e^{-x't} x'^{-2} dx' = t\,\Gamma(-1, tx). \tag{3.120}$$

Substituting all of these into Eq. (3.105), we find

$$\langle R_N^{(\max)} \rangle \simeq N \int_0^\alpha dt \int_0^\infty dx\, x\, e^{-xt} f(x) (1 - (-t \log t + C_1 t) - t\Gamma(-1, tx))^{N-1}$$

$$\simeq N \int_0^\alpha dt \int_0^\infty dx\, x\, e^{-xt} f(x)\, e^{N(t \log t - (C_1 + \Gamma(-1, tx))t)}$$

$$= N \int_0^\alpha dt \int_0^\infty \frac{dz}{z} e^{-z}\, e^{N(t \log t - (C_1 + \Gamma(-1, z))t)}$$

$$= N \int_0^{\tilde\alpha} d\tilde t \int_0^\infty \frac{dz}{z} e^{-z}\, e^{C_1 + \Gamma(-1, z)}\, e^{A\tilde t\, \log \tilde t}, \tag{3.121}$$

where we changed the integration variable in the x integration to $z := xt$, and used

$$\tilde{t} := te^{-(C_1+\Gamma(-1,z))}, \quad A := Ne^{C_1+\Gamma(-1,z)}. \tag{3.122}$$

Moreover, the upper limit of the t integration is changed from ∞ to a constant α. This is because the approximation formula (A.44) of $G_1(t)$ is true only for small t ($< 1/e$ for example) and as seen in the exponential factor, the only contribution comes from $t = O(1/N)$. Therefore, we may choose α ($\tilde{\alpha}$) to be an appropriate value ($1/e$ for example) and its choice does not affect the behavior for $N \to \infty$. The following formula then holds:

$$J(A) := \int_0^\alpha e^{-A\tilde{t}\log \tilde{t}} d\tilde{t} \simeq \frac{1}{A \log A} \quad (A \gg 1). \tag{3.123}$$

Using this for Eq. (3.121), we obtain the following:

$$\langle R_N^{(\max)} \rangle = \int_0^\infty \frac{e^{-z}}{\log N + C_1 + \Gamma(-1,z)} \frac{dz}{z}. \tag{3.124}$$

This integration converges for $z \to 0$ owing to $\Gamma(-1,z) \simeq 1/z$. Therefore, $\Gamma(-1,z)$ around $z = 0$ makes the dominant contribution. In fact,

$$\frac{1}{\log N + C_1 + \Gamma(-1,z)} = \frac{1}{\log N + 1/z + \log z + O(1)}$$

$$= \frac{1}{\log N + 1/z} - \frac{\log z}{(\log N + 1/z)^2} + \cdots \tag{3.125}$$

yields, from the first term in the expansion,

$$\int_0^\infty \frac{e^{-z}}{z \log N + 1} dz = \frac{1}{\log(N)} e^{\frac{1}{\log N}} \Gamma\left(0, \frac{1}{\log N}\right)$$

$$= \frac{\log(\log N) - \gamma_E}{\log N} + O\left(\frac{\log(\log N)}{(\log N)^2}\right). \tag{3.126}$$

The second term in Eq. (3.125) do not contribute to this dominant term. We arrive at the following:

$$\langle R_N^{(\max)} \rangle = \frac{\log(\log N) - \gamma_E}{\log N}. \tag{3.127}$$

Strictly speaking, the only dominant term is $\log(\log N)$. In reality, however, $N = 10^7$ yields $\log(\log N) = 2.7799...$, a small number, which is comparable to γ_E. (Mind that since $\log(10^7) = 16.118...$, the expansion in $1/\log N$ is meaningful.)

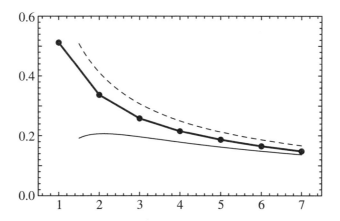

Fig. 3.5 Behavior of $\langle R_N^{(\max)} \rangle$ at $\mu = 1$. The ordinate shows $\langle R_N^{(\max)} \rangle$, the abscissa $\log_{10} N$. The dots connected with solid lines show the result of the Monte Carlo calculation, the broken line, the result of the numerical integration of Eq. (3.124) with $C_1 = -\gamma_E$, and the solid curve, Eq. (3.127).

Comparison of Eq. (3.127) and the result of the Monte Carlo calculation is given in Fig. 3.5, where we readily observe that the theoretical expression (3.127) correctly gives the asymptotic behavior for $N \to \infty$,

$\mu < 1$

In this case, $G_1(t)$ can be evaluated by partial integration as follows:

$$\begin{aligned}
G_1(t) &= \int_0^\infty (1 - e^{-x't}) \left(-\frac{d}{dx'} \bar{F}(x') \right) dx' \\
&= -(1 - e^{-tx}) \bar{F}(x) \Big|_0^\infty + t \int_0^\infty e^{-x't} \bar{F}(x') dx' \\
&\simeq 0 + t \int_0^\infty e^{-x't} x'^{-\mu} dx' \\
&= t^\mu \Gamma(1 - \mu).
\end{aligned} \tag{3.128}$$

Here, we used the fact that the x' integration in the second term of the second line diverges for $t \to 0$ due to $\mu \leq 1$ and is determined by the behavior of $\bar{F}(x')$ for $x' \to \infty$.

Next, $G_2(x, t)$. Using the asymptotic form for $f(x')$, we find

$$G_2(x, t) \simeq \int_x^\infty e^{-x't} \mu x'^{-\mu-1} dx' = t^\mu \mu \Gamma(-\mu, tx). \tag{3.129}$$

From this, we obtain the following:

$$\langle R_N^{(\max)} \rangle \simeq N \int_0^\infty dt \int_0^\infty x\, e^{-tx} \mu x^{-\mu-1} e^{-Nt^\mu(\Gamma(1-\mu)+\mu\Gamma(-\mu,tx))} dx. \quad (3.130)$$

By changing the integration variable (t,x) to $(\tilde{t}, \tilde{x})$ as defined here,

$$t = N^{-1/\mu}\tilde{t}, \quad x = N^{1/\mu}\tilde{x}, \quad (3.131)$$

we find that

$$\langle R_N^{(\max)} \rangle \simeq \int_0^\infty d\tilde{t} \int_0^\infty \tilde{x}\, e^{-\tilde{x}\tilde{t}} \mu \tilde{x}^{-\mu-1} e^{-\tilde{t}^\mu(\Gamma(1-\mu)+\mu\Gamma(-\mu,\tilde{t}\tilde{x}))}\, d\tilde{x} \quad (3.132)$$

N does not appear on the right-hand side of this equation. Therefore, $O(1)$ ranges of $(\tilde{t}, \tilde{x})$ contribute. This means that our previous assumption is correct and the aforementioned expression gives the correct evaluation for $N \to \infty$. Here, by changing the integration variable to $\rho = \tilde{t}^\mu$ and $z = \tilde{t}\tilde{x}$, we obtain the following:

$$\langle R_N^{(\max)} \rangle \simeq \int_0^\infty d\rho \int_0^\infty dz\, e^{-z} z^{-\mu} e^{-\rho(\Gamma(1-\mu)+\mu\Gamma(-\mu,z))} dz$$

$$= \int_0^\infty \frac{e^{-z} z^{-\mu}}{\Gamma(1-\mu) + \mu\Gamma(-\mu,z)} dz. \quad (3.133)$$

The denominator of this expression can be expressed as,

$$\Gamma(1-\mu) + \mu\Gamma(-\mu,z) = e^{-z} z^{-\mu} + \gamma(1-\mu,z) \quad (3.134)$$

where $\gamma(1-\mu,z)$ is the first-order incomplete gamma function. Therefore, the analytic expression for $N \to \infty$ becomes the following:

$$\langle R_\infty^{(\max)} \rangle = \int_0^\infty \frac{1}{1 + e^z z^\mu \gamma(1-\mu,z)} dz. \quad (3.135)$$

By taking $\mu = 0$ in the denominator of Eq. (3.135), from $\gamma(1,z) = 1 - e^{-z}$ we find that $\langle R_\infty^{(\max)} \rangle = 1$. Also, at $\mu = 1$, $\langle R_\infty^{(\max)} \rangle = 0$ from $\gamma(0,z) = \infty$.

Figure 3.6 shows the analytical result and the results of the Monte Carlo calculation, where we observe the latter approaches the former for $N \to \infty$.

3.2.2 The second largest value and its share

Using the method developed in the previous subsection, let us examine the second largest value $X^{(2)}$.

First, its average:

$$\langle X_N^{(2)} \rangle = \int_0^\infty f(x_1)dx_1 \int_0^{x_1} x_2 f(x_2) dx_2 N(N-1) \left(\int_0^{x_2} f(x') dx' \right)^{N-2}. \quad (3.136)$$

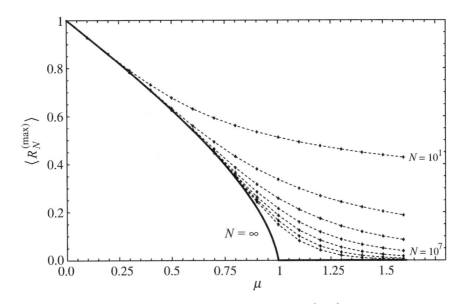

Fig. 3.6 The results of the Monte Carlo calculation of $\langle R_N^{(\max)} \rangle$ and the analytical result, Eq. (3.135) for $\langle R_\infty^{(\max)} \rangle$. The former is for $N = 10^{1,2,3,4,5,6,7}$ from top to bottom. As $N \to \infty$, the results of the Monte Carlo calculation converges to the analytical result for $\langle R_\infty^{(\max)} \rangle$. The dots are the results in the 95% confidence range as shown by error bars. For most points, the error bars are too short to be seen.

For $N \to \infty$, just as in Eq. (3.97), we evaluate as follows:

$$\left(\int_0^{x_2} f(x') dx' \right)^{N-2} \simeq e^{-Nx^{-\mu}} \tag{3.137}$$

and by exchanging the order of integrations (x_1, x_2), we obtain the following:

$$\langle X_N^{(2)} \rangle \simeq \int_0^\infty x_2 f(x_2) dx_2 N^2 e^{-Nx^{-\mu}} \int_{x_2}^\infty f(x_1) dx_1$$

$$\simeq \int_0^\infty \mu x_2^{-2\mu} N^2 e^{-Nx^{-\mu}} dx_2$$

$$= N^{1/\mu} \Gamma\left(2 - \frac{1}{\mu}\right). \tag{3.138}$$

This equation is good for $\mu > 1/2$ and for $\mu \leq 1/2$, we have $\langle X_N^{(2)} \rangle = \infty$.

Next, we look at the share: Let us examine the average of

$$\langle R_N^{(2)} \rangle := \frac{X^{(2)}}{\sum_{i=1}^N X_i}. \tag{3.139}$$

In this case, an evaluation similar to the previous case implies that $\langle R_N^{(2)} \rangle = 0$ for $\mu > 1$ at $N = \infty$, but it is finite for $\mu < 1$.

The expectation value is expressed as follows:

$$\langle R_N^{(2)} \rangle = N(N-1) \int_0^\infty f(x_1) dx_1 \int_0^{x_1} f(x_2) dx_2 \prod_{i=3}^{N} \int_0^{x_2} f(x_i) dx_i \frac{x_2}{\sum_{i=1}^N x_i}$$

$$= N(N-1) \int_0^\infty dt \int_0^\infty x_2 e^{-tx_2} f(x_2) dx_2 \int_{x_2}^\infty e^{-tx_1} f(x_1) dx_1$$

$$\times \left(\int_0^{x_2} e^{-tx'} f(x') dx' \right)^{N-2}. \tag{3.140}$$

In Eq. (3.140), the integration order of x_1 and x_2 are interchanged. By evaluating this in the same way as the previous case, we notice that this calculation involves an extra factor coming from the x_1 integration,

$$\int_{x_2}^\infty e^{-tx_1} f(x_1) dx_1 \simeq t^\mu \mu \Gamma(-\mu, tx_2) \tag{3.141}$$

By taking into account this factor, we obtain the following instead of Eq. (3.133):

$$\langle R_N^{(2)} \rangle \simeq \int_0^\infty d\rho \int_0^\infty dz\, e^{-z} z^{-\mu} \rho \mu \Gamma(-\mu, z) e^{-\rho(\Gamma(1-\mu) + \mu \Gamma(-\mu, z))}$$

$$= \int_0^\infty \frac{e^{-z} z^{-\mu} \mu \Gamma(-\mu, z)}{(\Gamma(1-\mu) + \mu \Gamma(-\mu, z))^2} dz. \tag{3.142}$$

By using Eq. (3.134), we find

$$\langle R_\infty^{(2)} \rangle = \int_0^\infty \frac{1 - e^z z^\mu \Gamma(1-\mu, z)}{(1 + e^z z^\mu \gamma(1-\mu, z))^2} dz. \tag{3.143}$$

This analytical result and the results of the Monte Carlo calculation are shown in Fig. 3.7 where the correctness of the expression (3.143) is apparent.

3.2.3 All about shares

Extending the analysis of this section, we can obtain the share of the kth largest value,

$$R_N^{(k)} := \frac{X^{(k)}}{\sum_{i=1}^N X_i} \tag{3.144}$$

for $N = \infty$. It should be noted that we take the limit $N \to \infty$ with k fixed.

Income and Firm-size Distributions

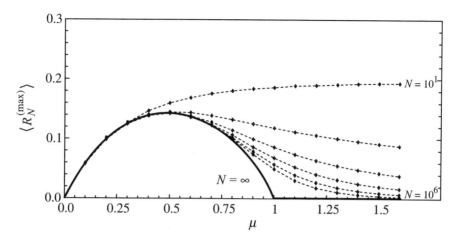

Fig. 3.7 The analytic result (3.143) for the share of the second largest value $\langle R_N^{(2)} \rangle$ for $N = \infty$ and the result of the Monte Carlo calculation for $N = 10^{1,2,3,4,5,6}$. The latter approaches the former as N increases.

First, corresponding to Eq. (3.140), we have

$$\langle R_N^{(k)} \rangle = N(N-1)\ldots(N-k+1) \int_0^\infty dt \int_0^\infty x_k e^{-tx_k} f(x_k) dx_k$$

$$\times \frac{1}{(k-1)!} \left(\int_{x_k}^\infty f(x'') dx'' \right)^{k-1} \left(\int_0^{x_k} e^{-tx'} f(x') dx' \right)^{N-k}. \tag{3.145}$$

By change of integrations, we find

$$\langle R_N^{(k)} \rangle \simeq \frac{1}{(k-1)!} \int_0^\infty d\rho \int_0^\infty dz \, e^{-z} z^{-\mu} (\rho \mu \Gamma(-\mu, z))^{k-1} e^{-\rho(\Gamma(1-\mu) + \mu \Gamma(-\mu, z))}$$

$$= \int_0^\infty \frac{e^{-z} z^{-\mu} (\mu \Gamma(-\mu, z))^{k-1}}{(\Gamma(1-\mu) + \mu \Gamma(-\mu, z))^k} dz. \tag{3.146}$$

Here the factor $(k-1)!$ that was initially in the denominator has been canceled by $\Gamma(k)$ coming from the ρ integration. Using Eq. (3.134) just as before, we find the following:

$$\langle R_\infty^{(k)} \rangle = \int_0^\infty \frac{(1 - e^z z^\mu \Gamma(1-\mu, z))^{k-1}}{(1 + e^z z^\mu \gamma(1-\mu, z))^k} dz. \tag{3.147}$$

Let us now calculate summation over $k = 1, 2, \ldots, m$, namely the share of the top m:

$$\sum_{k=1}^m \langle R_\infty^{(k)} \rangle = 1 - \frac{1}{\Gamma(1-\mu)} \int_0^\infty e^{-z} z^{-\mu} \left(\frac{1 - e^z z^\mu \Gamma(1-\mu, z)}{1 + e^z z^\mu \gamma(1-\mu, z)} \right)^m dz$$

This becomes 1 for $\mu < 1$ and $m = \infty$, which is a curious and interesting property.

$$\sum_{k=1}^{\infty} \langle R_{\infty}^{(k)} \rangle = \begin{cases} 1 & (\mu < 1), \\ 0 & (\mu \geq 1). \end{cases} \quad (3.148)$$

Let us talk in terms of firms whose sales obey power-distributions. The results we have obtained so far mean that for $\mu < 1$, a finite number of top firms have finite shares even if there are an infinite number of firms. This is a *monopolistic or oligopolistic* situation. On the other hand, for $\mu \geq 1$, any finite number of top firms have an infinitesimally small share among an infinite number of firms, which is a *non-oligopolistic* situation. Therefore, the Pareto index μ is important for economics. Figure 3.8 give the results of the expression (3.148) for $m = 1, 10, 20, \ldots, 100$ from bottom to top.

Table 3.4   Share of the top 10, 50, 100 firms in % given by Eq. (3.148)

μ	Top 10	Top 50	Top 100
0.90	40.0	49.8	53.5
0.95	24.2	30.5	33.0
0.98	11.8	14.7	15.9

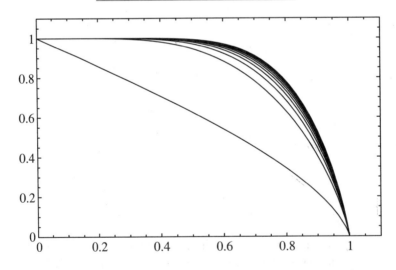

Fig. 3.8   Share of the sum of the top values. The curves show it for $m = 1, 10, 20, \ldots, 100$ from bottom to top.

Table 3.4 lists the share of the top 10, 50, and 100 firms for several values of μ in percentage. For example, at $\mu = 0.9$, the top 10 firms account for 40% of all (∞) firms.

This situation is also observed in Fig. 3.8. This kind of situation is not acceptable in developed economies and measures such as the Anti-Monopoly Act are taken, which has the effect of increasing μ. As μ approaches one, this pressure for an increase in μ weakens as the share of the top finite number of firms approaches zero. This we believe has the effect of universality of $\mu = 1$ in firm size distribution observed elsewhere. We refer readers to Section 2.6 of Aoyama et al. (2010a) for details on this issue.

3.3 Personal Income

In this section, we examine personal income distributions. First, we review a short history of the studies of this subject. Those by Pareto (1897) and Gibrat (1931) are historically key articles. Hence, we explain their essence in Section 3.3.1, Secondly, in Section 3.3.2, we explain the precise structure of personal income, i.e., a detailed balance and Gibrat's law, through an investigation of the income tax of high tax payers in Japan. Lastly, in Section 3.3.3, we propose a two-factor model to explain the personal income distribution.

3.3.1 Short history

Pareto (1895) opened the door to a new world where inequality is reckoned numerically. In his 1897 study, he investigated income distributions of some countries and cities in Europe in different years. For example, if we denote income as x and the number of taxpayers with an income greater than x as N, the double-logarithmic scale plot of the distributions of income in England and Ireland in 1893–94 is depicted in Fig. 3.2. Pareto (1897) explained this distribution by

$$\ln N = \ln A - \mu \ln x, \tag{3.149}$$

where $\ln A$ is the intercept in Fig. 3.9 and A is the normalization constant of PDF. The PDF is derived from Eq. (3.149) as

$$f(x; \mu) = -\frac{dN}{dx} = \frac{\mu A}{x^{\mu+1}}, \tag{3.150}$$

for $0 < b < x$ and zero otherwise. This is the type I Pareto distribution given by Eq. (3.67). Here, b is called the Pareto scale and $\mu > 0$, the Pareto index.

Pareto (1897) also pointed out another possibility that the distribution in Fig. 3.9 is explained by

$$\ln N = \ln A - \mu \ln(a + x) - \beta x. \tag{3.151}$$

The PDF corresponding to Eq. (3.151) is given by

$$f(x; a, \mu, \beta) = \frac{\mu A}{(x+a)^{\mu+1}} e^{-\beta x} + \frac{\beta A}{x+a} e^{-\beta x}. \tag{3.152}$$

for $0 < b < x$ and zero otherwise with $\mu > 0$ and $\beta > 0$. Pareto considered the case of $\beta \ll 1$, and approximated Eq. (3.152) as

$$f(x; a, \mu) \approx \frac{\mu A}{(x+a)^{\mu+1}}. \tag{3.153}$$

Pareto (1897) suggested that income and wealth universally obey the power-law distribution with $\mu \approx 1.5$, and this observation is called **Pareto's law**. The estimated values of μ for some countries and cities in Europe in different years are listed in Pareto (1897, p. 312). However, Pareto recognized that Pareto's law is only applicable to the high income and wealth range (Arnold, 2015).

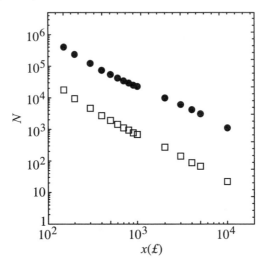

Fig. 3.9 Income distribution in Great Britain (filled circles) and Ireland (squares) 1893-94. This figure is drawn based on the table in Pareto (1897, p. 305).

After the publication of Pareto (1897), many studies criticized Pareto's law. For example, Shirras (1935) investigated the income and supertax statistics of India over a number of years and concluded: *There is indeed no Pareto law. It is time that it should be entirely discarded in studies on the distribution of income.* On the other hand, Macgregor (1936) and Johnson (1937) supported Pareto's law. At that time, the log-normal distribution volunteered itself as a candidate for understanding income distributions.

The log-normal distribution was first suggested by Galton (1879) and mathematically formulated by McAlister (1879). Their idea is explained as follows. We define random variables T_n as the multiple of independent positive variables $x_1, x_2, \ldots, x_n$:

$$T_n = \prod_{i=1}^{n} x_i. \tag{3.154}$$

If we take a logarithm of Eq. (3.154), we obtain

$$\ln T_n = \sum_{i=1}^{n} \ln x_i. \tag{3.155}$$

If $\ln x_i$ is such that a central limit theorem is applicable, then the normalized distribution of $\ln T_n$ obeys the normal distribution as n goes to infinity. Thus, T_n would obey the log-normal distribution.

This fact was re-discovered by Kapteyn et al. (1903). Gibrat (1931) was stimulated by Kapteyn's text, and proposed **the law of proportional effect**. This is expressed by

$$x_n = (1 + a_{n-1})x_{n-1}, \qquad (3.156)$$

where $x_0, x_1, \ldots, x_n$ is a sequence of random variables and $a_0, a_1, \ldots, a_{n-1}$ are mutually independent random variables. Here, Eq. (3.156) is also known as **Gibrat's law**. If we denote the growth rate of x_i as $R_i = x_i/x_{i-1}$ and introduce the conditional PDF as $f(R|x)$, Gibrat's law can be expressed as

$$f(R|x) = f(R), \qquad (3.157)$$

namely, R is independent of x.

We can also write Eq. (3.156) as

$$x_n = x_0 \prod_{i=0}^{n-1}(1 + a_i). \qquad (3.158)$$

If we assume $a_i \ll 1$ and take the logarithm of Eq. (3.158), we obtain

$$\ln x_n = \ln x_0 + \sum_{i=0}^{n-1} a_i. \qquad (3.159)$$

As is the case with Eq. (3.155), the additive central limit theorem is applicable. Thus, x_n obeys the log-normal distribution with two parameters (mean and variance).

If we believe the law of proportional effect, the distribution must obey the log-normal distribution. For this reason, Gibrat (1931) applied the log-normal distribution to many kind of data, such as income, wealth, rent, real estate, wages, firm sizes, city sizes, family sizes, etc. Gibrat's fittings are also summarized in Kleiber and Kotz (2003, p. 109).

The surge of research on income distribution that followed is listed in Kleiber and Kotz (2003, pp. 278–279; Table B.2). In the history of the study of income distribution, almost every study used classed and tabulated income data. Exceptions are Aoyama et al. (2000); Fujiwara et al. (2003); Souma and Nirei (2005); and Nirei and Souma (2006), because these articles investigated income data of individuals as explained here.

Aoyama et al. (2000) investigated two types of income data. These are self-declared income data and income tax of individuals in Japan for the fiscal year 1998. Here, the self-declared income data is the classed and tabulated data of individuals in Japan, and is publicly available on the web pages of the National Tax Agency of Japan (NTAJ) from 1951 (available from 1887 to the present, on paper). The income data for the fiscal year 1998 covers 6,224,254 individuals. The declaration of one's income tax was made individually. Under the Japanese tax system, if a worker has only one source of income (salary) and the income is less than 20 million yen, that person does not have to file a tax return (the tax deducted from the monthly salary is adjusted at the end of the year by the employer and that becomes the final amount of tax paid). This distribution is contained in the employment income data. Employment income data is also classed and tabulated data, and publicly available on the web pages of NTAJ from 1951. On the other hand, the income tax data covers *84,515 individuals* who are high income taxpayers. Here, a high income taxpayer is defined as a person who files income tax of more than 10 million yen. Aoyama et al. (2000) found that CCDFs of both data obey a power-law with a Pareto exponent very close to $\mu = 2$.

By investigating Japanese income data that is classed and tabulated data, Souma (2001), Souma and Nirei (2005), and Aoyama et al. (2010a) suggested that μ fluctuated around $\mu = 2$ from 1887 to 2003 in Japan. Souma and Nirei (2005), Nirei and Souma (2006), and Nirei and Souma (2007) suggested that the same property was observed in the U.S.

Table 3.5 The number of individuals and classes contained in employment income data, self-declared income data, and income tax data from 1987 to 2000 in Japan.

Fiscal Year	Employment income		Self-declared income		Income tax
	# individuals	# classes	# individuals	# classes	# individuals
1987	42,652,250	14	7,707,308	14	110,817
1988	43,686,503	14	7,797,019	14	111,765
1989	45,057,032	14	7,965,871	18	141,211
1990	46,549,065	14	8,547,375	18	172,183
1991	48,146,547	14	8,562,552	18	175,723
1992	49,090,672	14	8,577,661	18	125,066
1993	49,869,037	14	8,428,477	18	128,666
1994	50,157,991	14	8,223,171	18	95,683
1995	51,598,961	14	8,020,634	18	95,358
1996	52,009,683	14	8,239,858	18	99,284
1997	52,521,112	14	8,271,709	18	93,394
1998	52,682,839	14	6,224,254	18	84,571
1999	52,649,968	14	7,400,607	18	75,272
2000	52,691,963	14	7,273,506	18	79,999

3.3.2 Distributions and fluctuations

For an accurate grasp of the shape of Japanese personal income distribution, Nirei and Souma (2006) and Souma and Nirei (2005) investigated three data sets, i.e., employment income data, self-declared income data, and income tax data of high tax payers from 1987 to 2000 in Japan. The number of individuals covered by these data are detailed in Table 3.5.

Income and Firm-size Distributions

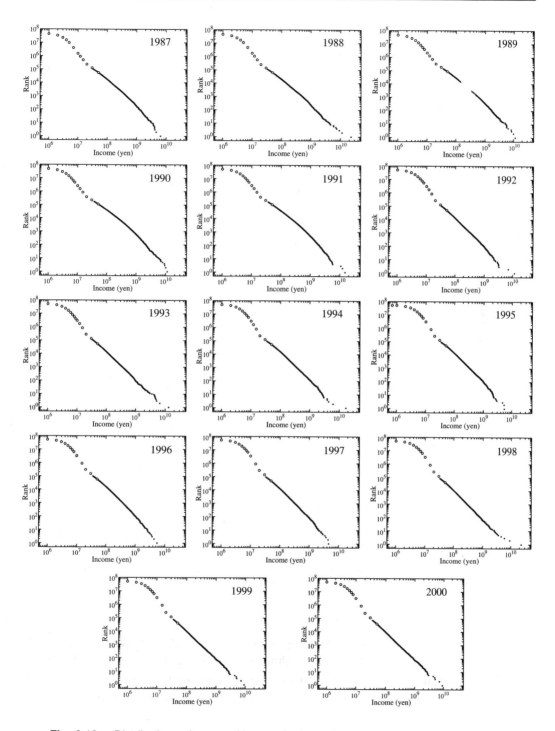

Fig. 3.10 Distributions of personal income in Japan from 1987 to 2000.

The distributions are depicted in Fig. 3.10. They represent the most accurate personal income distributions in the history of human civilization. The tail of the distributions are well explained by power-law. Exceptions are the distributions for 1990 and 1991. These figures show the *breakdown of Pareto's law*. It is generally believed that the tail of such a distribution always obeys Pareto's law. Based on this belief, many distributions have been proposed, such as the beta-type distribution aimed at reproducing the power-law behavior of the tail. Thus, we need to investigate new types of distributions that can explain the breakdown of Pareto's law.

The observation made earlier refers to the static structure of personal income; therefore, we must proceed to investigate the fluctuations in personal income to understand the nature of personal income. The income tax data of high tax payers contains the names, addresses, and telephone numbers of each person. Thus, we can identify each person, and calculate the growth rate of income tax of each high tax payer.

Fujiwara et al. (2003) and Aoyama et al. (2003) investigated the fluctuations in income tax of high tax payers. For example, the scatter plot of income tax for two consecutive years (1997 and 1998) is depicted in Fig. 3.11. In this figure, x_1 is income tax for the fiscal year 1997 and x_2 is that for the fiscal year 1998. Although the data on high tax payers in the fiscal year 1997 covers 93,394 people, some may have become excluded from the high tax bracket in the fiscal year 1998. Likewise, there may have been new entrants to this segment in the fiscal year 1998, leaving the data with 84,571 observations. As a result, 52,902 persons remained as high tax payers; thus, Fig. 3.11 is constructed from these 52,902 persons.

In Fig. 3.11, dots are symmetrically distributed with respect to the diagonal line. This is expressed by introducing the joint PDF $f(x_1, x_2)$ as

$$f_{12}(x_1, x_2) = f_{12}(x_2, x_1). \tag{3.160}$$

This is known as the **detailed balance** in statistical physics. By applying the two-dimensional Kolmogorov–Smirnov test, Fujiwara et al. (2003) and Aoyama et al. (2003) clarified that the detailed balance is preserved for the fiscal years 1997 and 1998.

Fujiwara et al. (2003) defined the growth rate R and the logarithms of it, r, as

$$R := \frac{x_2}{x_1}, \qquad r := \log_{10} R. \tag{3.161}$$

The distribution of r is depicted in the left panel of Fig. 3.12. In this figure, the bins $n = 1, \ldots, 5$ are used to correspond to five logarithmically equal intervals from $x_1 = 10^7 \sim 10^8$ yen. In the left panel of Fig. 3.12, the growth rate collapses into a single curve irrespective of the bins considered. This is Gibrat's law, i.e., the growth rate is statistically independent of the previous years income tax.

Income and Firm-size Distributions

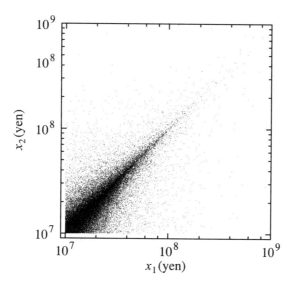

Fig. 3.11 Scatterplot for personal income tax for two consecutive years. Here, x_1 is the 1997 income tax and x_2 is the 1998 income tax (from Aoyama et al. (2010a)).

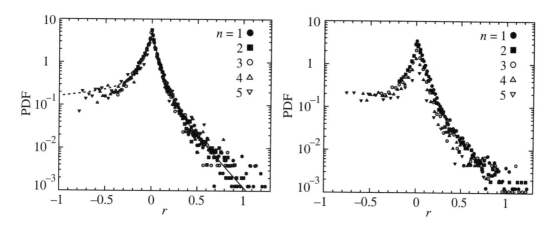

Fig. 3.12 Gibrat's law (left) and breakdown of Gibrat's law (right) (from Aoyama et al. (2010a)).

By using the detailed balance given by Eq. (3.160) and Gibrat's law given by Eq. (3.157), Fujiwara et al. (2003) derived the equation

$$\frac{f_1(x_1)}{f_2(x_2)} = \frac{1}{R}\frac{f_R(R^{-1})}{f_R(R)}. \tag{3.162}$$

Here, to emphasize the difference of the functional form of PDF, we introduce notation f_1, f_2, and f_R. This equation suggests that the left-hand side of Eq. (3.162) is a function of x_1 and $x_2 = Rx_1$, while the right-hand side of Eq. (3.162) is a function of R. This means that such a relation is preserved only if $f_1(x)$ and $f_2(x)$ are power-law distributions. Again, Aoyama et al. (2003) suggested that the detailed balance and Gibrat's law are preserved, and the growth rate obeys the equation called a **reflection law**:

$$f(R) = R^{-\mu-2} f(R^{-1}). \tag{3.163}$$

This line is drawn in the left panel of Fig. 3.12 as a dashed line.

In summary, when the detailed balance and Gibrat's law are maintained, the power-law distribution is realized. Conversely, when these are not preserved, the power-law distribution is not realized. If we calculate r for the period from 1991 to 1992, i.e., the collapse of the Heisei bubble in the Japanese economy, we obtain the right panel of Fig. 3.12. The growth rate does not collapse into a single curve in the right panel of Fig. 3.12. This means that Gibrat's law is broken, and the detailed balance is also broken (Fujiwara et al., 2003). Thus, income tax distribution does not obey the power-law distribution as already displayed in Fig. 3.10.

3.3.3 Two-factor model

A multiplicative process of wealth has been a standard explanation for the fat tail. This description makes good economic sense because the rate of return for an asset is a stationary process. Gibrat's celebrated law of proportional effect (Gibrat, 1931) first embodied the idea that the multiplicative process, i.e., Gibrat's process, generates a log-normal distribution (see Sutton (1997) for a review). Kalecki (1945) pointed out the problem with Gibrat's process, which is that the variance of the log-normal distribution derived from it increases continuously as time goes by. This apparently contradicts many economic phenomena.

After Gibrat (1931), many models have been proposed to explain the distribution of income, for example, the Markov chain model (Champernowne, 1953), the open population model (Rutherford,1955), and the Yule distribution (Simon, 1955) explained in Section 2.1.1, besides income determined by inherited wealth (Wold and Whittle, 1957), the law of Pareto–Lévy (Mandelbrot, 1960, 1961), etc.

The field has been made exciting recently by studies on a reflected multiplicative process (Solomon and Levy, 1996) or a closely related Kesten process (Kesten (1973); Solomon and Levy (1996); and Takayasu et al. (1997)) which have revealed the effect of a reflective lower bound on the tail of the stationary distribution. The reflective barrier model provides economists with an interestingly sharper structure in the multiplicative processes than previous models did. Gabaix (1999) constructed an economic model of city-size distributions by utilizing this structure and suggested its application to income-distribution tails. Levy (2003) also derived the power-law distribution of wealth in the same framework. We extend this section by incorporating the labor wage process in the wealth accumulation process.

As explained earlier, many models have been proposed to explain the distribution of personal income. However, almost all models have tried to explain the distribution by using a one-factor model. We propose a **two-factor model** incorporating the labor wage process in the wealth accumulation process (Nirei and Souma, 2007). Thus, to verify the validity of our model, we must investigate the income source (Souma and Nirei, 2005). Though the main income source of middle- and low-income earners is wages, high-income earners have a variety of sources. We can investigate the income sources of high income earners by using data reported by NTAJ.

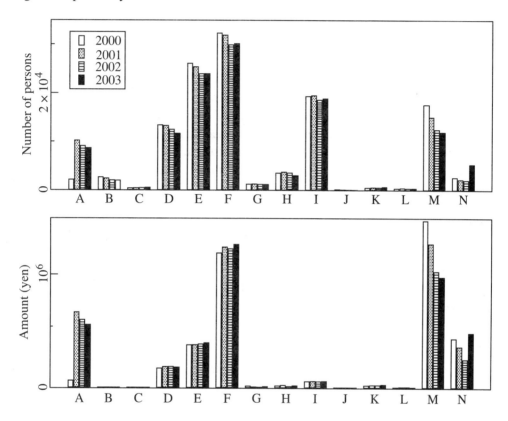

Fig. 3.13 Income sources of high-income earners from 2000 to 2003. The top panel represents the number of high-income earners, and the bottom panel represents the amount of income. In both panels, A stands for business income, B for farm income, C for interest income, D for dividends, E for rental income, F for wages & salaries, G for comprehensive capital gains, H for sporadic income, I for miscellaneous income, J for forestry income, K for retirement income, L for short-term separate capital gains, M for long-term separate capital gains, and N for capital gains from stocks (from Souma and Nirei (2005)).

The top panel of Fig. 3.13 depicts the number of high-income earners who earned income greater than 50 million yen in each year from 2000 to 2003. In this figure, income sources are divided into 14 categories. On the other hand, the bottom panel of Fig. 3.13 depicts the amount of income for each income source. By comparing these figures, we see that the main income sources of high-income earners are wages and capital gains. This fact suggests that the minimal model of personal income process is constructed from the sum of the wages process and the capital process (Nirei and Souma, 2007).

First we model the wage process, in which $w_i(t)$, $(i = 1, \ldots, N)$ denotes the wages of the ith person at time t. We define the wage process as

$$w_i(t+1) := \begin{cases} uw_i(t) + s\epsilon_i(t)\bar{w}(t) & \text{if } uw_i(t) + s\epsilon_i(t)\bar{w}(t) > \bar{w}(t), \\ \bar{w}(t) & \text{otherwise,} \end{cases} \quad (3.164)$$

where u is the trend growth of wage, and corresponds to an automatic growth in nominal wage. Here, we use $u = 1.0422$, which is an average inflation rate for the period from 1961 to 1999. In Eq. (3.164), $\epsilon_i(t)$ obeys the normal distribution with mean 0 and variance 1, i.e., $N(0, 1)$. Again, s determines the level of income for the middle class. We choose $s = 0.32$ to fit the middle part of the empirical distribution. In Eq. (3.164), $\bar{w}(t)$ is the reflective lower bound, which is interpreted as a subsistence level of income. We assume that $\bar{w}(t)$ grows deterministically,

$$\bar{w}(t) = v^t \bar{w}(0). \quad (3.165)$$

Here, we use $v = 1.0673$, which is the time average of the growth rate of the nominal income per capita.

Second, we model the asset process, in which $a_i(t)$, $(i = 1, \ldots, N)$ denotes the assets of the ith person at time t. We define the asset accumulation process as

$$a_i(t+1) = \gamma_i(t)a_i(t) + w_i(t) - c_i(t), \quad (3.166)$$

where the log return, $\ln \gamma_i(t)$, obeys the normal distribution with mean y and variance x^2, i.e., $N(y, x^2)$. We use $y = 0.0595$, which is the time average of the growth rate of the Nikkei average index from 1961 to 1999. We use $x = 0.3122$, which is the variance calculated from the distribution of the income growth rate for high-income earners. In Eq. (3.166), we assume that a consumption function, $c_i(t)$, is given by

$$c_i(t) = \bar{w}(t) + b\{a_i(t) + w_i(t) - \bar{w}(t)\}. \quad (3.167)$$

We chose $b = 0.059$ from the empirical range estimated from Japanese micro data.

We denote the income of the ith person at time t as $I_i(t)$, and define it as

$$I_i(t) = w_i(t) + \mathrm{E}[\gamma_i(t) - 1]a_i(t). \quad (3.168)$$

The results of the simulation for $N = 10^6$ are depicted in Fig. 3.14. The left panel of this figure is a double-logarithmic plot of the CCDF for income normalized by an average. The right panel of Fig. 3.15 is the simulation results for the Lorenz curve. These figures suggest that the two-factor model exactly explains the real distribution of personal income.

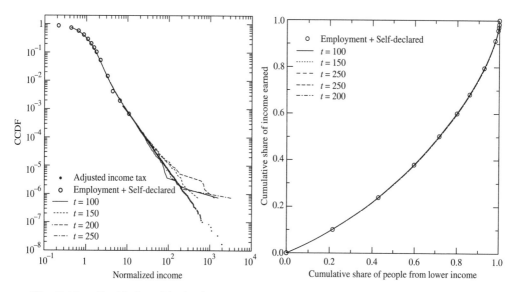

Fig. 3.14 Double-logarithmic plot of the cumulative distributions of normalized income in 1999 and simulation results (left), and the Lorenz curve in 1999 and simulation results (right) (from Souma and Nirei (2005)).

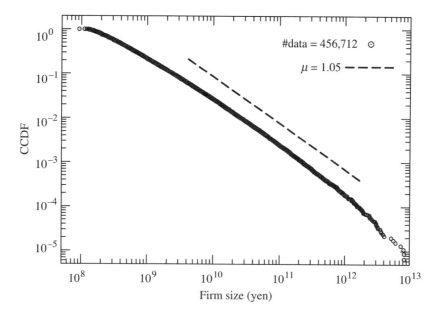

Fig. 3.15 CDF of Japanese firm size (sales) in 2002 (from Aoyama et al. (2010a)).

Nirei and Souma (2007) suggested that the exponent in the power-law part of the distribution, i.e., Pareto index, is given by

$$\mu = 1 - \frac{2\log(1 - z/g)}{x^2} \approx 1 + \frac{2z}{gx^2}, \tag{3.169}$$

where z is the steady state value of $[w(t) - c(t)]/\langle a(t)\rangle$, $\langle a(t)\rangle$ is the average assets, and g is the steady state value of the growth rate of $\langle a(t)\rangle$. Here, Eq. (3.169) suggests that μ fluctuates around $\mu = 2$, if $2z \sim gx^2$.

3.4 Firm Size

The old history of the study of firm sizes is summarized by Steindl (1965). Recently, the distribution of firm sizes have been vigorously investigated in the field of econophysics. There exists a huge literature on company, industry, and country growth in economics. For example, there are the studies by Stanley et al. (1996); Amaral et al. (1997); Sutton (1997); Takayasu et al. (1997); Ijiri and Simon (1977b); Amaral et al. (1998); Okuyama et al. (1999); Bottazzi and Secchi (2003); Fu et al. (2005); and Gabaix (2008).

Compared with personal income, data on firm sizes are easily available. However, when we investigate the data on firm sizes, we must be aware of the exhaustiveness of the data, as argued by Aoyama et al. (2010a).

Fujiwara et al. (2004a,b) investigated AMADEUS offered by the European electronic publishing company Bureau van Dijk. AMADEUS covers all firms with either an operating profit greater than 15 million euros, or a total capital greater than 30 million euros, or more than 150 employees. This means that AMADEUS contains large firms. Fujiwara et al. (2004a,b) suggested that distributions of firm sizes obey power-law distributions with $\mu = 1$, and this value of μ is sustainable. In addition, the detailed balance and Gibrat's law are also preserved (see also Aoyama et al. (2010a)).

Aoyama et al. (2010a) investigated sales data in Japan offered by the Tokyo Shoko Research (TSR). TSR provides exhaustive data and covers all firms with sales greater than 2 billion yen. This means that TSR contains large firms. The CDF of Japanese firm-size (sales) in 2002 is depicted in Fig. 3.15. This figure suggests that the distribution of sales obeys the power-law distribution with $\mu = 1$. The distribution of the growth rate is depicted in the left panel of Fig. 3.16. This figure shows the realization of Gibrat's law.

Again, Aoyama et al. (2010a) investigated sales data in Japan offered by Credit Risk Database (CRD). CRD provides data for small and medium firms. The distribution of the growth rate is depicted in the right panel of Fig. 3.16. This figure shows the *breakdown of Gibrat's law*. By comparing both figures depicted in Fig. 3.16, we can distinguish between large firms and small and medium firms. To do so, we calculate the variance of the distribution of the growth rate, and obtain Fig. 3.7. This figure suggests that the distribution is separated into the scaling regime and the Gibrat regime.

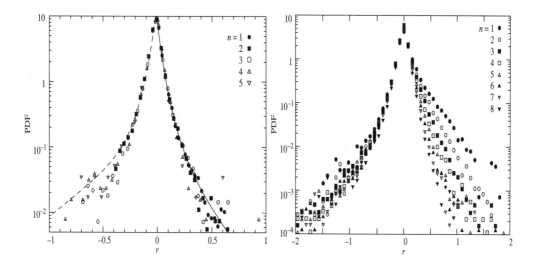

Fig. 3.16 Relation between firm sizes (sales) and variance of growth rate for large firms (Left) and small and medium firms (Right) for the years 2000/2001 (from Aoyama et al. (2010a)).

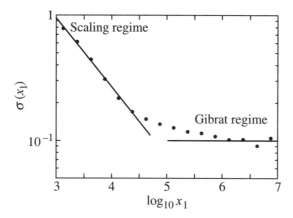

Fig. 3.17 Relation between company-size (sales) and variance of growth-rate for small and medium companies for the years 2000/2001 (from Aoyama et al. (2010a)).

3.5 Excercise

3.1 Solve Eq. (3.12), when $a = \langle x \rangle, c_0 = -\sigma^2, c_1 = 0$, and $c_2 = 0$.

3.2 Solve Eq. (3.12), when $a = 0, c_0 = -\lambda, c_1 = 0$, and $c_2 = 0$.

3.3 Prove Eq. (3.92) by induction on N.

3.4 When the PDF is

$$f(x) = \delta(x-a) \tag{3.170}$$

so that the variable X takes only the value a, answer the following questions.

1. Find the cumulative distribution $\bar{F}_>(x)$ and plot it.
2. Evaluate Eq. (3.93) and discuss its validity.

3.5 Derive the approximation formula (3.123).

3.6 Calculate the nonleading terms in Eq. (3.119) for the type-II Pareto distribution in Table 3.3.

4

Productivity Distribution and Related Topics

> A hand that's dirty with honest labor is fit to shake with any neighbor.
>
> **Proverb**

This chapter is concerned with the distribution of labor productivity and related topics that are standard toolkits in economics, and the production function. In Section 4.1, we demonstrate that although widely used, the production function does not actually fit the data. Instead, we propose a production copula. Section 4.2 presents both theoretical and empirical analyses of the distribution of labor productivity. In theory, we propose the concept of stochastic macro-equilibrium. Entropy plays a central role in this concept. As Yoshikawa (2003, 2014) explains, it provides the right micro-foundation for Keynesian macroeconomics. We also demonstrate that the model presented here fits the Japanese data well.

4.1 Production Function and Production Copula

Although size and growth rate are important characteristics of firms, it is also important to measure their activities. In this section, we examine the production activities of firms (Iyetomi et al., 2012).

4.1.1 Value added and labor productivity

The activities of firms are characterized by **value added** and **productivity**. There are many definitions of value added Y. For example, the definition proposed by the Bank of Japan is

$$Y = O + C + I + R + T + D, \tag{4.1}$$

where O, C, I, R, T, and D are ordinary gain, labor and welfare expenses, interest expense and discount premium, rent, taxes and public charges, and depreciation expenses during the year, respectively. On the other hand, Japan's Minister of Economy, Trade, and Industry defines Y as

$$Y = S + C + I + R + T + D, \tag{4.2}$$

where S is operating profit. We use Eq. (4.1) to define the productivity of firms.

Although many definitions of productivity have been proposed by researchers and organizations, the simplest definition is **labor productivity** c, defined by the value added per employee:

$$c := \frac{Y}{n}, \tag{4.3}$$

where n stands for the number of employees.

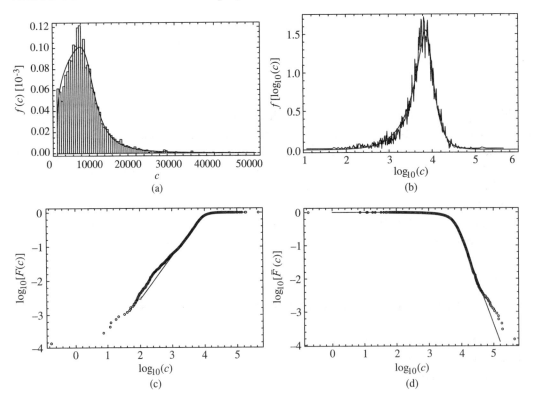

Fig. 4.1 The distributions of labor productivity for the electrical machinery sector in 2006 are shown as an example of distributions for the manufacturing sectors. The solid curve represents the results of fitting by GB2 (from Ikeda et al. (2010)).

We obtained c_i for each firm i ($i = 1, \ldots, N$) by using the NEEDS (Nikkei Economic Electronic Databank System) database (Nikkei Media Marketing, Inc., 2008). It has accumulated financial statements for Japanese listed firms for the past 40 years. We use the data set for 2006 throughout this section; the total number of firms is $N = 1360$. A selected example of the distribution of labor productivity for the manufacturing sector is shown in Fig. 4.1 (Ikeda et al., 2010). Fig. 4.1 (a) is the PDF of c, i.e., $f(c)$, Fig. 4.1 (b) is the PDF of $\log_{10}(c)$, Fig. 4.1 (c) is the CDF $F(c)$, and Fig. 4.1 (d) is the CCDF $\bar{F}(c)$. In these figures, we fit the data by GB2 defined by Eq. 3.47.

A selected example of the distribution of labor productivity for non-manufacturing sectors is shown in Fig. 4.2 The context of Fig. 4.2 is similar to that of Fig. 4.1. The distribution of labor productivity follows the GB2. However, there is another peak at the low side of productivity. This peak is observed for the entire non-manufacturing sector (Ikeda et al., 2010).

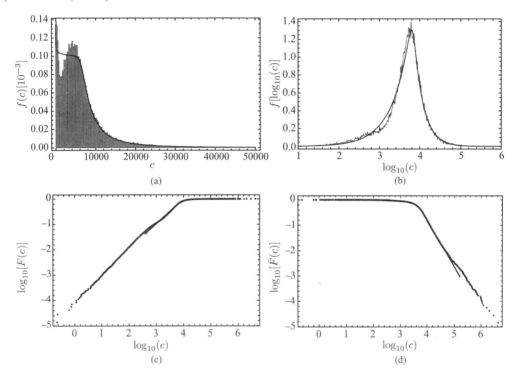

Fig. 4.2 The distributions of labor productivity for the service sector in 2006 are shown as an example of distributions for the non-manufacturing sector. The solid curve represents the results of fitting by GB2 (from Ikeda et al. (2010)).

4.1.2 Production function

The production function is one of basic ingredients in microeconomic theory (Wicksteed, 1894; Varian, 1992). It specifies output (value added) Y of a firm for given input factors such as labor L and capital K:

$$Y = \Phi(L, K), \tag{4.4}$$

where it is assumed that each firm produces goods optimally using its own production function.

Figure 4.3 shows scatter plots for all pairs of the three financial quantities, i.e., Y, L, K, in the manufacturing sector in 2006. We see that these variables are mutually correlated. Spearman's rank correlation coefficient ρ_S for the three pairs, K-Y, L-Y, and L-K, are 0.86, 0.95, and 0.83, respectively. Thus the L-Y pair has stronger correlation than the remaining two pairs K-Y and L-K have. Moreover the remaining ones have correlations of similar strength. The detailed correlation structure for each pair is described in terms of copulas later, with the implications of these results for the overall correlations.

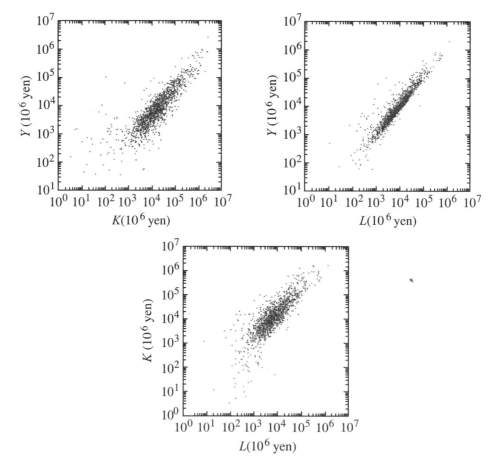

Fig. 4.3 Correlations for each pair of the three financial quantities, labor cost L, capital K, and value added Y, for the manufacturing sector in Japan. Note that these are double-logarithmic plots (from Iyetomi et al. (2012)).

As shown earlier, Y, L, K are mutually correlated; however, the inputs L and K are mathematically treated as independently controllable variables in the production function. Such dependence between K and L is beyond the scope of the production function itself.

A ridge theory of the production function elucidates that the dependence between the inputs arises from the profit maximization behavior of firms (Souma, 2007; Aoyama et al., 2010b).

Fitting to Cobb–Douglas and CES production function

Cobb and Douglas (1928) introduced the function given by

$$\Phi(L, K) = AK^\alpha L^\beta, \tag{4.5}$$

where A, α, and β are adjustable parameters to fit data. Equation (4.5) is called the Cobb–Douglas (CD) production function and has been extensively adopted.

We applied Eq. (4.5) to the data of Japanese listed firms ($N = 1360$) in 2006 and obtained the following values:

$$A = 2.160, \quad \alpha = 0.183, \quad \beta = 0.788. \tag{4.6}$$

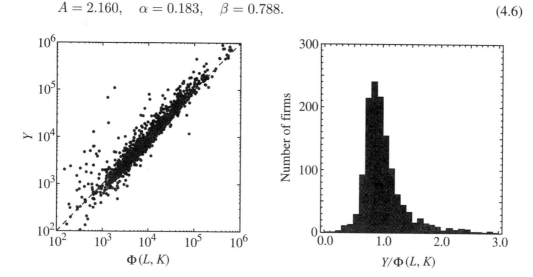

Fig. 4.4 Difference between the actual Y and the best-fit CD function $\Phi(L, K)$: (left) double-logarithmic scatter plot of the two variables, where the dashed line shows the diagonal line to gauge the accuracy of the CD function; (right) histogram of their ratio with bin size of 0.1 (from Iyetomi et al. (2012)).

Figure 4.4 shows the difference between the original values of Y and the corresponding results $\Phi(L, K)$ obtained by the best-fit CD production function with the parameters as given in Eq. (4.6). In the scatter plot (a) as well as in the histogram (b), we observe that the actual data is widely scattered around $\Phi(L, K)$. If the CD production function worked perfectly, the dots would be aligned along the diagonal line in Fig. 4.4 (a) and the distribution of $Y/\Phi(L, K)$ would be of a delta-function shape centered on $Y/\Phi(L, K) = 1$ in Fig. 4.4 (b). In the CD form, quantitatively, 41%(27%) of 540 firms have Y values which are more than 30%(50%) larger than the corresponding values of $\Phi(K, L)$ on the higher side $(Y/\Phi(K, L) > 1)$, and 18%(3.5%) of 820 firms have Y values more than 30%(50%) smaller than the $\Phi(K, L)$ values on the lower side $(Y/\Phi(K, L) < 1)$. The wide discrepancies between the actual data and the predictions due to the CD production function are more clearly visible in Fig. 4.5, which plots the complementary CDFs of $Y/\Phi(K, L)$ by separating its region into two sides with the boundary $Y/\Phi(K, L) = 1$. Both the CDFs are shown to have linear-like behavior in a double logarithmic plot, indicating that the distribution functions decay very slowly.

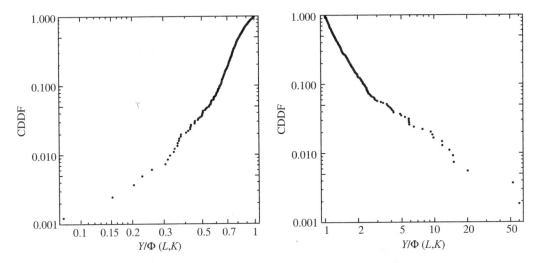

Fig. 4.5 Complementary CDFs of $Y/\Phi(L, K)$ on the lower side $[0, 1]$ (left) and the higher side $[1, \infty]$ (right), where the CD function is used. Note that both panels are double-logarithmic plots (from Iyetomi et al. (2012)).

Arrow et al. (1961) introduced the function given by

$$\Phi(L, K) = A(\gamma L^{cp} + (1 - \gamma) K^{cp})^{1/p}, \tag{4.7}$$

where A, γ, p and c are adjustable parameters to fit data. In the limit of $p \to 0$, Eq. (4.7) reduces to the CD form defined by Eq. (4.5). Equation (4.7) is referred to as the constant elasticity of substitution (CES) production function and has been extensively adopted.

Although the CES form is more flexible than the CD form, adoption of the CES form eliminates the aforementioned failure of the production function. We apply Eq. (4.7) to the data of the Japanese listed firms ($N = 1360$) in 2006 and obtain the following values:

$$A = 1.617, \quad \gamma = 0.932, \quad p = 1.172, \quad c = 1.004. \tag{4.8}$$

The optimized CES function shares the *constant returns to scale* property with the CD function, because the parameter c is nearly equal to 1. We iterate the same calculations as previously carried out for the CD function. The results confirm the limited capability of the production function. In the CES form, quantitatively, 40%(24%) of 576 firms are outside the less-than-30%(50%)-discrepancy range in the case of $Y/\Phi(K, L) > 1$ and 16%(4.0%) of 784 firms are outliers for the same discrepancy range when $Y/\Phi(K, L) < 1$.

Thus, we need to work out an alternative theoretical device that will allow us to take explicit account of the distribution itself. We will discuss this in the following sections.

4.1.3 Copula theory

The statistical distributions of the explanatory variables L and K, and the explained variable Y, in the production function are treated equally by using a PDF. The PDF is determined by

$$f(L, K, Y) = \frac{1}{N} \sum_{i=1}^{N} \delta(L - L_i) \delta(K - K_i) \delta(Y - Y_i), \tag{4.9}$$

where the quantities with a subscript i are those of a firm i and $\delta(x)$ denotes Dirac's delta function. Equation (4.9) is used as an estimator for a theoretical PDF. The CDF corresponding to Eq. (4.9) is given by

$$\bar{F}(L, K, Y) = \int_0^L \int_0^K \int_0^Y dL'dK'dY' f(L', K', Y'). \tag{4.10}$$

The marginal PDFs with two variables are deduced from $p(L, K, Y)$ by integrating over one of the variables, e.g.,

$$f(L, Y) = \int_0^\infty dK\, f(L, K, Y). \tag{4.11}$$

This relation is cast onto the CDFs such as

$$\bar{F}(L, Y) = \int_0^L \int_0^Y dL'dY' f(L', Y') = \bar{F}(L, K = \infty, Y). \tag{4.12}$$

The marginal PDF for each variable is likewise deduced from the binary PDFs as

$$f(Y) = \int_0^\infty dL\, f(L, Y), \tag{4.13}$$

for instance. Corresponding to Eq. (4.13), the marginal CDF is related to the binary CDF through

$$\bar{F}(Y) = \int_0^Y dY' f(Y') = \bar{F}(L = \infty, Y). \tag{4.14}$$

Unlike the production function given by Eq. (4.4), Y is not determined uniquely as a function of K and L, but obeys a certain PDF in current framework. The PDF of Y at a given set of L and K is calculated as a conditional probability given by

$$f(Y|L,K) = \frac{f(L,K,Y)}{f(L,K)}. \tag{4.15}$$

This describes the data exactly in a probabilistic way. Our task is thereby to adopt a mathematical measure suitable for describing $f(L,K,Y)$. In passing, we note that Eq. (4.15) corresponds to the generalized production function which incorporates the stochastic nature of the total factor productivity, the coefficient A in Eqs. (4.5) and (4.7).

Definition of copula

We take advantage of **copulas** to devise an analytic model for $f(L,K,Y)$. Copulas measure the dependence among stochastic variables. In other words, copulas extract correlations among stochastic variables free from the marginal CDFs. A number of different types of copulas have been proposed.

Sklar's theorem (Sklar, 1959; Nelsen, 2006) guarantees that $\bar{F}(L,K,Y)$ is a unique function of the marginal CDFs associated with L, K, and Y:

$$\bar{F}(L,K,Y) = C(u_L, u_K, u_Y), \tag{4.16}$$

where

$$u_s := \bar{F}(s) = 1 - F(s) \quad (s = L, K, Y) \tag{4.17}$$

are assumed to be continuous functions of s. The function $C(u_L, u_K, u_Y)$ is called a copula. The PDF, $f(L,K,Y)$, is then derived from $\bar{F}(L,K,Y)$ by the partial differentiation with respect to each of the variables:

$$f(L,K,Y) = \frac{\partial^3 \bar{F}(L,K,Y)}{\partial L \partial K \partial Y} = f(L)f(K)f(Y)c(u_L, u_K, u_Y), \tag{4.18}$$

where $c(u_L, u_K, u_Y)$ is the copula density defined by

$$c(u_L, u_K, u_Y) := \frac{\partial^3 C(u_L, u_K, u_Y)}{\partial u_L \partial u_K \partial u_Y}. \tag{4.19}$$

The copula density reduces to 1 if the variables are statistically independent of each other. Hence, the copula density is referred to as the correlation function in many-body physics (Hansen and McDonald, 2006). The copula describes genuine correlations among the variables.

The copulas have the boundary conditions:

$$C(u_L, u_K = 0, u_Y) = 0, \tag{4.20}$$

$$C\left(u_L, u_Y\right) = C\left(u_L, u_K = 1, u_Y\right). \tag{4.21}$$

Equation (4.20) is apparent from the definition of CDF given by Eq. (4.10); Eq. (4.21) just rephrases Eq. (4.12) (Nelsen, 2006). In addition, we note that the copulas read

$$C(u_s, u_{s'}) = u_s u_{s'}, \tag{4.22}$$

$$C(u_L, u_K, u_Y) = u_L u_K u_Y, \tag{4.23}$$

if the financial quantities are independent of each other.

Archimedean copulas

Archimedean copulas (Nelsen, 2006) are a well-known family of copula. They have been widely used because of their ease of mathematical handling together with the diversity of their correlation properties. We first explain Archimedean copulas with two variables and then proceed to those with many variables.

A bivariate Archimedean copula can be readily constructed from a generator function $\eta(u)$, which satisfies

$$C_A(u_1, u_2) = \eta^{-1}\left(\eta(u_1) + \eta(u_2)\right), \tag{4.24}$$

where we assume that $\eta(u)$ is a continuous and strictly decreasing function mapping $[0,1]$ to $[0,\infty]$ with $\eta(1) = 0$ so that its inverse $\eta^{-1}(t)$ is well-defined for $0 \leq t \leq \infty$. We also impose convexity on $\eta(u)$ so that Eq. (4.24) is a valid copula.

We will test three typical Archimedean copulas to model the real data. The first is the **Frank copula** defined by

$$\eta_F(u; \theta) = -\log\left(\frac{e^{-\theta u} - 1}{e^{-\theta} - 1}\right) \quad \text{with} \; -\infty \leq \theta \leq \infty \; (\theta \neq 0), \tag{4.25}$$

$$C_F(u_1, u_2; \theta) = -\frac{1}{\theta} \log\left[1 + \frac{\left(e^{-\theta u_1} - 1\right)\left(e^{-\theta u_2} - 1\right)}{\left(e^{-\theta} - 1\right)}\right]. \tag{4.26}$$

The second is the **Gumbel copula**:

$$\eta_G(u; \theta) = (-\log u)^\theta \quad \text{with} \; 1 \leq \theta \leq \infty, \tag{4.27}$$

$$C_G(u_1, u_2; \theta) = \exp\left[-\left\{(-\log u_1)^\theta + (-\log u_2)^\theta\right\}^{1/\theta}\right]. \tag{4.28}$$

The last is the **Clayton copula** given[1] by

$$\eta_C(u;\theta) = \frac{1}{\theta}\left(u^{-\theta} - 1\right) \quad \text{with } -1 \leq \theta \leq \infty \, (\theta \neq 0), \tag{4.29}$$

$$C_C(u_1, u_2; \theta) = \left[\max\left(u_1^{-\theta} + u_2^{-\theta} - 1, 0\right)\right]^{-1/\theta}. \tag{4.30}$$

These three copulas possess different correlation properties as shown in Fig. 4.6. The strength of dependence in the Frank copula is almost flat and spans the lower part ($u_1, u_2 \simeq 0$) to the upper part ($u_1, u_2 \simeq 1$). The Gumbel copula has stronger correlation in the upper part than in the lower part, while the Clayton copula shows a reversed correlation structure.

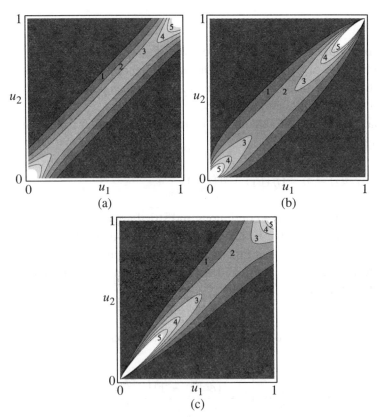

Fig. 4.6 Contour plots of the bivariate copula densities. Panel (a) is for Frank with $\theta = 14.14$; (b), for Gumbel with $\theta = 4$; and (c), for Clayton with $\theta = 6$. The parameter θ is chosen to yield the same value of the Kendall τ ($= 0.75$) for each copula. Level values of the contours are annotated on the figures (from Iyetomi et al. (2012)).

---

[1] For $\theta < 0$, $\eta_C^{-1}(t)$ should be replaced by the pseudo-inverse $\eta_C^{[-1]}(t)$ which is equal to $\eta_C^{-1}(t)$ in $0 \leq t \leq \eta_C(0)$ but set to be 0 beyond $t = \eta_C(0)$.

This asymptotical correlation structure of copulas is, however, superficial. In fact, we derive a survival copula from a given copula, by changing the variables, $u_i \to 1 - u_i$. For instance, the survival Clayton copula (s-Clayton) is given by

$$\hat{C}_C(u_1, u_2; \theta) = u_1 + u_2 - 1 + C_C(1 - u_1, 1 - u_2; \theta), \tag{4.31}$$

i.e., $\hat{C}_C(u_1, u_2; \theta)$ is related to the original Clayton copula (Nelsen, 2006). As can be easily appreciated, $\hat{C}_C(u_1, u_2)$ has stronger correlation in the upper part ($u_1, u_2 \simeq 1$).

The bivariate Archimedean copulas can be generalized to multivariate copulas with n variables:

$$C_A(u_1, \cdots, u_n) = \eta^{-1}(\eta(u_1) + \cdots + \eta(u_n)). \tag{4.32}$$

The generator function, however, is required to satisfy a more stringent mathematical constraint that its inverse $\eta^{-1}(x)$ is completely monotonic, i.e.,

$$(-1)^m \frac{d^m \eta^{-1}(x)}{dx^m} \geq 0 \quad (m = 0, 1, 2, \cdots). \tag{4.33}$$

We refer the readers to the textbook by Nelsen (2006) for details. The generators of Frank, Gumbel, and Clayton are completely monotonic with the following conditions for θ: $\theta > 0$ (Frank), $\theta > 1$ (Gumbel), and $\theta > 0$ (Clayton). We will call the multivariate Archimedean copula defined by Eq. (4.32) in a more specific way after its generator; namely, the copula given by Eq. (4.32) generated by Eq. (4.25) is referred to as the n-variate Frank copula and so forth.

Generalization of Archimedean copulas

The Archimedean copulas are sometimes too restrictive to accommodate real data because they are symmetric for the exchange of variables. However, for bivariate copulas, this constraint is relaxed by

$$C_A(u_1, u_2; \theta, \alpha, \beta) = u_1^{1-\alpha} u_2^{1-\beta} C_A(u_1^\alpha, u_2^\beta; \theta), \tag{4.34}$$

where two additional parameters with $0 \leq \alpha, \beta \leq 1$ are introduced. When $\alpha = \beta$, the generalized copula is exchangeable. Hence, the difference between α and β measures the degree of asymmetry in correlations. Taking a special limit of $\alpha = \beta = 1$, Eq. (4.34) recovers the original Archimedean copula.

Moreover, we recall that the trivariate Archimedean copula, i.e., Eq. (4.32) with $n = 3$, is characterized by a single parameter θ. This means that all marginal CDFs have the same correlation structure. This is not true for real data as has been already indicated; so we generalize the original form as

$$C_{A\text{-nex}}(u_1, u_2, u_3; \theta_1, \theta_2) = C_A(C_A(u_1, u_2; \theta_2), u_3; \theta_1)$$

$$= \eta_1^{-1}\left(\eta_1\left(\eta_2^{-1}(\eta_2(u_1) + \eta_2(u_2))\right) + \eta_1(u_3)\right), \tag{4.35}$$

with different generating functions $\eta_1(u) = \eta(u;\theta_1)$ and $\eta_2(u) = \eta(u;\theta_2)$. The copula given by Eq. (4.35) has two characteristic parameters θ_1 and θ_2 leading to the margin $C_A(u_1, u_2; \theta_2)$, generated by $\eta_2(u)$ and the remaining margins, $C_A(u_1, u_3; \theta_1)$ and $C_A(u_2, u_3; \theta_1)$ generated by $\eta_1(u)$. If we set $\theta_1 = \theta_2$ in Eq. (4.35), the exchangeable Archimedean copula given by Eq. (4.32) with $n = 3$ is recovered. We note that the condition $\theta_1 \leq \theta_2$ should be satisfied for Eq. (4.35) to be a copula. We refer to the generalized form given by Eq. (4.35) as a non-exchangeable Archimedean copula (McNeil et al., 2005).

4.1.4 Construction of a production copula

In this section, we model real data by using copulas. We refer to the resulting copula as a **production copula**.

Bivariate

Let us focus our attention on the bivariate correlations for all pairs of L, K, and Y as depicted in Fig. 4.3. Three Archimedean copulas, i.e., Frank, Gumbel, and Clayton, are examined as has been mentioned. Actually, we use the survival copula of the Clayton copula instead of the original one. This is because the former is better suited to a description of the correlation structure in real data than the latter.

Table 4.1 Maximized log-likelihood ℓ in fitting bivariate Archimedean copulas (Frank, Gumbel, s-Clayton) to the pair correlations in real data as shown in Fig. 4.3. The Kendall rank correlation coefficient τ corresponding to the value of θ is listed for each copula (from Iyetomi et al. (2012)).

Copula		K-Y	L-Y	L-K
Frank	θ	11.0	21.2	9.23
	ℓ	940.1	1612.4	790.2
	τ	0.691	0.826	0.644
Gumbel	θ	3.21	5.30	2.73
	ℓ	1081.8	1694.0	892.1
	τ	0.688	0.811	0.634
s-Clayton	θ	3.43	6.11	2.59
	ℓ	992.1	1483.7	787.3
	τ	0.632	0.753	0.564

By using the maximum likelihood estimation, we fitted those copulas to the pair correlations shown in Fig. 4.3. The results are summarized in Table 4.1. Here, the optimized values of the correlation parameter θ and the maximized log-likelihood ℓ are listed. We also calculated the Kendall rank correlation coefficient τ from the θs. From Table 4.1, we observe that the Gumbel copula gives the best fit among the three copulas tested as manifested by the largest value for ℓ. Hence, we compare the optimized results obtained with the Gumbel copula with the corresponding empirical copulas derived from the real data in Fig. 4.7. These two results are in good agreement with each other. The comparison for copula densities are also shown in Fig. 4.8.

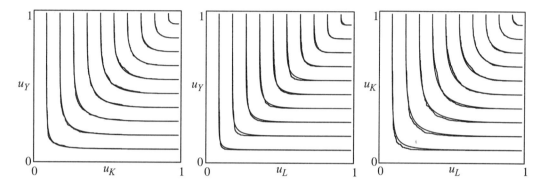

Fig. 4.7 Modeling of the pair correlations in Fig. 4.3 in terms of the Gumbel copula. The fitted results (smooth solid curves) are compared with the corresponding real data (jagged solid curves); the difference between the two curves is almost invisible. Ten contours are drawn at equal spacing ranging from 0 (on the bottom horizontal axis and the left vertical axis) to 1 (at the top right corner) on each panel (from Iyetomi et al. (2012)).

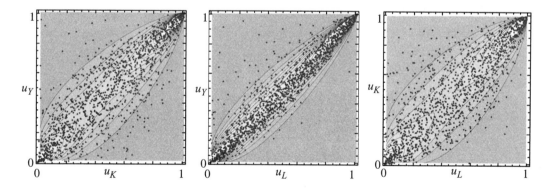

Fig. 4.8 Contour plots of the copula densities corresponding to the fitted copulas in Fig. 4.7. The dots refer to the real data. The contours are equally spaced with the interval ranging from 0.5 (dark side) to 5 (bright side) (from Iyetomi et al. (2012)).

The overall strength of the pair correlations is represented by θ and τ. Comparison of θs and τs in Table 4.1 confirms the conclusion derived from Fig. 4.3, i.e., the L-Y correlation is significantly stronger than the K-Y and L-K correlations. In contrast, the K-Y and L-K correlations resemble each other. Furthermore, we address to what extent the bivariate correlations among L, K, and Y are asymmetric. By using the asymmetric Gumbel copula given by Eq. (4.34) with Eq. (4.27), we repeat the maximum likelihood estimate and obtain Table 4.2. This result demonstrates that none of the pairs has notable asymmetry in its correlation structure.

Table 4.2 Degree of asymmetry in the pair correlations in Fig. 4.3 estimated using the asymmetric Gumbel copula (from Iyetomi et al. (2012)).

	K-Y	L-Y	L-K
θ	3.34	6.59	2.86
α	0.994	0.999	1.000
β	0.984	0.934	0.957
ℓ	1090.1	1779.5	896.0

Trivariate

We are now ready to discuss the construction of the production copula. The models we consider here are enumerated as:

Model (I) Frank copula;

$$C^{(\mathrm{I})}(u_L, u_K, u_Y) = C_\mathrm{F}(u_L, u_K, u_Y; \theta). \quad (4.36)$$

Model (II) Gumbel copula;

$$C^{(\mathrm{II})}(u_L, u_K, u_Y) = C_\mathrm{G}(u_L, u_K, u_Y; \theta). \quad (4.37)$$

Model (III) s-Clayton copula;

$$C^{(\mathrm{III})}(u_L, u_K, u_Y) = \hat{C}_C(u_L, u_K, u_Y; \theta)$$
$$= u_L + u_K + u_Y - 1 + C_\mathrm{C}(1 - u_L, 1 - u_K, 1 - u_Y; \theta). \quad (4.38)$$

Model (IV) Non-exchangeable Gumbel copula;

$$C^{(\mathrm{IV})}(u_L, u_K, u_Y) = C_\mathrm{G}(C_\mathrm{G}(u_L, u_Y; \theta_2), u_K; \theta_1). \quad (4.39)$$

Here, the trivariate Archimedean copulas $C_{\mathrm{F,G,C}}(u_L, u_K, u_Y; \theta)$ are constructed from their η-functions given by Eq. (4.25), Eq. (4.27), and Eq. (4.29), respectively, by using Eq. (4.32). The first three models, i.e., $C^{(\mathrm{I})}, C^{(\mathrm{II})}, C^{(\mathrm{III})}$, are specified by a single parameter result in all the marginal CDFs and PDFs with the same correlation structure. The last model $C^{(\mathrm{IV})}$ is based on Eq. (4.35) having Eq. (4.28) plugged into $C_\mathrm{A}(u_i, u_j; \theta_1)$ and $C_\mathrm{A}(u_i, u_j; \theta_2)$. The variables u_1, u_2, and u_3 in Eq. (4.35) should read u_L, u_Y, and u_K, respectively. The last model can account for the difference in correlations among the marginals with different values for θ_1 and θ_2, which is a desirable property as we will see.

Productivity Distribution and Related Topics

Table 4.3 Maximum likelihood optimization in the models for the production copula (from Iyetomi et al. (2012)).

	Model (I)	Model (II)	Model (III)	Model (IV)
θ	11.5	3.27	3.34	2.89 (θ_1), 5.26 (θ_2)
ℓ	2207.5	2428.2	2164.3	2701.1
AIC	−4413.0	−4854.4	−4326.6	−5398.2

We applied the four models to the real data using the maximum likelihood method. The optimization results are listed in Table 4.3. As expected, from modeling the bivariate correlations, the Gumbel copula used in Model (II) outperforms the fit when compared with the other Archimedean copulas in Models (I) and (III). This is the motivation behind choosing Model (IV) in which the Gumbel copula is specially selected. The result of the optimization based on the generalized model is also listed in Table 4.3. The numbers of parameters are different for Models (II) and (IV); therefore, we have to replace the maximum log-likelihood ℓ by Akaike's information criterion, AIC for model selection (Akaike, 1974b); the model with the smallest AIC is adopted. Comparison of the AIC proves that Model (IV) significantly improves the fit over even Model (II).

We delve deeper into the performance of Model (IV). To make a detailed comparison with the real data, we introduce a trivariate copula cumulant[2] defined by

$$\Omega(u_L, u_K, u_Y) = C(u_L, u_K, u_Y) - u_K C(u_L, u_Y) - u_L C(u_K, u_Y) \\ - u_Y C(u_L, u_K) + 2 u_L u_K u_Y. \tag{4.40}$$

Here, the contributions owing to the bivariate correlations are subtracted from the trivariate copula. We will consider two special cases to understand this. First, assume all the variables are independent of each other. Replacement of the copulas on the right-hand side of Eq. (4.40) by the corresponding independent copulas, Eq. (4.22) and Eq. (4.23) leads to $\Omega(u_L, u_K, u_Y) = 0$. Next, assume that only one pair of L and K are correlated among the three variables; the copula $C(u_L, u_K, u_Y)$ can then be decomposed as

$$C(u_L, u_K, u_Y) = u_Y C(u_L, u_K). \tag{4.41}$$

Thus, the bivariate copulas involving u_Y as a variable can be replaced by the corresponding independent copulas. Again, we see that $\Omega(u_L, u_K, u_Y)$ vanishes in this case. Such subtraction is also manifested by the boundary condition that the copula cumulant vanishes on the marginal boundaries:

$$\Omega(1, u_K, u_Y) = \Omega(u_L, 1, u_Y) = \Omega(u_L, u_K, 1) = 0. \tag{4.42}$$

Here, Fig. 4.9 shows the fitted results for $\Omega(u_L, u_K, u_Y)$ in Model (IV) together with the corresponding empirical data on the typical cross-sections A-D depicted in Fig. 4.10. We

---
[2] Cumulant functions are usually defined in terms of PDFs in place of CDFs (Hansen and McDonald, 2006).

observe that the copula model reproduces the empirical results in an almost indistinguishable manner. In addition, Fig. 4.11 gives more in-depth comparison of the results between the copula model and real data for $\Omega(x,x,x)$ along the diagonal specified by $x = u_L = u_K = u_Y$. Again, switching from Model (II) to (IV) leads to significant improvement in reproducing the empirical copula cumulant.

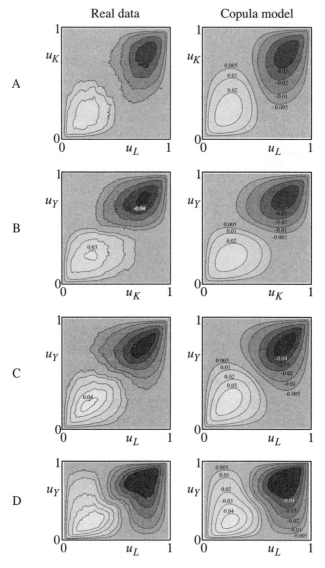

Fig. 4.9 Contour plots of the copula cumulant Eq. (4.40) on the cross sections in the u_L-u_K-u_Y space as specified in Fig. 4.10. The results obtained from real data (left-hand side) are compared with those derived from the production copula in Model (IV) (right-hand side); the contours are drawn at the same levels on both sides (from Iyetomi et al. (2012)).

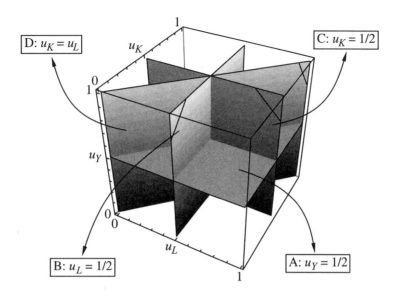

Fig. 4.10 Typical cross sections in u_L-u_K-u_Y space (from Iyetomi et al. (2012)).

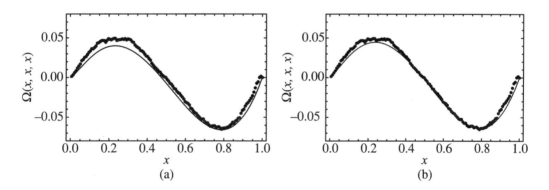

Fig. 4.11 Comparison of the results for $\Omega(x,x,x)$ between the copula model (solid curve) and real data (dots). Panel (a) shows Model (II) and panel (b), Model (IV) (from Iyetomi et al. (2012)).

Now that the copula model has been established, it can be used for various economic investigations of firms. We demonstrate one such an example. We carried out a simulation for the production activity of the Japanese listed firms in 2006 and display the result in Fig. 4.12. The simulated points ($N = 1360$) were generated using a rejection method (Press et al., 2007) with the trivariate PDF obtained by combining the production copula and the marginal PDFs according to Eq. (4.18). As shown in Fig. 4.12, for reference, we did a simulation based on the random model where no correlations among the financial quantities were taken into account. The comparison confirms the effectiveness of the production copula.

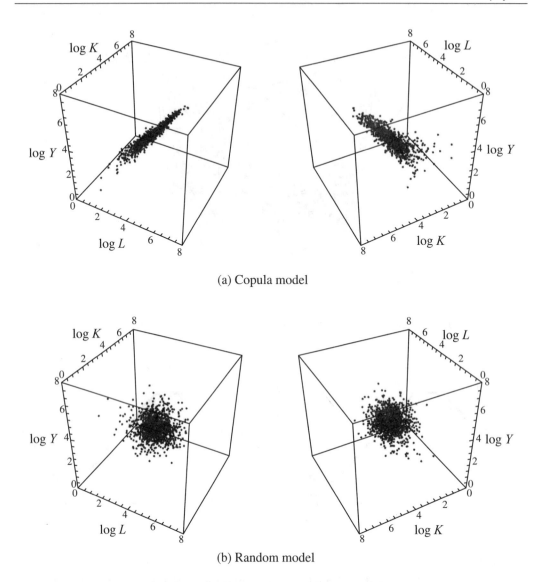

Fig. 4.12 A simulated result for the production activity of listed Japanese firms obtained with the production copula in Model (IV), accompanied by that based on the random model without any correlations among the financial quantities (from Iyetomi et al. (2012)).

We calculated the deviation from the CD production function for the copula and random models, as has been already done for the real data in Fig. 4.4. We can also express the CCDF of the ratio $\xi = Y/\Phi(L, K)$ on both sides in terms of the trivariate copula. For instance, the CCDF on the upper side is given by

$$\bar{F}(\xi) = f(\xi)/f(1) \quad (1 \leq \xi < \infty), \tag{4.43}$$

where

$$f(\xi) = 1 - \int_0^1 du_L \int_0^1 du_K \frac{\partial^2 C(u_L, u_K, u_{Y_\xi})}{\partial u_L \partial u_K}, \qquad (4.44)$$

with

$$Y_\xi = \xi\Phi(L, K) = \xi\Phi(\bar{F}^{-1}(u_L), \bar{F}^{-1}(u_K)). \qquad (4.45)$$

For the random model, Eq. (4.44) is replaced by

$$f(\xi) = 1 - \int_0^1 du_L \int_0^1 du_K u_{Y_\xi}. \qquad (4.46)$$

A detailed comparison of the results based on the copula and random models with the corresponding real data is shown in Fig. 4.13. It shows that the correlations involved in the real data are well reproduced by the copula model, except for outliers on the upper side occupying about 5% of the data; their functional behavior is rather close to that of the results in the random model.

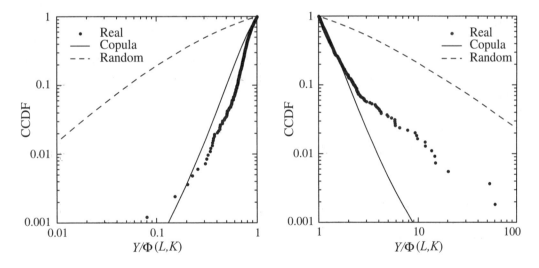

Fig. 4.13 The CCDFs of $Y/\Phi(L, K)$ calculated in Model (IV), compared with the corresponding results based on the real data and those based on the random model as demonstrated in Fig. 4.12 (from Iyetomi et al. (2012)).

In fact, we are required to determine totally 14 parameters, 12 for the three marginal distributions and two for the copula to calculate $f(L, K, Y)$ in the current copula model. The production copula may appear to contain many more parameters than the CD production function, which has just three. However, this is not true because we have to specify $f(L, K)$ to take into account correlations between the input variables for the production function. This is a fair way to compare the two ideas. If the GB2 and the Gumbel copula are employed, totally nine additional parameters appear, that is, eight for the two marginal distributions and one for the copula. Thus, we can claim that the production copula is very successful in reproducing the real data and is workable, considering it has only two extra parameters.

Finally, we remark that economists have studied microscopic foundation of the macroscopic production function for a long time. Houthakker (1955) derived the Cobb–Douglas production function on the basis of the assumption that labor and capital coefficients obey power distributions; later, his approach was elaborated by Sato (1975). With the production copula, it is not necessary to solve such a long-standing aggregation problem. The production copula bridges the gap between micro- and macro-economics in a seamless manner.

4.2 Stochastic Macro-equilibrium

4.2.1 The basic idea

In this section, we consider the distribution of workers across different levels of labor productivity. Workers are always interested in better job opportunities and occasionally do change their jobs. While workers search for suitable jobs, firms also search for suitable workers. A firm's job offer is, of course, conditional on its economic performance. This analysis focuses on the firm's labor productivity, which increases because of capital accumulation and technical progress or innovations. However, those job sites with high productivity remain only *potential* unless firms face high enough demand for their products; firms may not post job vacancy signs and may even discharge existing workers when demand is low.

The explanations of statistical physics are as follows. Although we assume that firms with higher productivity make more attractive job offers to workers, we do not know how attractive they are to which workers. Whenever possible, workers move to firms with higher productivity, but we never know the particular reasons for such moves. For workers to move to firms with higher productivity, it is necessary that those firms must decide to fill the vacant job sites, and post a high enough number of vacancy signs and/or make enough hiring efforts. They will post such vacancy signs and make hiring efforts only when they face an increase in demand for their products, and decide to raise the level of production. It also goes without saying that high-productivity firms will retain their existing workers only when they face a high enough demand.

The question to ask is, what is the distribution of employed workers across firms with different productivities. Since microeconomic shocks to both workers and firms are so complex and vague, optimization exercises based on representative agent assumptions are not of much help. In particular, we never know how the aggregate demand is distributed across firms. Besides, among other things, the job arrival rate, the job separation rate, and the probability distribution of wages (or, more generally, the measure of the desirability of a job) differ across workers and firms. This recognition is the starting point of the fundamental method of statistical physics. Foley (1994), in his seminal application of this approach to general equilibrium theory, called this the "statistical equilibrium theory of markets." Following the lead of Foley (1994), Yoshikawa (2003) applied the concept to macroeconomics. At first, one might think that allowing too large a dispersion of individual characteristics leaves so many degrees of freedom that almost anything can happen. However, it turns out that the methods of statistical physics provide us with not only qualitative results but also with quantitative predictions.

In the current model, the fundamental constraint on the economy as a whole is aggregate demand D. Accordingly, for each firm facing a downward-sloping kinked individual demand curve, the level of demand is a fundamental constraint. The problem is to ascertain how the aggregate demand D is allocated to these monopolistically competitive firms. Our model offers a solution. The method used is standard in statistical physics. The basic idea behind the analysis can be explained using the simplest case. We focus on productivity dispersion here.

The statistical equilibrium theory of an ideal gas of *distinguishable* particles has already been given in Sec. 2.2. We begin by assuming that n_k workers are working at firms with productivity c_k ($k = 1, 2, \cdots, K$), where there are K levels of productivity in the economy arranged in the ascending order: $c_1 < c_2 < \cdots < c_K$. The analogy between physics and economics for a statistical physics formulation of productivity distribution is then established through the following mapping:

$$
\begin{array}{ccc}
\text{particles} & \leftrightarrow & \text{workers} \\
\text{microscopic states} & \leftrightarrow & \text{firms} \\
\text{energy level } \epsilon_k & \leftrightarrow & \text{productivity level } c_k \\
\text{total energy } E & \leftrightarrow & \text{total output } Y
\end{array}
$$

The first macro-constraint is concerned with labor endowment:

$$N = \sum_{k=1}^{K} n_k. \tag{4.47}$$

The second macro-constraint is concerned with effective demand. We assume that given that aggregate demand D is balanced by the total output Y:

$$D = Y = \sum_{k=1}^{K} c_k n_k. \tag{4.48}$$

The Lagrangian, Eq. (2.81), now reads:

$$L = -N \sum_{k=1}^{K} \left(\frac{n_k}{N}\right) \ln \left(\frac{n_k}{N}\right) - \alpha N - \beta D, \tag{4.49}$$

Corresponding to Eq. (2.87), the number of workers in firms with productivity c_k is exponentially distributed:

$$\frac{n_k}{N} = e^{-\alpha - \beta c_k} = \frac{e^{-\beta c_k}}{\sum_{k=1}^{K} e^{-\beta c_k}}. \tag{4.50}$$

Here arises a crucial difference between economics and physics. Whenever possible, particles tend to move toward the *lowest* energy level. On the contrary, in economics, workers always strive for better jobs offered by firms with higher productivity c_k. As a result of optimization under unobservable respective constraints, workers move to better jobs. In fact, if allowed all the workers would move up to the job sites with the highest productivity, c_K. This situation corresponds to the textbook Pareto optimal Walrasian equilibrium with no frictions and uncertainty. However, this state is actually impossible unless the level of aggregate demand D is so high as to equal the maximum level $D_{\max} = c_K N$. When D is lower than $D_{\max}$, the story is quite different. Some workers — a majority of workers, in fact, must work at job sites with productivity lower than c_K.

How are workers distributed over job sites with different productivities? Obviously, that depends on the level of aggregate demand. When D reaches its lowest level, $D_{\min}$, workers are distributed evenly across all the firms with different levels of productivity, $c_1, c_2, \cdots, c_K$. Here, $D_{\min}$ is defined as $D_{\min} = N(c_1 + c_2 + \cdots + c_k)/K$. It is easy to see that the lower the level of D is, the greater the combinatorial number of distribution $(n_1, n_2, \cdots, n_K)$ which satisfies the aggregate demand constraint, Eq. (4.48), becomes.

Entropy S increases when D decreases. For example, in the extreme case where D is equal to the maximum level $D_{\max}$, all the workers work at job sites with the highest productivity. In this case, the entropy S becomes zero, its lowest level because $n_K/N = 1$ and $n_k/N = 0$ ($k \neq K$). At the other extreme, where aggregate demand is equal to the minimum level $D_{\min}$, we have $n_k = N/K$, and entropy S reaches its maximum level, $\ln K$. The relation between the entropy S and the level of aggregate demand D is, therefore, as schematically represented in Figure 4.14.

At this stage, we recall that β in Eq. (4.49) is determined according to

$$\beta = \frac{\partial L}{\partial D} = \frac{\partial S}{\partial D}, \tag{4.51}$$

where β is the slope of the tangent of the curve as shown in Figure 4.14 and is, therefore, *negative*.

In physics, β is normally positive. This difference arises because workers strive for job sites with higher productivity, not the other way round (Iyetomi, 2012). In physics, β is equal to the inverse of temperature, or more precisely, temperature is defined as the inverse of $\partial S/\partial D$ when S is the entropy and D, energy. Thus, a negative β means temperature is negative. It may sound odd, but the notion of negative temperature is perfectly legitimate in such systems as the one in the current analysis; see Section 73 of Landau and Lifshitz

(1980) and Appendix E of Kittel and Kroemer (1980). With a negative β, the exponential distribution, Eq. (4.50) is upward-sloping. However, unless the aggregate demand is equal to (or greater than) the maximum level, $D_{\max}$, workers' efforts to access job sites with the highest productivity c_K must be frustrating because firms with the highest productivity do not employ a large number of workers and are less aggressive in recruitment, and accordingly, it becomes harder for workers to find such jobs. This means that workers are distributed over all the job sites with different levels of productivity.

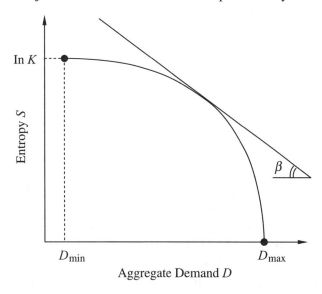

Fig. 4.14 Relationship between entropy S and aggregate demand D, where β is the Lagrangian multiplier in Eq. (4.49) in the text. (From Aoyama et al. (2015b))

The maximization of entropy under the aggregate demand constraint (4.48), in fact, balances two forces. On the one hand, whenever possible, workers move to better jobs identified with job sites with higher productivity. This is the outcome of successful job matching resulting from the worker's search and the firm's recruitment. When the level of aggregate demand is high, this force dominates. However, when D is lower than $D_{\max}$, there are, in general, a number of different allocations $(n_1, n_2, \cdots, n_K)$ that are consistent with D.

As we argued earlier, micro shocks facing both workers and firms are truly unspecifiable. We simply do not know which firms with what productivity face how much of a demand constraint and need to employ how many workers with what qualifications. We do not know which workers are seeking what kind of jobs with how much productivity, either. Here comes the maximization of entropy. It gives us the distribution $(n_1, n_2, \cdots, n_K)$ which corresponds to the maximum combinatorial number consistent with the given D.

Standard economic theory, and search theory in particular, emphasizes non-trivial job matching in labor markets with frictions and uncertainty. Our analysis shows that the matching of high-productivity jobs is ultimately conditioned by the level of aggregate demand. That is, uncertainty and frictions emphasized by the standard search theory are *not* exogenously given, but depend crucially on aggregate demand. In a booming gold-rush town, one would not waste a minute to find a good job! The opposite holds in a depressed city.

It is essential to understand that the current approach does *not* regard economic agents' behaviors as random. Certainly, firms and workers maximize their profits and utilities. This analysis, in fact, presumes that workers always strive for better jobs characterized by higher productivity. The randomness underlying entropy maximization comes from the fact that both the objective functions of and constraints facing a large number of economic agents are constantly subject to *unspecifiable* micro shocks. We must recall that the number of households is of the order 10 million, and the number of firms, 1 million. Therefore, outside observers, namely economists analyzing the macro-economy, must regard a particular allocation *under macro-constraints* as equi-probable. Thus, it is most likely that the allocation of aggregate demand and workers which maximizes the probability P_n or Eq. (2.80) *under macro-constraints* is realized.

This method has time and again been successful in the natural sciences in the analysis of objects comprising many micro elements. Economists might be afraid that optimization of economic agents is essentially different from optimization of inorganic atoms and molecules comprising gas. However, there is not much difference between the behavior of the Ramsey consumer and that of particles; both are described by the Euler equation for a problem of calculus of variation. The method of statistical physics can be usefully applied not because motions of micro units are "mechanical," but because the object or system under investigation comprises the many micro units, individual movements of which we are unable to know.

The aforementioned analysis shows that the distribution of workers at firms with different productivities depends crucially on the level of aggregate demand. As Yoshikawa (2003, 2014) explains, this analysis provides the right micro-foundations for Keynesian macroeconomics. Although a simple model is useful to explain the basic idea, it will not be applicable to the empirically observed distribution of labor productivity (Aoyama et al., 2015a,b).

4.2.2 Empirical distribution of productivity

In this study, we take advantage of the Orbis database (Bureau van Dijk Electronic Publishing, Brussels, Belgium) that has information on more than 120 million firms across the globe. We focus on Japanese firms with non-empty entries in annual operating revenue Y and the number n of employees, and calculate the labor productivity c of firms according to Eq. (4.3). We count 31,512 manufacturing firms and 304,006 non-manufacturing firms in 2012. The data source is different from that of our previous studies (Souma et al., 2009; Iyetomi, 2012; Aoyama et al., 2015a), which were based on the data set constructed by unifying two domestic databases, the NEEDS database (Nikkei

Digital Media, Inc., Tokyo, Japan) for large firms and the CRD database (CRD Association, Tokyo, Japan) for small to medium firms. The Orbis database used here may enable us to extend our study to labor productivity in other countries and do international comparisons of its dynamics.

Figure 4.15 shows the PDF of firms and workers with respect to $\log c$, empirically determined from the Orbis database in 2012. Here, c is measured in units of 1000 USD per person. The fact that the major peak of the latter shifts to the right compared to that of the former indicates that the average number $\bar{n}$ of workers per firm increases in this region. In fact, Figure 4.16 shows the functional dependence of $\bar{n}$ on the labor productivity c of firms. We observe that as productivity increases, it first goes up to about $n \simeq 100$ and then decreases. Iyetomi (2012) explained the upward-sloping distribution in the low productivity region by introducing the negative temperature theory. The downward-sloping part in the high productivity region is close to linear (denoted by the dotted line) in this double-log plot. This indicates that it obeys the power law:

$$\bar{n} \propto c^{-\gamma}. \tag{4.52}$$

We thus see that the number of workers exponentially increases as c increases up to a certain level of productivity, and then decreases with the power-law form, Eq. (4.52), in the high productivity region. The latter behavior of $\bar{n}$ may be somewhat counter-intuitive, because firms that have achieved higher productivity through innovation and high-quality management would continue to grow larger and larger leading to a monotonically increasing $\bar{n}$ with c.

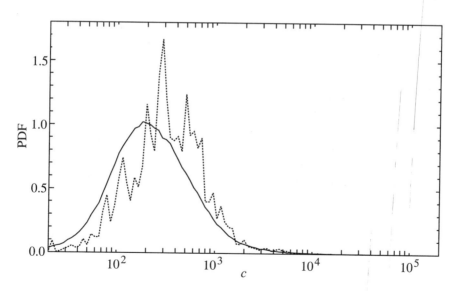

Fig. 4.15 Probability density function (PDF) of firms' $\log c$ (solid line) and workers' $\log c$ (dotted line) in 2012. The labor productivity c is measured in units of 1000 USD/person. (From Aoyama et al. (2015b))

In the previous section, we demonstrated that entropy maximization under macro constraints leads to an exponential distribution. This distribution with negative β can explain the broad pattern of the left-hand side of the distribution shown in Fig. 4.16, namely an upward-sloping exponential distribution (Iyetomi, 2012). However, we cannot reproduce the downward-sloping power distribution for high-productivity firms. To explain this, we need to make an additional assumption that the number of potentially available high-productivity jobs is limited and it decreases as the level of productivity rises (Aoyama et al., 2015a).

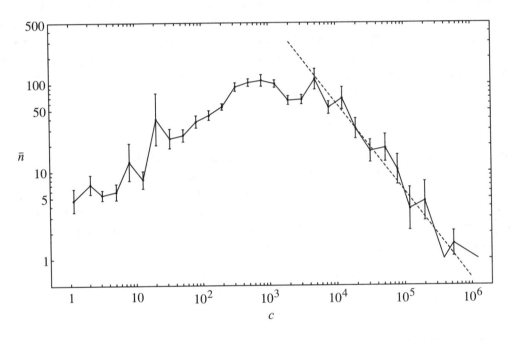

Fig. 4.16 Dependence of the average number $\bar{n}$ of workers of individual firms on the labor productivity c (dots with their error bars connected by thick lines) in 2012. The dotted straight line has a gradient -1, that is, $\bar{n} \propto 1/c$. (From Aoyama et al. (2015b))

Potential jobs are created by firms by accumulating capital and/or introducing new technologies, particularly new products. On the other hand, they are destroyed by firms' losing demand for their products permanently. Schumpeterian innovations by way of creative destruction raise the levels of some potential jobs, but at the same time lower the levels of others. In this way, the number of potential jobs with a particular level of productivity keeps changing. Note, however, that they remain *potential* because firms do not necessarily attempt to fill all the job sites with workers. To fill them, firms either keep the existing workers on the job or post job vacancy signs and make enough hiring efforts, but these are economic decisions and depend crucially on the economic conditions facing firms. The number of potential job sites, therefore, is not exactly equal to, but rather imposes a ceiling on, the sum of the number of filled job sites, or employment, and the unfilled jobs.

Under reasonable assumptions, distribution of potential job sites with high productivity becomes a downward-sloping power law. Adapting the model of Marsili and Zhang (1998b), we can derive a power-law distribution such as the one for the tail of the empirically observed distribution of labor productivity; see Yoshikawa (2014) for details. However, the determination of employment by firms with various levels of productivity is another matter. To fill potential job sites with workers is a firm's economic decision. The most important constraining factor is the level of demand facing the firm in the product market. To fill potential job sites, the firm must either retain its existing workers, or make enough hiring efforts including posting vacancy signs aimed at successful job matching. Such actions of the firms and job search of workers are purposeful. However, micro shocks affecting firms and workers are just unspecifiable. So how are workers actually employed at firms with various levels of productivity? This is the problem we considered in the previous section. In what follows, we will examine this in a more general framework.

The number of workers working at firms with productivity c_k, namely n_k is

$$n_k \in \{0, 1, \cdots, g_k\} \quad (k = 1, 2, \cdots K). \tag{4.53}$$

Here, g_k is the number of potential jobs with productivity c_k, and puts a ceiling on n_k. We assume that in the low productivity region, g_k is so large that n_k is virtually unconstrained by g_k. In contrast, in the high productivity region, g_k constrains n_k and actually diminishes in a power form as we have analyzed earlier. When the number of potential jobs with high productivity is limited, the behavior of economic agents necessarily becomes correlated; if good jobs are taken by some workers, it becomes more difficult for others to find such jobs. The current analysis represents this precisely by introducing ceilings on n_j.

4.2.3 Universal statistics

One way to analyze the equilibrium distribution of labor productivity based on statistical physics is to maximize entropy. Instead, Scalas and Garibaldi (2010) suggest that we can usefully apply the Ehrenfest–Brillouin model, a Markov chain to analyze the problem. We first present such a model (Aoyama et al., 2015a) and then a statistical physics formulation leading to the same results (Aoyama et al., 2015b).

Ehrenfest–Brillouin Model

The macro-economy consists of many firms with different levels of productivity. Differences in productivity arise from different capital stocks, levels of technology and/or demand conditions facing firms. We call a group of firms with the same level of productivity, a *cluster*. Workers randomly move from one cluster to another for various reasons at various times. Despite these random changes, the distribution of labor productivity as a whole remains stable because these incessant random movements balance out each other. This balancing must be achieved for each cluster, and is called the detail balance. In the following section, we present a general treatment of this detail balance using the particle-correlation theory of Costantini and Garibaldi (1989). The empirical results lead us to make the assumption that the number of workers who belong to clusters with high productivity is constrained; see also (Yoshikawa, 2011).

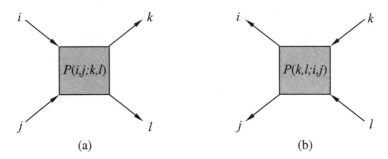

Fig. 4.17 (a) Elementary binary process where a worker at firm i and a worker at firm j move to firms k and ℓ, with probability $P(i, j; k, \ell)$ and (b) the reverse process, with probability $P(k, \ell; i, j)$. (From Aoyama et al. (2015b))

For the productivity distribution of workers specified by $\{n_1, n_2, n_3, \ldots\}$ to be in equilibrium, the number of workers who move *out* of cluster i per unit time must be equal to that of workers who move *into* this cluster per unit time. We consider the minimal process that satisfies the condition that the total output in the economy as a whole is conserved with Eq. (4.48). In this process, two workers move simultaneously (Garibaldi and Scalas, 2010; Scalas and Garibaldi, 2009). This is illustrated in Fig. 4.17. (a): A worker in a cluster with productivity c_i and a worker in a cluster with productivity c_j move to clusters with productivities c_k and c_ℓ, respectively. For the total output Y to remain constant, the following condition must be satisfied:

$$c_i + c_j = c_k + c_\ell. \qquad (4.54)$$

Such job switching occurs for various unspecifiable reasons. The best we can do is to consider a Markov chain defined by transition rates, $P(i, j; k, \ell)$. They have the following trivial symmetries:

$$P(i, j; k, \ell) = P(j, i; k, \ell), \quad P(i, j; k, \ell) = P(i, j; \ell, k). \qquad (4.55)$$

We also assume that the reverse process, illustrated in Fig. 4.17 (b) occurs with the same probability:

$$P(i, j; k, \ell) = P(k, \ell; i, j). \qquad (4.56)$$

The equilibrium condition then requires that the number of workers moving from (i, j) to (k, ℓ) per unit time denoted by $N(i, j; k, \ell)$ must be equal to the number of those from (k, ℓ) to (i, j) denoted by $N(k, \ell; i, j)$:

$$N(i,j;k,\ell) = N(k,\ell;i,j). \tag{4.57}$$

The flux $N(i,j;k,\ell)$ is proportional to n_i and n_j, the numbers of workers in clusters i and j, and the corresponding transition rate $P(i,j;k,\ell)$:

$$N(i,j;k,\ell) \propto P(i,j;k,\ell)n_i n_j. \tag{4.58}$$

The fundamental assumption we make is that a cluster with productivity c can accommodate $g(c)$ workers at most. This means that

$$N(i,j;k,\ell) = P(i,j;k,\ell)n_i n_j L(c_k, n_k) L(c_\ell, n_\ell), \tag{4.59}$$

where the factor $L(c,n)$ limits the number n of workers in a cluster with productivity c:

$$1 \geq L(c,n) \geq 0 \quad \text{for} \quad n \leq g(c), \tag{4.60}$$

$$L(c,n) = 0 \quad \text{for} \quad n > g(c). \tag{4.61}$$

One can obtain a general solution in terms of $L(c,n)$ for the detail-balance equation (4.57) as follows. Thanks to Eq. (4.56), substituting Eq. (4.59) into Eq. (4.57) gives,

$$H(c_i)H(c_j) = H(c_k)H(c_\ell), \tag{4.62}$$

where $H(c) := n/L(c,n)$. From Eq. (4.54), we obtain

$$H(c_i)H(c_j) = H(c_i + c_j)H(0), \tag{4.63}$$

or, by denoting $G(c) := \log(H(c)/H(0))$,

$$G(c_i) + G(c_j) = G(c_i + c_j). \tag{4.64}$$

This proves that $G(c)$ is linear in c if the continuity of $G(c)$ with respect to c is assumed. It thus leads us to derive

$$n = L(c,n)e^{-\beta(c-\mu)}, \tag{4.65}$$

where μ and β are real free parameters. Once the function $L(c,n)$ is given, the equilibrium distribution $\bar{n}$ can be obtained by solving Eq. (4.65).

To model the distribution of labor productivity, we need to allow g to be any integer number. Furthermore, we find it most natural to choose $L(c,n)$ so that it is continuous at $n = g(c)$. Here we adopt a simple linear model,

$$L(c,n) = \begin{cases} \dfrac{g(c) - n}{g(c)} & \text{for } n \leq g(c), \\ 0 & \text{for } n \geq g(c), \end{cases} \tag{4.66}$$

as depicted in Fig. 4.18. We can reasonably assume that $L(c,0) = 1$, because there would be no restrictions on hiring workers if the firm has none.

By substituting Eq. (4.66) into Eq. (4.65) and solving for n, we finally obtain

$$\bar{n} = \frac{g(c)}{g(c)e^{\beta(c-\mu)} + 1}. \tag{4.67}$$

This equation is a simple extension of the Fermi–Dirac (FD) statistics; the FD distribution is obtained by setting $g(c) = 1$. We note that the partition function Z that yields Eq. (4.67) is

$$Z = \left(1 + \frac{1}{g(c)} e^{-\beta(c-\mu)}\right)^{g(c)}. \tag{4.68}$$

It is a reasonable extension of the FD statistics in the sense that the partition function has the expansion

$$Z = 1 + e^{-\beta(c-\mu)} + \cdots, \tag{4.69}$$

and yet allows existence of $g(c)$ levels.

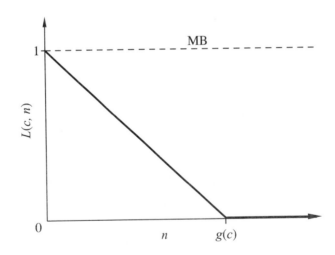

Fig. 4.18 A model of worker limitation $L(n,c)$ in Eq. (4.66); $L(n,c) = 1$ corresponds to the Maxwell–Boltzmann (MB) distribution. (From Aoyama et al. (2015b))

Grand canonical formulation

We have developed the Ehrenfest–Brillouin model to incorporate the assumption that a firm can accommodate a limited number of workers based on its productivity. Here, we derive an identical model within a framework of statistical physics, taking advantage of the idea of a grand canonical ensemble.

Productivity Distribution and Related Topics

We begin with the grand canonical modelling for $\bar{n}(c)$ by going back to the situation where firms can accommodate as many workers as they wish. In this case, the partition function Z_N can be written as

$$Z_N = \sum_{n_1=0}^{N} \sum_{n_2=0}^{N} \cdots \sum_{n_K=0}^{N} W_N(n) \exp\left[-\beta \sum_{k=1}^{K} c_k n_k\right], \tag{4.70}$$

where the multiple summation with respect to the number of workers at each productivity level is carried out subject to the constraint, Eq. (2.71). The adoption of a grand canonical formulation does away with such a troublesome restriction. We thus think about the situation in which the system under study is connected to a worker reservoir with "chemical potential" μ_0. The grand partition function Ξ is then defined as

$$\Xi = \sum_{N=0}^{\infty} e^{\beta \mu_0 N} \binom{N_\infty}{N} Z_N, \tag{4.71}$$

where N_∞ is the total number of workers in the extended system and $\binom{N_\infty}{N}$ denotes the binomial coefficient. Note that the combinatorial factor in Eq. (4.71) stems from the statistical assumption that workers are *distinguishable*. Since one can assume $N \ll N_\infty$, the combinatorial factor is well approximated by

$$\binom{N_\infty}{N} \simeq \frac{N_\infty^N}{N!}. \tag{4.72}$$

With this approximation, we can calculate Eq. (4.71) as

$$\Xi = \prod_{k=1}^{K} \Xi(c_k), \tag{4.73}$$

where we set

$$\Xi(c) = \sum_{n=0}^{\infty} \frac{1}{n!} e^{\beta(\mu-c)n} = \exp\left[e^{\beta(\mu-c)}\right], \tag{4.74}$$

and

$$e^{\beta\mu} = N_\infty e^{\beta\mu_0}. \tag{4.75}$$

The equilibrium distribution $\bar{n}(c)$ is then obtained from $\Xi(c)$ according to

$$\bar{n}(c) = q(c) \frac{\partial \ln \Xi(c)}{\partial q(c)}, \tag{4.76}$$

with

$$q(c) = e^{\beta(\mu-c)}. \tag{4.77}$$

The substitution of Eq. (4.74) in this formula gives the Maxwell–Boltzmann (MB) distribution.

Next, we impose the restriction on the distribution of productivity by putting a ceiling on the number of job positions at each productivity level c:

$$\Xi(c) = \sum_{n=0}^{\infty} f(n; g(c)) \frac{1}{n!} q(c)^n, \qquad (4.78)$$

and introduce the ceiling function $f(n; g(c))$ which is characterized by

$$1 \geq f(n; g(c)) \geq 0 \quad \text{for } n \leq g(c), \qquad (4.79)$$

$$f(n; g(c)) = 0 \quad \text{for } n > g(c). \qquad (4.80)$$

These two conditions, Eq. (4.79) and Eq. (4.80), imposed on $f(n; g(c))$ correspond to Eq. (4.60) and Eq. (4.61) on $L(c, n)$, respectively. Equation (4.78) is compared with Eq. (23) in (Yoshikawa, 2014), where workers are treated as being *indistinguishable*.

To reproduce Eq. (4.67), in fact, we only have to choose $f(k; g(c))$ in the following form:

$$f(n; g(c)) = \binom{g(c)}{n} \frac{n!}{g(c)^n} = \frac{g(c)!}{(g(c) - n)! g(c)^n}, \qquad (4.81)$$

supplemented with Eq. (4.80). Then, the grand partition function, Eq. (4.78), is explicitly calculated as

$$\Xi(c) = \sum_{k=0}^{g(c)} \binom{g(c)}{k} \left(\frac{q(c)}{g(c)}\right)^k = \left(1 + \frac{q(c)}{g(c)}\right)^{g(c)}. \qquad (4.82)$$

We can derive the desired result, Eq. (4.67), from this grand partition function through Eq. (4.76). The formula, Eq. (4.82), has already been derived heuristically by Aoyama et al. (2015a).

It should be remarked that Eq. (4.81) is well defined even at $g = -1$:

$$f(n; g = -1) = \binom{-1}{n} \frac{n!}{(-1)^n} = n!, \qquad (4.83)$$

with

$$\binom{-1}{n} = \frac{(-1)(-2) \cdots (-n)}{n!} = (-1)^n. \qquad (4.84)$$

This leads to the Bose–Einstein (BE) distribution as expected. We thus see that the ceiling function, Eq. (4.81) is capable of representing the three typical distributions, MB, FD, and BE, in a unified way. This is within the framework of the grand canonical formulation for *distinguishable* particles. On the other hand, the grand canonical formulation for *indistinguishable* particles, as in the case of quantum physics, can accommodate only the FD and the BE distributions; the MB distribution is obtained by taking the classical limit of $\bar{n} \ll 1$ in either of the two quantum distributions.

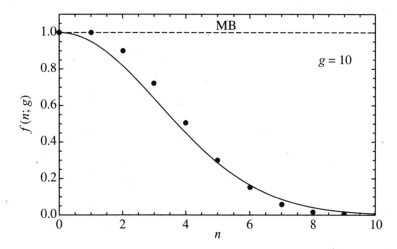

Fig. 4.19 Ceiling function $f(n; g)$, Eq. (4.81), calculated in the exact functional form (dots) and the approximated Gaussian form (solid curve) at $g = 10$. Setting $f(n; g) = 1$ results in the MB distribution. (From Aoyama et al. (2015b))

In passing, we note that the Gaussian approximation to $f(n; g(c))$ in Eq. (4.81),

$$f(n; g(c)) \simeq \exp\left[-\frac{n^2}{2g(c)}\right], \tag{4.85}$$

is valid for $g(c) \gg 1$. Although we assumed $n \ll g(c)$ besides to derive Eq. (4.85) from Eq. (4.81), the additional condition is automatically satisfied by the resulting Gaussian form; its relevant range of n is $n \lesssim \sqrt{2g(c)} \ll g(c)$ for $g(c) \gg 1$. Figure 4.19 compares the results calculated in the exact functional form for $g = 10$ with the corresponding results in the Gaussian form. We see that the approximation works well even for a not so large value of g.

4.2.4 Fitting of the model to the Japanese data

First, we note that when there is no limit to the number of workers, i.e., $g \to \infty$, Eq. (4.67) boils down to the Maxwell–Boltzmann distribution,

$$\bar{n}_{\text{MB}}(c) = e^{-\beta(c-\mu)}. \tag{4.86}$$

When we apply Eq. (4.86) to a low to intermediate range of c, where $\bar{n}$ is an exponentially increasing function of c as observed in Figure 4.16 we must have

$$\beta < 0. \tag{4.87}$$

The negative β is tantamount to the negative temperature (Landau and Lifshitz, 1980; Kittel and Kroemer, 1980). The current model thus accommodates the Boltzmann statistics model with negative temperature advanced by Iyetomi (2012) as a special case.

Secondly, we recall that the power law, Eq. (4.52), holds for $\bar{n}$ in the high productivity side. We can use this empirical fact to determine the functional form for $g(c)$. Equation (4.67) implies that when the temperature is negative, $\bar{n}$ approaches $g(c)$ in the limit $c \to \infty$. These arguments lead us to adopt the following ansatz for $g(c)$,

$$g(c) = Ac^{-\gamma}. \tag{4.88}$$

Given the present model, explaining the empirically observed distribution of productivity is equivalent to determining four parameters, β, μ, A, and γ in Eq. (4.67) and Eq. (4.88). We estimate these four parameters by the χ^2 fit to the empirical results. Figure 4.20 demonstrates the results of the best fit for three data sets of firms, namely, those for all the sectors, the manufacturing sector, and the non-manufacturing sector (we note that the extraction of the empirical distributions, especially for the manufacturing sector, may be hampered considerably by the missing material costs in the data). The fitted parameters are listed in Table 4.4, together with the crossover productivity c_p separating low-to-medium and high productivity regimes, where c_p is defined according to

$$\frac{\partial \bar{n}(c)}{\partial c} = \bar{n}^2 \left(\frac{1}{g(c)} \frac{\partial \ln g(c)}{\partial c} + \frac{1}{q(c)} \frac{\partial \ln q(c)}{\partial c} \right) = 0. \tag{4.89}$$

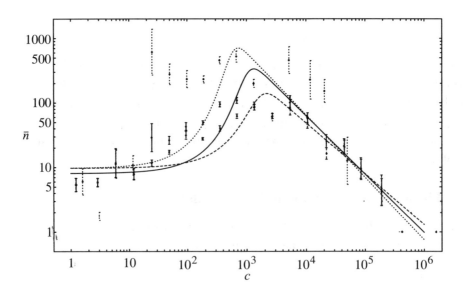

Fig. 4.20 The best fit to the empirical data in 2012 as demonstrated in Fig. 4.16; the solid curve is for all sectors, the dotted curve for the manufacturing sector, and the dashed curve for the non-manufacturing sector. (From Aoyama et al. (2015b))

The current model is quite successful in unifying the two opposing functional behaviors of the average number of workers with low-to-medium and high productivities.

In the aforementioned discussion, we treated the economy as a whole and also regarded it as consisting of two sectors. Economic systems are generally non-homogeneous, since they can be decomposed into various components such as industrial sectors, regional sectors and business groups. Here, we derive necessary conditions for such economic subsystems to be in equilibrium with each other. We must modify the standard derivation (Landau and Lifshitz, 1980) of those conditions in physics, because we have to take into account the distinguishability of workers.

Let us assume that an economic system consists of two subsystems A and B with N_A and N_B workers and demands of D_A and D_B, respectively. The total number W_{A+B} of the microscopic states of the whole system is calculated as

$$W_{A+B}(N_A, N_B; D_A, D_B) = \frac{N!}{N_A! N_B!} W_A(N_A, D_A) W_B(N_B, D_B), \qquad (4.90)$$

where we assume that the total number of workers, $N = N_A + N_B$, and the total demand, $D = D_A + D_B$, are conserved during contact between the two subsystems. The pre-factor in the right-hand side of Eq. (4.90) counts the number of ways of distributing *distinguishable* workers between the two subsystems. This counting is unnecessary in the case of statistical physics because identical particles are indistinguishable in nature. From Eq. (4.90), we obtain the entropy of the whole system as

$$S_{A+B} \simeq S_A + S_B + N \ln N - N_A \ln N_A - N_B \ln N_B. \qquad (4.91)$$

It should be noted that the last three terms arising from the extra counting factor destroy the additivity of the entropy; the entropy of the whole system is not separable into components of the subsystems.

The entropy maximum principle determines the most probable distribution of workers and demands between the two subsystems. Since both N and D are assumed to be constant, the variational condition is explicitly written down with Eq. (2.82) as

$$\delta S_{A+B} = \delta S_A + \delta S_B - \delta N_A \ln N_A - \delta N_B \ln N_B$$
$$= (\alpha_A - \alpha_B - \ln N_A + \ln N_B) \delta N_A + (\beta_A - \beta_B) \delta D_A \qquad (4.92)$$
$$= 0.$$

where δN_A and δD_A denote the infinitesimal variations of N_A and D_A which are independent of each other. We thus obtain the following equilibrium conditions:

$$\beta_A = \beta_B, \qquad (4.93)$$

$$N_A e^{-\alpha_A} = N_B e^{-\alpha_B}. \qquad (4.94)$$

The first condition, Eq. (4.93), shares the same form as in physics. It is required for the system to be in equilibrium against an exchange of demands. The second condition, Eq. (4.94), guarantees that there is no macroscopic flow of workers between the two subsystems. The equation, on each side of which the number of workers appears, is unfamiliar to physicists; distinguishability of workers brings the extra factor to the normal form. Demand flows from subsystem A to subsystem B if $\beta_A < \beta_B$ and in the reversed direction if $\beta_A > \beta_B$. Moreover, workers flow from A to B if $N_A e^{-\alpha_A} > N_B e^{-\alpha_B}$ and vice versa in the opposite case.

Table 4.4 Estimated parameters in Eq. (4.67) and the position of the peak c_p determined by Eq. (4.89) for the results given in Fig. 4.20. The units are 10^{-3} for β, 10^3 for μ, 10^5 for A, and 10^3 for c_p. (From Aoyama et al. (2015b))

	β	μ	A	γ	c_p	$\beta\mu$
All	-4.32	-0.48	2.57	0.90	1.32	2.09
Manufacturing	-8.68	-0.26	4.75	0.97	0.73	2.27
Non-manufacturing	-2.13	-1.07	0.68	0.76	2.15	2.28

Table 4.4 shows that the temperature of the non-manufacturing sector is significantly lower than that of the manufacturing sector. This implies that there is a much wider demand gap in the non-manufacturing sector. The system as a whole is thus far from equilibrium with respect to exchanges of product demand between the two sectors. In contrast, β times μ takes almost the same value for the two sectors, indicating that the subsystems seem to be balanced against flow of workers. Note that by comparing Eq. (2.87) and Eq. (4.86), the equilibrium condition, Eq. (4.94), can be rewritten as

$$\beta_A \mu_A = \beta_B \mu_B . \tag{4.95}$$

These empirical findings on equilibration of the Japanese economic system with respect to exchanges of demand and workers have been established by Iyetomi (2012) and Aoyama et al. (2015a) on the basis of the alternative data set for the years spanning from 2000 through 2009.

4.3 Exercise

4.1 Derive the elasticity of substitution σ for the Cobb–Douglas production function.

4.2 Derive the elasticity of substitution σ for CES production function, and show that

$$\sigma = \frac{1}{1-cp}. \tag{4.96}$$

4.3 Prove the equation:

$$\lim_{p \to 0} \Phi_{\text{CES}}(L, K; \gamma, c, p) = \Phi_{\text{CD}}(L, K; \alpha = (1-\gamma)c, \beta = \gamma c). \tag{4.97}$$

4.4 Prove the formula in Eq. (4.65).

4.5 Derive the Gaussian approximation form, Eq. (4.85), for the ceiling function $f(n; g(c))$.

5

Multivariate Time Series Analysis

The only reason for time is so that everything doesn't happen at once.

Albert Einstein

Most economic variables tend to move more or less together. In many case, we reasonably expect to identify a small number of factors describing general comovements of variables. The standard method is the principal component analysis (PCA). It is used not only in economics but also in sociology, political science, and natural sciences. The problem of this method is that it fails when there are significant leads and lags among variables. This case is norm, not exceptional in economics. The complex Hilbert PCA (CHPCA) takes care of this problem.

This method has been originally developed in Rasmusson et al. (1981), Barnett (1983), Horel (1984), Stein et al. (2011), Hannachi et al. (2007), using the Hilbert transformation developed by Hilbert (1912), Gabor (1946a), Granger and Hatanaka (1964), Bendat and Piersol (2011), Feldman (2011), Ikeda et al. (2013b) and Ikeda et al. (2013c) among others, and has been successful in many areas of natural sciences. We have further introduced improvements on CHPCA and combined it with the rotational random shuffling method (Arai and Iyetomi, 2013) to establish this tool set for analysis of a set of time series data and applied it to several economic data. In this chapter, we first explain the ordinary PCA and then proceed to CHPCA and its applications.

5.1 Principal Component Analysis (PCA)

The PCA consists of the following steps:

1. Calculating correlation coefficients $\mathcal{C}_{\alpha\beta}$ for all the pairs of the time series α and β.

2. Obtaining eigenvalues and eigenvectors of the correlation matrix $\mathcal{C}$, which comprises the correlation coefficients $C_{\alpha\beta}$.

3. Using the random matrix theory (RMT) or rotational random simulation (RRS) to identify which part of the eigen system signals the comovements.

In this section, we examine the first two steps.

Let us review how to analyze the correlation of a set of N time series $r_\alpha(t)$, where the suffix $\alpha = 1, 2, 3, \cdots, N$ denotes the time series and $t = 1, 2, 3, \cdots, T$ denotes time in the suitable unit, which can be day, month, year depending on the data set.[1]

The standardized (canonical) time series is defined to have an average equal to zero and a standard deviation equal to one:

$$\widehat{r}_\alpha(t) := \frac{r_\alpha(t) - \langle r_\alpha \rangle}{\sigma_\alpha}, \tag{5.1}$$

where $\langle \cdot \rangle$ denotes the average over time, for example,

$$\langle r_\alpha \rangle := \frac{1}{T} \sum_{t=1}^{T} r_\alpha(t), \tag{5.2}$$

and $\sigma_\alpha (> 0)$ denotes the standard deviation;

$$\sigma_\alpha^2 := \frac{1}{T} \sum_{t=1}^{T} |r_\alpha(t) - \langle r_\alpha \rangle|^2 = \frac{1}{T} \sum_{t=1}^{T} \left(\langle r_\alpha^2 \rangle - \langle r_\alpha \rangle^2 \right). \tag{5.3}$$

The correlation coefficient $C_{\alpha\beta}$ between the time series α and β is defined by the following.

$$C_{\alpha\beta} := \langle \widehat{r}_\alpha \widehat{r}_\beta \rangle. \tag{5.4}$$

This evidently satisfies

$$C_{\alpha\alpha} = 1, \tag{5.5}$$

by definition of the standardized time series, Eq. (5.1), and

$$C_{\beta\alpha} = C_{\alpha\beta}. \tag{5.6}$$

The correlation matrix $\boldsymbol{C}$ is the symmetric $N \times N$ matrix whose (α, β) component is $C_{\alpha\beta}$:

$$\boldsymbol{C} := (C_{\alpha\beta}). \tag{5.7}$$

---

[1] For example, if we are analyzing daily data, starting January 1, 2015, the unit is a day. Thus, $t = 1$ is the first day, Jan. 1, 2015; $t = 2$, Jan. 2, 2015, and so on.

The eigenvalues $\lambda^{(n)}$ and the corresponding eigenvectors $\boldsymbol{V}^{(n)}$ ($n = 1, 2, 3, \cdots, N$) are defined by the following equation:

$$\boldsymbol{C}\boldsymbol{V}^{(n)} = \lambda^{(n)}\boldsymbol{V}^{(n)}, \tag{5.8}$$

The eigenvalues are real and the eigenvectors can be chosen to be real, owing to the fact that the matrix $\boldsymbol{C}$ is real and symmetric. We choose their suffix (n) in descending order of the eigenvalue;

$$\lambda^{(1)} \geq \lambda^{(2)} \geq \lambda^{(3)} \cdots \geq \lambda^{(N)}. \tag{5.9}$$

They satisfy the orthogonality;

$$\boldsymbol{V}^{(n)} \cdot \boldsymbol{V}^{(m)} = \delta_{nm}, \tag{5.10}$$

where δ_{nm} is Kronecker's delta;

$$\delta_{nm} = \begin{cases} 1 & \text{if } n = m, \\ 0 & \text{if } n \neq m, \end{cases} \tag{5.11}$$

the trace property;

$$\sum_{n=1}^{N} \lambda^{(n)} = N, \tag{5.12}$$

and

$$\boldsymbol{C} = \sum_{n=1}^{N} \lambda^{(n)} \boldsymbol{V}^{(n)} \boldsymbol{V}^{(n)\text{t}}, \tag{5.13}$$

which means that

$$C_{\alpha\beta} = \sum_{n=1}^{N} \lambda^{(n)} V_{\alpha}^{(n)} V_{\beta}^{(n)}. \tag{5.14}$$

The time series are expanded in terms of the eigenvectors:

$$\widehat{r}_{\alpha}(t) = \sum_{n=1}^{N} a^{(n)}(t) V_{\alpha}^{(n)}, \quad a^{(n)}(t) = \sum_{\alpha=1}^{N} V_{\alpha}^{(n)} \widehat{r}_{\alpha}(t). \tag{5.15}$$

We call the coefficients $a^{(n)}(t)$ the "mode signals", as they give how much the nth eigenmode is present at time t. They satisfy,

$$\langle a^{(n)} a^{(m)} \rangle = \delta_{nm} \lambda^{(n)}, \tag{5.16}$$

which can be proven by the use of Eqs. (5.4), (5.7), (5.8), and (5.10). This proves that the eigenvalues $\{\lambda^{(n)}\}$ are positive. Equation (5.16) shows that the larger the eigenvalue is, the larger is the eigenvector's presence, with the mean strength proportional to the square root of the eigenvalues. The next question then is which eigenmodes (with large eigenvalues) are significant, being free from noise.

5.2 Complex Hilbert PCA (CHPCA)

Readers may encounter the Hilbert transform in the context of continuous time in the literatures. Therefore, we first leave our time series and discuss it for functions on continuous time. Then we go back to our time series to proceed to complexify all the PCA context and arrive at the full CHPCA methodology.

5.2.1 Continuous time

For a function $f(t)$ defined for $t \in [-\infty, \infty]$, its Hilbert transform $\mathcal{H}[f]$ is a linear transformation defined as follows,

$$\mathcal{H}[f](t) = \frac{1}{\pi} \mathrm{P} \int_{-\infty}^{\infty} \frac{f(z)}{z-t} dz. \tag{5.17}$$

The symbol P means the Cauchy principal value. We further define the complexification of a real function, $\mathcal{C}[\cdot]$, by adding the Hilbert transform as an imaginary part:

$$\mathcal{C}[f] := f + i\mathcal{H}[f], \tag{5.18}$$

which may be evaluated as follows:

$$\mathcal{C}[f](t) = f(t) + i\mathcal{H}[f](t) \tag{5.19}$$

$$= \frac{i}{\pi} \int_{-\infty}^{\infty} \frac{f(z)}{z - t + i\epsilon} dz. \tag{5.20}$$

In the previous equation, we used an identity,

$$\mathrm{P}\frac{1}{z-t} - \pi i \delta(z-t) = \frac{1}{z - t + i\epsilon}, \tag{5.21}$$

with positive infinitesimally small ϵ ($\epsilon \to +0$). Alternatively, it may be written as

$$\mathcal{C}[f](t) = \frac{i}{\pi} \int_C \frac{f(z)}{z-t} dz, \tag{5.22}$$

whose integration contour C is drawn in Fig. 5.1.

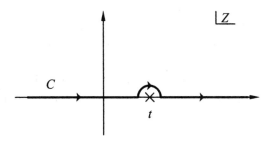

Fig. 5.1 The integration contour C for the integration in Eq. (5.20) on the complex z plane.

Let us examine what this does to trigonometric functions: From Eq. (5.22), one can show easily the following for $\omega > 0$:

$$\mathcal{C}[e^{i\omega t}] = 0, \tag{5.23}$$

$$\mathcal{C}[e^{-i\omega t}] = e^{-i\omega t}. \tag{5.24}$$

Using these and the fact that complexification is a linear operation, we find that

$$\mathcal{C}[\cos \omega t] = e^{-i\omega t}, \tag{5.25}$$

$$\mathcal{C}[\sin \omega t] = i\, e^{-i\omega t}. \tag{5.26}$$

Therefore, the complexification, Eq. (5.18), converts oscillations to *clockwise rotation* in the complex plane.[2]

It is useful to note that Eqs. (5.25) and (5.26) imply the following:

$$\mathcal{H}[\cos \omega t] = -\sin \omega t, \tag{5.27}$$

$$\mathcal{H}[\sin \omega t] = \cos \omega t, \tag{5.28}$$

which may be proved directly from Eq. (5.17).

5.2.2 Discrete time series in a finite time range

Hilbert transformation and complexification are basically performing the transformations, Eqs (5.25), (5.26), (5.27), (5.28) to each Fourier component of the time series in question, respectively. A straightforward way to obtain the complex time series, therefore, is to

(i) obtain the discrete Fourier constants of the time series,

(ii) do the transformation, Eq. (5.25) and Eq. (5.26) to each Fourier component, and

(iii) sum over the transformed Fourier series.

There is, however, a redundancy in this process: There are two summing processes, one in the Fourier transformation in (i) and the other in the inverse Fourier transformation in (iii). A combination of these two summing processes can be done independently from the original time series. Therefore, we carry out that summation shown in the following equation to obtain a single transform, which not only saves on calculation time, but also offers several important insights into the resulting correlation coefficients and the eigen system.

---

[2] Readers may change the sign of the imaginary unit i everywhere in the previous equations to obtain the *counterclockwise rotation*. The authors' notation was chosen to keep the direction clockwise.

The discrete Fourier transform of the time series r is as follows:[3]

$$r(t) = \frac{1}{\sqrt{T}} \sum_{k=1}^{T} \widetilde{r}(k) \, e^{-i\omega_k t}, \tag{5.29}$$

where

$$\omega_k := \frac{2\pi}{T} k \tag{5.30}$$

and its inverse Fourier transformation;

$$\widetilde{r}(k) = \frac{1}{\sqrt{T}} \sum_{t=1}^{T} r(t) \, e^{i\omega_k t}, \tag{5.31}$$

from which it follows that

$$\widetilde{r}(k) = \widetilde{r}(T-k)^*, \tag{5.32}$$

$$\widetilde{r}(T) = \frac{1}{\sqrt{T}} \sum_{t=1}^{T} r(t) = \sqrt{T} \langle r \rangle, \tag{5.33}$$

which guarantees that the degrees of freedom in the Fourier components $\{\widetilde{r}(k)\}$ is the same as that of the original time series $\{r(t)\}$, T. Let us see how this is so by looking at even T and odd T separately.

[Even T]

For even T, the independent Fourier components are complex $\widetilde{r}(k)$ for $k = 1, 2, \cdots T/2 - 1$, and $\widetilde{r}(T/2)$ and $\widetilde{r}(T)$, both of which are real because of Eq. (5.32) and Eq. (5.33), respectively, totaling $2 + 2 \times ((T/2) - 1) = T$ (real) degrees of freedom.

The Fourier expansion is written as follows,

$$r(t) = \langle r \rangle + (-1)^t \langle r \rangle_- + \frac{2}{\sqrt{T}} \sum_{k=1}^{(T/2)-1} \Re \left[\widetilde{r}(k) e^{-i\omega_k t} \right], \tag{5.34}$$

where $\Re$ takes the real part of the variable and we introduce the notation:

$$\langle r \rangle_- := \sum_{t'=1}^{T} (-1)^{t'} r(t') \tag{5.35}$$

---

[3] In this subsection, we simply drop the index (α, β, etc.) to save the notation, as the transformation is done on each time series.

The first two terms in the left-hand side of Eq. (5.34) comes from the $k = T$ and $k = T/2$ terms. To examine the complexification process, we write the last term in real quantities:

$$\Re\left[\widetilde{r}(k)e^{-i\omega_k t}\right] = \Re[\widetilde{r}(k)]\cos(\omega_k t) + \Im[\widetilde{r}(k)]\sin(\omega_k t), \tag{5.36}$$

where $\Im$ refers to the imaginary part. Using Eqs. (5.27) and (5.28), we find that complexification of Eq. (5.36) is as follows:

$$\mathcal{C}\left[\Re[\widetilde{r}(k)]\cos\omega_k t + \Im[\widetilde{r}(k)]\sin\omega_k t\right] = \Re[\widetilde{r}(k)]e^{-i\omega_k t} + \Im[\widetilde{r}(k)]\, i\, e^{-i\omega_k t} \tag{5.37}$$

$$= \widetilde{r}(k)e^{-i\omega_k t}. \tag{5.38}$$

The second term on the right-hand side of Eq. (5.34) needs careful discussion: Since it is the $k = T/2$ term with oscillation $e^{-i\pi t} = \cos(\pi t) = (-1)^t$, one might think that its Hilbert transform is $\mathcal{H}(\cos(\pi t)) = \sin(\pi t) = 0$. However, the following is also true:

$$(-1)^t = \sqrt{2}\cos\left(\pi t - \frac{\pi}{4}\right). \tag{5.39}$$

By going through the continuous-time analysis as in Subsection 5.2.1, we find that

$$\mathcal{H}\left[\sqrt{2}\cos\left(\pi t - \frac{\pi}{4}\right)\right] = -\sqrt{2}\sin\left(\pi t - \frac{\pi}{4}\right) = (-1)^t, \tag{5.40}$$

which, when added to the original $(-1)^t$ behavior as an imaginary part, produces a "clockwise" rotation

$$\sqrt{2}\, e^{-i(\pi t - \frac{\pi}{4})} = (-1)^t(1+i), \tag{5.41}$$

as depicted in Fig. 5.2. One might claim that this is only a matter of interpretation, not a logical step, and that is true. However, by adding this term in the complexification, one finds that it acquires several desirable features, namely Eq. (5.82), which it shares with the odd T case, as we will see later. Furthermore, an alternative to Eq. (5.41) is $(-1)^t(1-i)$, which can be induced by a slightly modified argument. Although this choice also satisfies Eq. (5.82), we choose to use the convention, Eq. (5.41).[4]

Therefore, we find that the complexified time series is given by the following:

$$\mathcal{C}[r](t) = \langle r \rangle + (1+i)(-1)^t \langle r \rangle_- + \frac{2}{\sqrt{T}}\sum_{k=1}^{T/2-1}\widetilde{r}(k)\,e^{-i\omega_k t}. \tag{5.42}$$

---

[4]Note that this modification was not adopted in the general definition of the Hilbert transformation in Vodenska et al. (2016) or Yoshikawa et al. (2015); the actual analysis in those papers were for odd T and this modification does not affect any of the results presented in those papers.

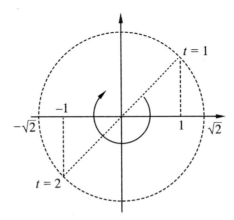

Fig. 5.2 Rotating behavior of Eq. (5.41) on its complex plane.

Let us now examine the summation on the right-hand side of Eq. (5.42). By substituting the inverse Fourier transform, Eq. (5.31) in this term, we obtain the following:

$$\frac{2}{\sqrt{T}} \sum_{k=1}^{(T/2)-1} \tilde{r}(k)\, e^{-i\omega_k t} = \frac{2}{T} \sum_{k=1}^{(T/2)-1} \sum_{t'=1}^{T} r(t')\, e^{-i\omega_k(t-t')} \tag{5.43}$$

The sum over k is independent of the time series $\{r(t)\}$ and can be carried out as follows, as mentioned at the beginning of this subsection: First, for $\ell := t - t' = 0$,

$$\frac{2}{T} \sum_{k=1}^{(T/2)-1} e^{-i\omega_k \ell} = 1 - \frac{2}{T}. \tag{5.44}$$

For $\ell = t - t' \neq 0$,

$$\frac{2}{T} \sum_{k=1}^{(T/2)-1} e^{-i\omega_k \ell} = \frac{2}{T} \frac{1 - e^{-i\frac{2\pi}{T}\left(\frac{T}{2}-1\right)\ell}}{1 - e^{-i\frac{2\pi}{T}\ell}} e^{-i\frac{2\pi}{T}\ell} = \frac{2}{T} \frac{1 - (-1)^{\ell} e^{i\frac{2\pi}{T}\ell}}{1 - e^{-i\frac{2\pi}{T}\ell}} e^{-i\frac{2\pi}{T}\ell}$$

$$\tag{5.45}$$

$$= \begin{cases} -\dfrac{2}{T} & \text{for } \ell = \text{even}, \\ -i\dfrac{2}{T} \cot\left(\dfrac{\pi}{T}\ell\right) & \text{for } \ell = \text{odd}. \end{cases} \tag{5.46}$$

Substituting these back into Eq. (5.42), we arrive at the following simple expression:

$$\mathcal{C}[r](t) = r(t) - i\frac{2}{T} \sum_{t'=1}^{T} \left(\delta_{t-t'}^{(\text{o})}\, r(t') \cot\left(\frac{\pi}{T}(t-t')\right) - \frac{1}{2}(-1)^{t-t'} \right) \tag{5.47}$$

where $\delta^{(o)}$ is defined as follows:

$$\delta_\ell^{(o)} = \begin{cases} 1 & \text{for } \ell = \text{odd}, \\ 0 & \text{for } \ell = \text{even}. \end{cases} \tag{5.48}$$

Later we need the even version,

$$\delta_\ell^{(e)} = \begin{cases} 1 & \text{for } \ell = \text{even}, \\ 0 & \text{for } \ell = \text{odd}. \end{cases} \tag{5.49}$$

The imaginary part is the discrete version of the Hilbert transform in continuous time, Eq. (5.17). For $t' - t \to 0$, the cot term approaches $T/\pi(t' - t)$ while the restriction $t' - t = \text{odd}$ guarantees the avoidance of the divergence at $t' = t$, reverting to the continuous case.[5]

[Odd T]

In case T is odd, independent components are complex $\widetilde{r}(k)$ for $k = 1, 2, \cdots (T-1)/2$ and real $\widetilde{r}(T)$, totaling $2 \times (T-1)/2 + 1 = T$ degrees of freedom.

The Fourier transformation is as follows:

$$r(t) = \langle r \rangle + \frac{2}{\sqrt{T}} \operatorname{Re}\left[\sum_{k=1}^{(T-1)/2} \widetilde{r}(k)\, e^{-i\omega_k t} \right], \tag{5.50}$$

Then, as in the even T case, we obtain the complexified time series as follows:

$$\mathcal{C}[r](t) = \langle r \rangle + \frac{2}{\sqrt{T}} \sum_{k=1}^{(T-1)/2} \widetilde{r}(k)\, e^{-i\omega_k t} \tag{5.51}$$

$$= \langle r \rangle + \frac{2}{T} \sum_{k=1}^{(T-1)/2} \sum_{t'=1}^{T} r(t')\, e^{-i\omega_k (t-t')}. \tag{5.52}$$

The summation over k is done as follows, where $\ell := t - t'$. For $\ell = 0$,

$$\frac{2}{T} \sum_{k=1}^{(T-1)/2} e^{-i\omega_k \ell} = 1 - \frac{1}{T}, \tag{5.53}$$

---

[5] This is a heuristic argument. To conduct an exact analysis, we need to define the time spacing Δt to be a variable, not as done earlier, and let $\Delta t \to 0$ at the end.

and for $\ell \neq 0$,

$$\frac{2}{T}\sum_{k=1}^{(T-1)/2} e^{-i\omega_k \ell} = \frac{2}{T}\frac{1-e^{-i\frac{2\pi}{T}\frac{T-1}{2}\ell}}{1-e^{-i\frac{2\pi}{T}\ell}} e^{-i\frac{2\pi}{T}\ell} = \frac{2}{T}\frac{1-(-1)^\ell e^{i\frac{\pi}{T}\ell}}{1-e^{-i\frac{2\pi}{T}\ell}} e^{-i\frac{2\pi}{T}\ell} \quad (5.54)$$

$$= \begin{cases} -\dfrac{1}{T} + \dfrac{i}{T}\tan\left(\dfrac{\pi}{2T}\ell\right) & \text{for } \ell = \text{even,} \\ -\dfrac{1}{T} - \dfrac{i}{T}\cot\left(\dfrac{\pi}{2T}\ell\right) & \text{for } \ell = \text{odd.} \end{cases} \quad (5.55)$$

By substituting this back into Eq. (5.52), we arrive at the following:

$$\mathcal{C}[r](t) = r(t) + \frac{i}{T}\left[-\sum_{\substack{t'=1 \\ t-t'=\text{odd}}}^{T} r(t') \cot\left(\frac{\pi}{2T}(t-t')\right) + \sum_{\substack{t'=1 \\ t-t'=\text{even}}}^{T} r(t') \tan\left(\frac{\pi}{2T}(t-t')\right)\right]. \quad (5.56)$$

To summarize, the complexified time series $\mathcal{C}[r]$ of $\{r(1), r(2), \cdots, r(T)\}$ is expressed as follows:

$$\mathcal{C}[r](t) = r(t) + i\sum_{t'=1}^{T} H(t-t';T)\, r(t'), \quad (5.57)$$

$$H(\ell;T) := \begin{cases} -\dfrac{2}{T}\left[\delta_\ell^{(o)} \cot\left(\dfrac{\pi}{T}\ell\right) - \dfrac{1}{2}(-1)^{t-t'}\right] & \text{for } T = \text{even,} \\ \dfrac{1}{T}\left[-\delta_\ell^{(o)} \cot\left(\dfrac{\pi}{2T}\ell\right) + \delta_\ell^{(e)} \tan\left(\dfrac{\pi}{2T}\ell\right)\right] & \text{for } T = \text{odd.} \end{cases} \quad (5.58)$$

5.2.3 Absolute values and phases of the complexified time series

Let us now discuss the amplitudes (absolute values) and the phases of the complexified time series in relation to the counting of the number of degrees of freedom. We denote amplitudes by A_t and phases by θ_t:

$$\mathcal{C}[r](t) := A_t e^{i\theta_t}, \quad A_t := |\mathcal{C}[r](t)|. \quad (5.59)$$

Since the real part of $\mathcal{C}[r](t)$ is $r(t)$, we have

$$r(t) = A_t \cos\theta_t. \quad (5.60)$$

Although there are T amplitudes, A_t and T phases θ_t totaling $2T$ quantities, there are only T original degrees of freedom. This means that there are some redundancies in the amplitudes and phases. Our aim here then is to pick up the independent variables among these $2T$ quantities.

First, let us express the imaginary part of Eq. (5.57) in simpler notation:

$$\mathcal{H}[r](t) = \sum_{t'=1} H_{tt'} r(t). \tag{5.61}$$

Therefore, the amplitudes and the phases satisfy

$$A_t \sin \theta_t = \sum_{t'=1}^{T} H_{tt'} A_t \cos \theta_t. \tag{5.62}$$

Let us express this in vector-matrix notation as follows:

$$\mathcal{B} A = 0. \tag{5.63}$$

Here, A is a column vector of amplitudes with N components:

$$A = \begin{pmatrix} A_1 \\ A_2 \\ \vdots \\ A_T \end{pmatrix}, \tag{5.64}$$

and $\mathcal{B}$ is a $T \times T$ square matrix $\mathcal{B} = (\mathcal{B}_{tt'})$,

$$\mathcal{B}_{tt'} = \sin \theta_t \delta_{tt'} - A_{tt'} \cos \theta_t. \tag{5.65}$$

Therefore, the matrix $\mathcal{B}$ has at least one zero eigenvalue;

$$\operatorname{Det} \mathcal{B} = 0. \tag{5.66}$$

Since $\mathcal{B}$ depends only on the phases, this is a constraint on the T phases. Therefore, only $T - 1$ phases are independent among $\{\theta_1, \theta_2, \cdots, \theta_T\}$,

As for the amplitudes, Eq. (5.63) specifies that the amplitude vector A is proportional (parallel) to the eigenvector of B with zero eigenvalue. Therefore, only one variable, say,

$$|A| := \sqrt{\sum_{t=1}^{T} A_t^2} \tag{5.67}$$

is an independent quantity.

Therefore, we can choose independent quantities as $T - 1$ phases and an overall amplitude $|A|$, totaling T quantities as it should.

Let us look at an example to explicitly confirm the aforementioned procedure for $T = 3$. In this case, Eq. (5.61) is as follows:

$$\mathcal{H}[r](1) = \frac{1}{\sqrt{3}}\left(-r(2) + r(3)\right), \tag{5.68}$$

$$\mathcal{H}[r](2) = \frac{1}{\sqrt{3}}\left(-r(3) + r(1)\right), \tag{5.69}$$

$$\mathcal{H}[r](3) = \frac{1}{\sqrt{3}}\left(-r(1) + r(2)\right). \tag{5.70}$$

From this, we find that

$$\mathcal{B} = \begin{pmatrix} \sin\theta_1 & \dfrac{\cos\theta_2}{\sqrt{3}} & -\dfrac{\cos\theta_3}{\sqrt{3}} \\ -\dfrac{\cos\theta_1}{\sqrt{3}} & \sin\theta_2 & \dfrac{\cos\theta_3}{\sqrt{3}} \\ \dfrac{\cos\theta_1}{\sqrt{3}} & -\dfrac{\cos\theta_2}{\sqrt{3}} & \sin\theta_3 \end{pmatrix} \tag{5.71}$$

and

$$\mathrm{Det}\,\mathcal{B} = \sin\theta_1 \sin\theta_2 \sin\theta_3 \left[1 + \frac{1}{3}\left(\cot\theta_1 \cot\theta_2 + \cot\theta_2 \cot\theta_3 + \cot\theta_3 \cot\theta_1\right)\right] \tag{5.72}$$

On the other hand, from Eqs. (5.68)–(5.70),

$$\cot\theta_1 = \frac{-r(2) + r(3)}{\sqrt{3}\,r(1)}, \tag{5.73}$$

$$\cot\theta_2 = \frac{-r(3) + r(1)}{\sqrt{3}\,r(2)}, \tag{5.74}$$

$$\cot\theta_3 = \frac{-r(1) + r(2)}{\sqrt{3}\,r(3)}. \tag{5.75}$$

Plugging these into the right-hand side of Eq. (5.72), we find that it is in fact identical to zero.

5.2.4 Complex correlation matrix

Let us now go back to the multivariate time series t_α with $\alpha = 1, 2, \cdots, N$. To simply the notation, let us denote the complex time series as c in this subsection:

$$\mathcal{C}[r_\alpha] := c_\alpha \quad (\Re(c_\alpha) = r_\alpha) \tag{5.76}$$

Its average has the following property:

$$\langle c_\alpha \rangle = \frac{1}{T} \sum_{t=1}^{T} c_\alpha(t) = \langle r_\alpha \rangle. \tag{5.77}$$

This is because Eq. (5.58) satisfies the property,

$$\sum_{t=1}^{T} H(t-t',T) = 0 \tag{5.78}$$

for any T and t', which can be proven from the fact that H is the imaginary part of the sum on the left-hand sides of Eq. (5.45) or (5.54).

The standard deviation for the complex time series is

$$\tilde{\sigma}_\alpha^2 := \frac{1}{T} \sum_{t=1}^{T} |c(t)_\alpha - \langle c_\alpha \rangle|^2 = \frac{1}{T} \sum_{t=1}^{T} \left(\langle |c_\alpha(t)|^2 \rangle - \langle c_\alpha \rangle^2 \right). \tag{5.79}$$

This standard deviation is related to the standard deviation of the real time series:

$$\tilde{\sigma}_\alpha = \sqrt{2}\, \sigma_\alpha \tag{5.80}$$

This is because of the following: By substituting Eq. (5.57) into Eq. (5.79), we find that

$$\tilde{\sigma}_\alpha^2 = \sigma_\alpha^2 + \frac{1}{T} \sum_{t=1}^{T} \sum_{t'=1}^{T} \sum_{t''=1}^{T} H(t-t';T) H(t-t'';T)\, r_\alpha(t')\, r_\alpha(t''). \tag{5.81}$$

using the identity, Eq. (5.77). The H function has the property that

$$\sum_{t=1}^{T} H(t-t';T) H(t-t'';T) = \delta_{tt'} - \frac{1}{T}. \tag{5.82}$$

By using this in Eq. (5.81), we arrive at the identity, Eq. (5.80).

The complex correlation coefficient $\tilde{C}_{\alpha\beta}$ for the time series α and β is defined by the following.

$$\tilde{C}_{\alpha\beta} := \langle \hat{c}_\alpha^* \hat{c}_\beta \rangle, \tag{5.83}$$

where $\hat{c}_\alpha$ is the standardized time series;

$$\hat{c}_\alpha(t) := \frac{c_\alpha(t) - \langle c_\alpha \rangle}{\tilde{\sigma}_\alpha}. \tag{5.84}$$

This satisfies

$$\tilde{C}_{\alpha\alpha} = 1, \quad \tilde{C}_{\beta\alpha} = \tilde{C}_{\alpha\beta}^*. \tag{5.85}$$

The correlation matrix C is the Hermitian $N \times N$ matrix whose (α, β) component is $\tilde{C}_{\alpha\beta}$:

$$\tilde{C} := (\tilde{C}_{\alpha\beta}). \tag{5.86}$$

The eigenvalues $\tilde{\lambda}^{(n)}$ and the corresponding eigenvectors $\tilde{V}^{(n)}$ ($n = 1, 2, 3, \cdots, N$) are defined by the following equation:

$$\tilde{C}\tilde{V}^{(n)} = \tilde{\lambda}^{(n)}\tilde{V}^{(n)}, \tag{5.87}$$

The eigenvalues are real because of the hermiticity of $\mathcal{C}$. As before, we choose their suffix (n) in descending order of the eigenvalue:

$$\tilde{\lambda}^{(1)} \geq \tilde{\lambda}^{(2)} \geq \tilde{\lambda}^{(3)} \cdots \geq \tilde{\lambda}^{(N)}. \tag{5.88}$$

They satisfy the orthogonality:

$$\tilde{V}^{(n)\dagger} \cdot \tilde{V}^{(m)} = \delta_{nm}, \tag{5.89}$$

the trace property:

$$\sum_{n=1}^{N} \tilde{\lambda}(n) = N, \tag{5.90}$$

and

$$\tilde{C} = \sum_{n=1}^{N} \tilde{\lambda}^{(n)} \tilde{V}^{(n)} \tilde{V}^{(n)\dagger}. \tag{5.91}$$

Mode signals are defined as for the real PCA case, Eq. (5.15),

$$\tilde{\hat{r}}_\alpha(t) = \sum_{n=1}^{N} \tilde{a}^{(n)}(t) \tilde{V}_\alpha^{(n)}, \quad \tilde{a}^{(n)}(t) = \sum_{\alpha=1}^{N} \tilde{V}_\alpha^{(n)} \tilde{\hat{r}}_\alpha(t). \tag{5.92}$$

and satisfies

$$\langle \tilde{a}^{(n)*}(t) \tilde{a}^{(m)}(t) \rangle = \delta_{nm} \tilde{\lambda}(n), \tag{5.93}$$

This proves that the eigenvalues $\{\tilde{\lambda}(n)\}$ are positive.

Let us now prove an identity[6]

$$\Re\left(\tilde{C}\right) = C. \tag{5.94}$$

From Eq. (5.77),

$$\hat{c}_\alpha(t) = \frac{1}{\tilde{\sigma}_\alpha} \left[r_\alpha(t) - \langle r_\alpha \rangle + i \sum_{t'=1}^{T} H(t-t'; T) r_\alpha(t') \right]. \tag{5.95}$$

---

[6] We are not aware of any studies where this identity is proven, or even mentioned.

We find that

$$\Re\left(\tilde{C}_{\alpha\beta}\right) = \frac{1}{T}\frac{1}{2\sigma_\alpha^2}\left[\sum_{t=1}^{T}(r_\alpha(t)-\langle r_\alpha\rangle)(r_\beta(t)-\langle r_\beta\rangle)\right.$$

$$\left.+\sum_{t=1}^{T}\sum_{t'=1}^{T}\sum_{t''=1}^{T} H(t-t';T)H(t-t'';T)\,r_\alpha(t')\,r_\beta(t'')\right]$$

$$= \frac{1}{T}\frac{1}{2\sigma_\alpha^2}\left[\sum_{t=1}^{T}(r_\alpha(t)-\langle r_\alpha\rangle)(r_\beta(t)-\langle r_\beta\rangle) + \sum_{t'=1}^{T} r_\alpha(t')r_\beta(t')\right.$$

$$\left.- T\langle r_\alpha\rangle\langle r_\beta\rangle\right] \tag{5.96}$$

$$= \frac{1}{T}\frac{1}{\sigma_\alpha^2}\sum_{t=1}^{T}(r_\alpha(t)-\langle r_\alpha\rangle)(r_\beta(t)-\langle r_\beta\rangle) = C_{\alpha\beta}, \tag{5.97}$$

where we used Eq. (5.82).

5.2.5 Examples

Let us know look at an example, which will show the usefulness of the CHPCA approach elaborated in the earlier subsections.

Let us take the following two time series:

$$r_1(t) = \sin\left(\frac{\pi t}{23}\right), \quad r_2(t) = \cos\left(\frac{\pi t}{23}\right), \tag{5.98}$$

for $t = 0, 1, 2, \cdots, 99$, which is plotted in Fig. 5.3.

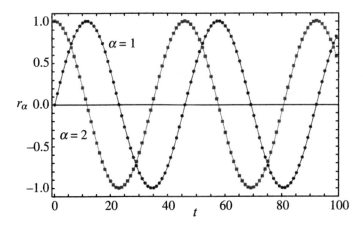

Fig. 5.3 The sample time series of sin and cos defined in Eq. (5.98).

The PCA correlation matrix is,

$$C = \begin{pmatrix} 1 & 0.049 \\ 0.049 & 1 \end{pmatrix}, \tag{5.99}$$

which fails to detect correlation with a time-lag between these two time series. This is as it should be, as there is no equal-time correlation between them if they are continuous in time. The small off-diagonal element, 0.049, is simply reflective of the fact that these time series are discretized in time.

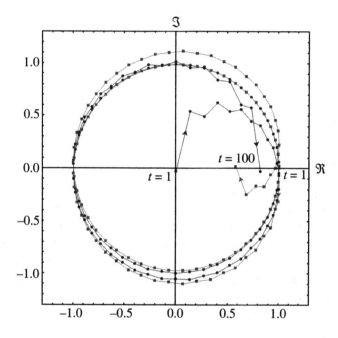

Fig. 5.4 The time series $\widetilde{r}_{1,2}(t)$ complexified as defined in Eq. (5.42).

When complexified, these time series behaves as shown in Fig. 5.4. Note that the period of these sinusoidal curves is equal to 46, which is *not* a divider of the whole time range $T = 100$. Therefore, these time series do not have just one Fourier component (see Eq. (5.31)), which explains the fact that the beginning part, say, $t \lesssim 10$ and the ending part, $t \gtrsim 90$. Nonetheless, the overall rotation of the time series in the complex plane is reproduced remarkably accurately, except for these edge regions.

The CPCA correlation matrix defined in Eq. (5.7) is now as follows:

$$\widetilde{C} = \begin{pmatrix} 1 & 0.981 e^{0.484\pi i} \\ 0.981 e^{-0.484\pi i} & 1 \end{pmatrix}. \tag{5.100}$$

And the eigenvalues and eigenvectors are the following:

$$\lambda^{(1)} = 1.982, \quad \boldsymbol{V}^{(1)} = \frac{1}{\sqrt{2}} \begin{pmatrix} 1 \\ e^{-0.484\pi} \end{pmatrix}, \tag{5.101}$$

$$\lambda^{(2)} = 0.019, \quad \boldsymbol{V}^{(2)} = \frac{1}{\sqrt{2}} \begin{pmatrix} 1 \\ e^{+0.484\pi} \end{pmatrix}. \tag{5.102}$$

The phase $\delta_{12} = 0.48\pi$ shows that the time series r_1 lags behind r_1 by 0.484π (see Fig. 5.3), which is very close to the actual value, 0.5π. Furthermore, its absolute value $|\widetilde{C}_{12}| = 0.981$ implies that the correlation with this time-lag is *very strong*, or *almost perfect*, which is the desired result. The eigenmode 1 is indeed the comovement of the sine and cosine with a time-lag.

This demonstrates the strength of CHPCA for detecting correlations with lead/lag.

5.3 Identification of Comovements

5.3.1 Random matrix theory

We must study which eigenmodes are significant, i.e., which signals represent systemic comovements in this system and not noises. This is the central issue that one always encounters in applying PCA or CHPCA to multivariate time series data.

The random matrix theory (RMT) provides us with a sound null hypothesis for such a statistical significance test. A set of random independent and identically distributed (i.i.d.) time series has a non-trivial correlation matrix and the eigenvalue spectrum $\rho(\lambda)$ is explicitly calculated as

$$\rho(\lambda) = \frac{Q}{2\pi} \frac{\sqrt{(\lambda_+ - \lambda)(\lambda - \lambda_-)}}{\lambda}, \tag{5.103}$$

with

$$\lambda_\pm = \left(1 \pm \frac{1}{\sqrt{Q}}\right)^2, \tag{5.104}$$

where $Q = T/N (>1)$ and $\lambda_- \leq \lambda \leq \lambda_+$. This formula was first derived by Marčenko and Pastur (1967). If $Q < 1$, we have to add a contribution of eigenvalues condensed at zero with fraction $1 - Q$ to the right-hand side of Eq. (5.103). Since the eigenvalues predicted by RMT are confined to $[\lambda_-, \lambda_+]$, the eigenvalues for the actual correlation matrix larger than λ_+ can be regarded as representing statistically meaningful correlations. For the CHPCA, Q in Eqs. (5.103) and (5.104) should be replaced by $Q/2$ as has been noted by Arai and Iyetomi (2013); this is because the imaginary part of a complexified time series is not independent of its real part, being related by the Hilbert transformation.

The RMT-based criterion is a refinement of Kaiser's selection rule, $\lambda > 1$. One can improve it further by comparing the actual eigenvalues with the corresponding eigenvalues of RMT, rank by rank, in place of λ_+. This method is basically the same as the one originally proposed by Horn (1965), known as **parallel analysis** (PA) in statistics (Zwick and Velicer, 1986; Buja and Eyuboglu, 1992; Franklin et al., 1995). However, PA does not take advantage of RMT; instead it carries out numerical simulations for its null model. The RMT has been extensively used by physicists to elucidate correlation structures in financial markets since the seminal works of Laloux et al. (1999a) and Plerou et al. (1999a). It is now attracting more and more interest by statisticians (Paul and Aue, 2014).

Figure 5.5 confirms the RMT formula Eq. (5.103) by comparing it with a numerical simulation; they agree quite well with each other. Figure 5.6 shows a PCA result obtained

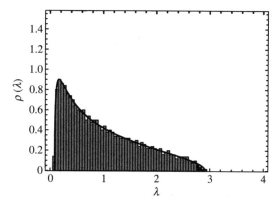

Fig. 5.5 Probability density function of the eigenvalues obtained numerically for a completely random correlation matrix with $N = 1000$ and $T = 2000$ ($Q = 2$), compared with the corresponding RMT result (solid curve), Eq. (5.103).

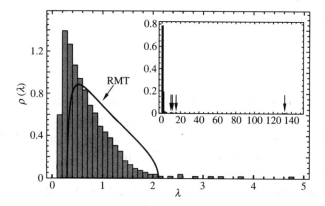

Fig. 5.6 Probability density function of the PCA eigenvalues, calculated for daily data of $N = 557$ stocks in the TSE market from 1996 to 2006 ($T = 2707$). In our case, $Q = \frac{T}{N} = 4.86$. If the returns are completely random, the eigenvalues must reside in the interval specified by the RMT, $0.30 \leq \lambda \leq 2.11$.

for daily prices of $N = 557$ stocks in the Tokyo Stock Exchange (TSE) collected over 11 years from 1996 ($T = 2706$ daily log returns). We see that the eigenvalue distribution $\rho(\lambda)$ is remarkably different from the RMT prediction, depicted by the solid curve. There are 13 eigenvalues larger than λ_+. The largest eigenvalue $\lambda \simeq 132$, corresponding to the collective behavior of the whole market, is 60 times larger than the maximum eigenvalue predicted for uncorrelated time series.

5.3.2 Autocorrelation effects

Here, we examine the autocorrelation effects on the RMT eigenvalues (Arai and Iyetomi, 2012). We first generate time series data $\{x_i(t)\}$ using the AR(1) model, which is given by

$$x_i(t+1) = \phi x_i(t) + \epsilon(t), \tag{5.105}$$

where $\epsilon(t)$ is supposed to be a random variable following the normal distribution $N(0, 1)$ and $|\phi| \leq 1$. We take the time difference of $\{x_i(t)\}$ as

$$w_i(t) = x_i(t+1) - x_i(t) \quad (i = 1, \ldots, N; t = 1, \ldots, T), \tag{5.106}$$

and then standardize $\{w_i(t)\}$ as

$$\omega_i(t) := \frac{w_i(t) - \langle w_i \rangle}{\sigma_i}, \tag{5.107}$$

with

$$\sigma_i := \sqrt{\frac{1}{T}\sum_{t=1}^{T}(w_i(t) - \langle w_i \rangle)^2}. \tag{5.108}$$

Each time series of the obtained data matrix $\{\omega_i(t)\}$ has autocorrelation characterized by the autocorrelation function,

$$\psi(\tau) = -\frac{1-\phi}{2}\phi^{\tau-1} \quad (\tau = 1, 2, \ldots), \tag{5.109}$$

and $\psi(0) = 1$.

Setting $\phi = 1$ in Eq. (5.105) reduces the AR(1) model to Brownian motion, so that the data matrix $\{\omega_i(t)\}$ is simply a collection of random numbers leading to the eigenvalue distribution of RMT. For $\phi = 0$ in Eq. (5.105), on the other hand, the AR(1) model generates just a random number series. Therefore, $\{\omega_i(t)\}$ possesses the strongest autocorrelation of $-1/2$ between its adjacent elements in the time direction. Figure 5.7 shows the eigenvalue probability density $\rho(\lambda)$ based on the AR(1) model at various ϕ values. The range of the eigenvalues is gradually extended as compared with that of the RMT as ϕ is decreased from 1 to 0. This example illustrates that some degree of deviation of $\rho(\lambda)$ from the RMT may be ascribable to autocorrelations involved in the time series data of finite length, which mimic cross correlations.

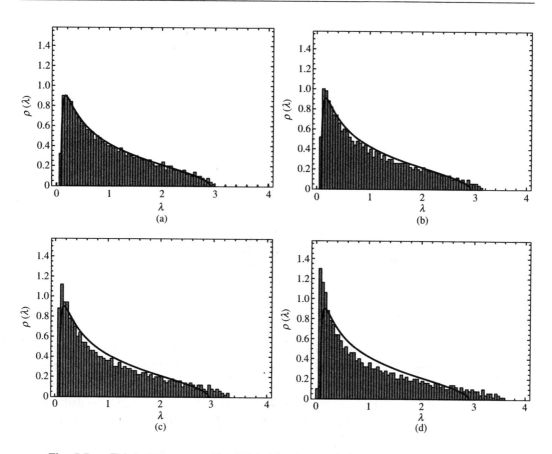

Fig. 5.7 This is the same as Fig. 5.5, but for the correlation matrices based on the AR(1) model at various ϕ: (a) $\phi = 0.75$, (b) $\phi = 0.5$, (c) $\phi = 0.25$, and (d) $\phi = 0$.

5.3.3 Rotational random shuffling

While the RMT-based method is clearly superior to other methods from both theoretical and practical points of view, Iyetomi et al. (2011a,b) demonstrates that it requires the following to be satisfied: 1) there is no autocorrelation in each time series, and 2) the time series are infinite in the sense that $N, T \to \infty$ with $Q = T/N$ kept finite.

To be free of these restrictions on the applicability of RMT to PCA and CHPCA, we resort to the rotational random shuffling (RRS) simulation proposed by (Iyetomi et al. (2011a,b) and Arai and Iyetomi (2013). In this simulation, we first randomly rotate each time series as follows;

$$r_\alpha(j) \to r_\alpha(\text{Mod}(j - \tau_\alpha, T) + 1) \tag{5.110}$$

where $\tau_\alpha \in [0, N]$ and $\text{Mod}(n, m)$ is the modulus function that gives the remainder of the division of n by m. It should be noted here that no autocorrelation is lost in each time series, as they are "rotated". This is necessary to keep the length of the time series intact.

On the other hand, since each time series is rotated differently, the comovements between them are destroyed. Therefore, the resulting eigenvalues λ_ℓ should reflect the same set of time series with the comovements destroyed. This, in turn, means that by comparing the resulting eigenvalue spectrum with the actual one, one can identify the true comovements in the data. The idea of RRS came from a combination dial lock as shown in Fig. 5.8.

Fig. 5.8 A combination dial lock.

Here, we give two examples of the RRS preprocessing of real data. One is the same stock prices data as used in Fig. 5.6. The other is the monthly sea level pressure (SLP) anomaly data observed worldwide for the period 1910 to 1990 ($T = 959$ monthly differences), where we sampled $N = 948$ observation points. The results for the eigenvalue distribution are shown in Fig. 5.9. For the TSE stock data, we see that the RMT works well to identify statistically meaningful eigenmodes. In contrast, the SLP anomaly data possess significant autocorrelations, so that the RMT should be replaced by the RRS for a better null hypothesis for the significance test.

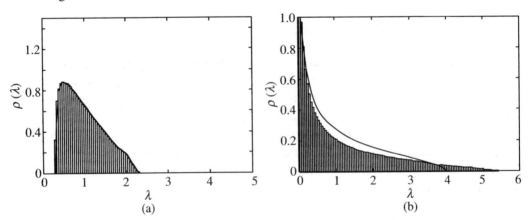

Fig. 5.9 Probability density function of the PCA eigenvalues for (a) the TSE stock data and (b) the SLP anomaly data preprocessed with the RRS. The solid curve in each panel shows the corresponding eigenvalue distribution of the RMT. (From Arai and Iyetomi (2012))

5.4 Case Study: Equity and Currency in the World

In this section, we will apply the toolkit expanded in the previous sections to the equity and currency time series of 48 countries in the world and extract significant lead–lag relationships among these markets. (Vodenska et al., 2016). As a result of this analysis, readers will see that in general, the foreign exchange market has predictive power for the global stock market performances. In addition, the United States, German and Mexican markets have forecasting power for the performances of other global equity markets. Later in this book, we shall construct a coupled synchronization network and carry out community analysis, to discover the formation of four distinct network communities.

Table 5.1 gives the list of the 48 counties together with their stock indices chosen. We study these *daily data* for the 14-year period from Jan. 1, 1999 to Dec. 31, 2012 (5,114 calendar days with 3,652 trading days).

A note on the currency is in order. The currency quotes that we analyze are expressed as currency/SDR, where SDR, special drawing right, is defined as a basket of four major currencies (the Japanese Yen, US Dollar, Pound Sterling, and the Euro). The composition of the SDR basket is reviewed every five years by the Executive Board of the International Monetary Fund (IMF) to ensure that it reflects the relative importance of currencies in the global financial and trading systems. Therefore, when it increases, it means that the currency has depreciated and more "currency" is needed for one SDR. For example, if the USD/SDR price increases, this means that the US Dollar has depreciated. For this reason, we adopt the negative of the logarithmic growth rate as the time series for the currency.

It should be noted that the definition of "day" differs from country to country, as these 48 counties are at various points of the globe. If we are studying a limited part of the world, say, only the European countries, this does not matter. However, that is not true in our case. When the New York Stock Exchange closes at 4 p.m., for example, it is 6 a.m. the following day in Japan (5 a.m. if the US is on daylight time). Thus, if something occurs in the US markets that has an immediate effect on Japanese markets, the event will not occur on the same recorded business day in the two countries. As we have elaborated in this chapter, our CHPCA is a perfect tool to overcome this point, as it can identify comovement with lead/lag.

One may believe that applying PCA to this set of time series with time-shifts could also help deal with this point. However, this is not true because of the large number of calculations it requires. For example, allowing a six-day delay between a pair will require us to examine $6N(N-1)/2 \simeq 2.7 \times 10^4$ cases, where $N = 96$ is the number of time series in our data. In addition, there is the issue of consistency of time-shift obtained in such a manner: what if the time-shifts of three time series are all of the same sign? For example, if time series A is ahead of B by one day, with B ahead of C by two days, and C ahead of A by a day, how can one look at the comovement of the three bodies (A,B,C)? None of these problems arise for CHPCA: with one calculation, it yields a set of (independent) comovements as significant eigenmodes and the strength of each can be ascertained by their eigenvalues.

Table 5.1 List of 48 countries with their stock-market indices with markers used in the figures. For the top 10 countries (2012 GDP, excluding China), the markers are given individually with their national flag motifs, while for others, the markers reflect their regionality. The currencies are marked with their stock-market marker on a gray hexagonal base in Fig. 5.12.

No.	Country	Stock	Index	No.	Country	Stock	Index
1	UK		UKX	25	Canada		SPTSX
2	Austria		ATX	26	Mexico		MEXBOL
3	Belgium		BEL20	27	Brazil		IBOV
4	Finland		HEX25	28	Argentina		MERVAL
5	France		CAC	29	Chile		IPSA
6	Germany		DAX	30	Peru		IGBVL
7	Ireland		ISEQ	31	Venezuela		IBVC
8	Italy		FTSEMIB	32	India		SENSEX
9	Netherlands		AEX	33	Sri Lanka		CSEALL
10	Portugal		PSI20	34	Indonesia		JCI
11	Spain		IBEX	35	Japan		NKY
12	Greece		ASE	36	South Korea		KOSPI
13	Malta		MALTEX	37	Malaysia		FBMKLCI
14	Slovakia		SKSM	38	Thailand		SET
15	Norway		OBX	39	Philippine		PCOMP
16	Sweden		OMX	40	Hong Kong		HSI
17	Iceland		ICEXI	41	Australia		AS51
18	Switzerland		SMI	42	Israel		TA-25
19	Czech		PX	43	Pakistan		KSE100
20	Denmark		KFX	44	Saudi Arabia		SASEIDX
21	Hungary		BUX	45	South Africa		TOP40
22	Poland		WIG	46	Oman		MSM30
23	Russia		INDEXCF	47	Qatar		DSM
24	USA		SPX	48	Mauritius		SEMDEX

The eigenvalues (the large dots with numbers) and the RRS results (the small dots connected by the dashed line) are plotted in Fig. 5.10 and displayed in Table 5.2 The first six eigenvalues, $\lambda^{(1)} \ldots \lambda^{(6)}$, lie outside the RRS 99% error range and thus are clearly identifiable. These six largest eigenvalues can be used to explain the significant comovements in the system.[7]

---

[7] Some readers might argue that (perhaps with knowledge from RMT) only the upper bound matters since the largest eigenvalue from RRS is above the third actual eigenvalue and therefore only the 1st and the 2nd eigenmodes are significant from Fig. 5.10. However, this is not so. The sixth eigenmode is a signature of the Iceland crisis and IS significant. We will come back to this point a little later.

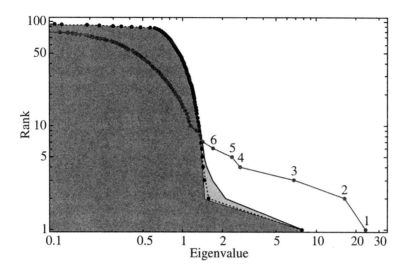

Fig. 5.10 Significant eigenvalues identified by CHPCA using the RRS method. The dot denoted by n shows the nth eigenvalue $\lambda^{(n)}$ (x-axis) and the eigenvalue rank (y-axis). The gray small dots and the lighter gray area show the average RRS and the 99% range. The six largest eigenvalues are clearly outside of their RRS ranges, and show significant relationships in the interdependent network (from Vodenska et al. (2016)) [see Color Plate at the end of book].

Table 5.2 List of CHPCA eigenvalues, their 99% RRS range, and the mean contribution rate $\sqrt{\lambda_n/\lambda_1}$. Although the seventh eigenmode is outside the RRS range, we exclude this mode as insignificant as it is very close to the boundary (from Vodenska et al. (2016)).

n	λ_n	99% RRS range	$\sqrt{\lambda_n/\lambda_1}$
1	23.74	$7.79^{+0.17}_{-0.02}$	1
2	16.4	$1.56^{+0.54}_{-0.12}$	0.83
3	6.8	$1.46^{+0.22}_{-0.05}$	0.53
4	2.7	$1.42^{+0.09}_{-0.03}$	0.37
5	2.3	$1.40^{+0.05}_{-0.03}$	0.31
6	1.7	$1.38^{+0.03}_{-0.02}$	0.27
7	1.4	$1.36^{+0.03}_{-0.02}$	0.24
8	1.3	$1.35^{+0.02}_{-0.02}$	0.24

As noted earlier in the chapter, CHPCA is superior to PCA as it is capable of detecting cross-correlations with lead–lag relations in multivariate time series. Figure. 5.11 shows that the cumulative sum of the CHPCA eigenvalues is consistently larger than that of the

PCA eigenvalues, which means that CHPCA is in fact superior to PCA in identifying the significant modes.

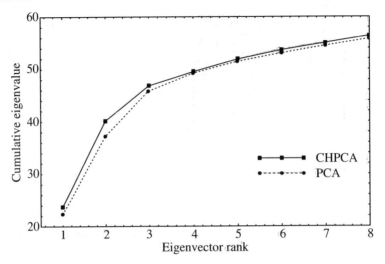

Fig. 5.11 Comparison between the PCA and CHPCA eigenvalues. The partial sum of the eigenvalues, $\sum_{n=1}^{K} \lambda^{(n)}$ (y-axis) versus K (x-axis), for PCA (the lower curve) and for CHPCA (the upper curve) are shown. The fact that the CHPCA sums of eigenvalues are always above the PCA-based eigenvalue sums shows that CHPCA is a stronger analytic tool in identifying important comovements (from Vodenska et al. (2016)).

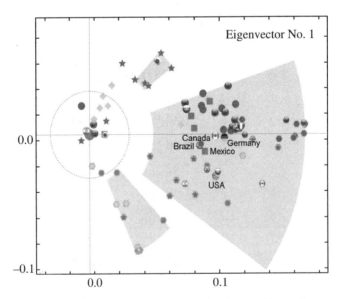

Fig. 5.12 The first eigenvector components. The flags without the gray background represent the equity markets, while the dark gray hexagons behind the flags represent the currencies as shown in Table 5.1 (from Vodenska et al. (2016)).

Let us now look at some of the eigenvectors. Figure. 5.12 shows the components of the first eigenvector in their complex plane, where the nodes with absolute value larger than 0.04 are grouped so that nodes with a phase gap smaller than 0.2 [rad] are in the same group, in order to visualize the group of nodes that move relatively close to each other. From this figure, we see clearly that in general, currencies lead the stock markets and that US stock market leads the stock markets of Mexico, Brazil and others. For other eigenmodes and interpretations, see Vodenska et al. (2016).

Let us now come back to the sixth mode. Figure. 5.13 shows the sixth eigenvector, where we observe that it is dominated by the Icelandic Krona and the Icelandic, Polish, and Russian stock markets on one side and the Middle Eastern stock markets on the other.

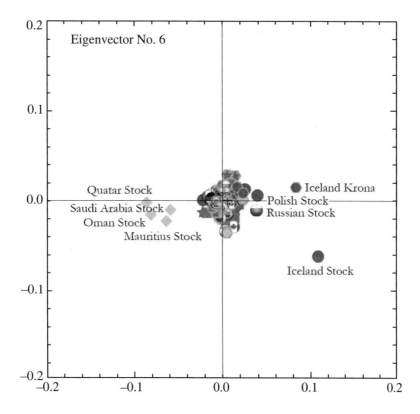

Fig. 5.13 The sixth eigenvector components of the world equity and foreign exchange markets (from Vodenska et al. (2016)) [see Color Plate at the end of book].

Figure 5.14 (a) shows the presence of 1–6 modes in time. We readily observe that on October 13, 2008, at the onset of the Icelandic banking crisis, this sixth eigenmode was quite strong: Although its average presence is weaker than the first to the fifth modes owing to the smallness of its eigenvalue, it was important at this point in time. Figure 5.14 (b) shows that the Icelandic stock market (the line with the shaded gray area) collapsed at this point. In fact, on this day, the Icelandic stock market lost over 90% of its market capitalization which had a strong influence on some of the other countries as seen in the eigenvector plot.

This demonstrates the power and the usefulness of CHPCA and RRS.

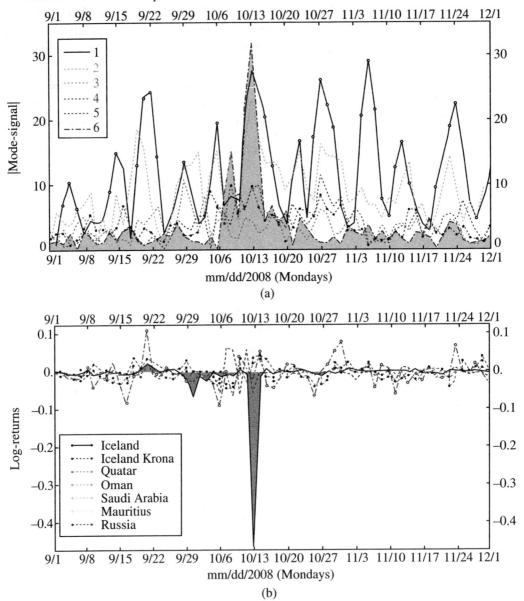

Fig. 5.14 (a) Absolute values of the mode-signals for the eigenmodes 1 to 6 from September 1st to December 1st of 2008. The red dots connected with red dash-dot lines show the sixth eigenmode. (b) Behavior of the log-returns of the time series that have large absolute values in the sixth eigenvector (from Vodenska et al. (2016)) [see Color Plate at the end of book].

5.5 Exercise

5.1 By using Eq. (5.20), prove Eq. (5.25) and Eq. (5.26) for $\omega > 0$.

5.2 Integrate Eq. (5.103) over λ to confirm that it is normalized to unity or Q depending on $Q > 1$ or $Q < 1$, respectively.

5.3 Derive Eq. (5.109).

6

Business Cycles

> No, no, you're not thinking; you're just being logical.
>
> Niels Bohr

What causes business cycles? A look at Haberler (1964) shows that all kinds of theories had already been advanced by the end of the 1950s. In recent years, macroeconomists had been so confident of their skillful policy management as to hail it as the "Great Moderation". Lucas Jr. (2003), in his presidential address to the meeting of the American Economic Association, declared:

> Macroeconomics was born as a distinct field in the 1940's, as a part of the intellectual response to the Great Depression. The term then referred to the body of knowledge and expertise that we hoped would prevent the recurrence of that economic disaster. My thesis in this lecture is that macroeconomics in this original sense has succeeded: Its central problem of depression prevention has been solved, for all practical purposes, and has in fact been solved for many decades (Lucas Jr., 2003, [p.1]).

Ironically, when Lucas delivered this address, the American economy had been on a steady road to the worst post-war recession, perhaps even depression. Business cycles are still with us and await further investigation.

Describing the characteristics of business cycles has been one of the most difficult problems in the study of economies because of the presence of various economic shocks. In this chapter, we analyze the characteristics of business cycles more directly than in the standard approach using a method motivated by physics. First, we identify the causes of business cycles using the random matrix theory and principal component analysis.

We then identify evidence of synchronization in Japanese business cycles and the international business cycle. Finally, we discuss the mechanism of this synchronization using the coupled limit-cycle oscillator model.

6.1 What Causes Business Cycles

Business cycles are defined as a type of fluctuation in the aggregate economic activity of nations that organize their work mainly around business enterprises. A cycle consists of expansions occurring at roughly the same time in many economic activities, followed by similar general recessions, contractions, and revivals that merge into the expansion phase of the next cycle; it is a sequence of change that is recurrent but not periodic.

Business cycles have a long history of theoretical studies (Granger and Hatanaka, 1964; Aoki and Yoshikawa, 2006). Samuelson (1939) showed that the second-order ordinary linear differential equation based on a multiplier and accelerator could generate a cycle in the gross domestic product (GDP). Hicks, by introducing a "ceiling" and a "floor" into such a model (Hicks, 1950a), demonstrated that the cycle is sustainable. In contrast, some other theorists focused on non-linearity. Goodwin (1951) introduced a non-linear accelerator to generate a sustainable cycle. Kaldor (1940) saw business cycles as limit cycles. Subsequently, the idea of non-linearity evolved into the application of chaos theory (Lorenz, 1993; Sterman and Mosekilde, 1994).

In neoclassical economics, the predominant causes of business cycles are seen as real changes on the supply side, such as innovation in production technology, instead of change in aggregate demand. The real business cycle (RBC) theory pioneered by Kydland and Prescott (1982), arguably the most dominant school today, regards changes in total factor productivity as the primary cause of business cycles. If we consider the Cobb–Douglas production function, Eq. (4.5), the Solow residual,

$$\frac{\Delta b}{b} = \frac{\Delta Y}{Y} - \alpha \frac{\Delta K}{K} - (1-\alpha)\frac{\Delta L}{L} \qquad (6.1)$$

is considered as a shock because of technological innovations. Here, Y, K, L, α, and b are total production, capital input, labor input, capital elasticities, and total factor productivity, respectively. If technological innovations occur randomly, the Solow residual can be emulated as a moving summation of random shocks. According to Slutzky (1937a), such averaging of random shocks generates "cyclical" business fluctuations. The Slutsky effect is illustrated in Fig. 6.1

Since the RBC theory is basically the neoclassical equilibrium theory, it minimizes the role of the government's stabilization policy. The Keynesian theory, in contrast, attributes business cycles to fluctuations of real aggregate demand (Summers, 1986; Mankiw, 1989; Tobin, 1993), and assigns the government and the central bank the substantial role of stabilization. The fundamental problem, namely, what causes business cycles, remains unresolved.

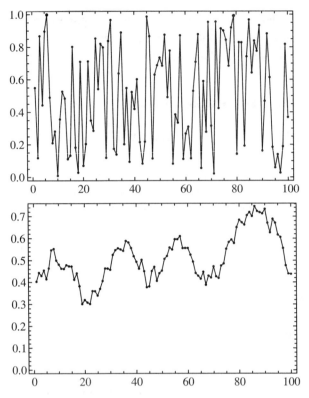

Fig. 6.1 Demonstration of Slutsky's effect. The upper panel shows a sequence of random numbers uniformly distributed between 0 and 1. The lower panel shows their 10 term moving average.

The standard approach to study this problem is to explore the relative importance of technology shocks on the one hand, and demand shocks on the other, as the sources of business cycles by making various identifying restrictions to macro time series models; see Shapiro and Watson (1988), Blanchard and Quah (1989), Francis and Ramey (2005), and works cited therein. Although the variables used and the methods differ in details, they all share the same key identifying restriction, namely, that the level of output (or real GDP) is determined in the long run only by supply shocks such as technology shocks. Conversely, the long-run effect of demand shocks on output is assumed to be nil. On this assumption, disturbances that are not directly observable are sorted into technology shocks and demand shocks.

A majority of works routinely make this assumption without questioning its validity. However, we maintain that the assumption is not actually as tenable as is commonly believed. The problem is that in the study of business cycles, technology shocks are usually taken to be exogenous, as typified by the RBC theory. Here, we must observe that modern micro-founded macroeconomics is curiously inconsistent in that while new growth theory puts much emphasis on the endogeneity of technical progress, the business cycle theory is naively content with taking advances in technology as exogenous. New technology or technical progress is certainly not a free lunch! First of all, it must be generated by R&D investment. Once new technology is discovered, it can augment production only by way of investment which embodies such new technology. Now, investment including R&D investment is, of course, much affected by cyclical fluctuations. Monetary factors as well as real factors affect investment. Thus, to the extent that new technology is embodied in investment, demand shocks are bound to affect the pace of technical progress. Therefore, one cannot naively accept the standard identifying restriction that demand shocks do not leave a permanent effect on the level of output.

One way to proceed without this questionable assumption is to take advantage of disaggregated data. The representative works on business cycles cited earlier all use macro variables. However, business cycles as the conspicuous macro phenomena are characterized by the fact that output movements across broadly defined sectors and types of goods move together (Mitchell, 1951). In Mitchell's terminology, they exhibit high conformity. It is then most natural to look for possibly unobserved common factors underlying sectoral comovements as the basic factors causing business cycles. Moreover, the separate effects of such unobserved factors on production, shipments and inventory are expected to provide one with valuable information on the causes of business cycles.

Factor models fit our purpose precisely (Sargent and Sims, 1977; Stock and Watson, 1998, 2002). Stock and Watson (1998, 2002), based on this method, came up with the diffusion index to improve the forecasting of cyclical fluctuations. Here, we take the same approach, but for a different purpose (Iyetomi et al., 2011b,a). By examining the relationship between production, shipments, and inventory of 21 types of goods, we explore what causes business cycles. We also show that the model estimated based on the pre-Lehman crisis (September 15th, 2008) can reasonably explain the post-Lehman (2008–09) recession of the Japanese economy. Our findings favor the old Keynesian view (Tobin, 1993).

6.1.1 Indices of industrial production

We use the monthly data of the Indices of Industrial Production (IIP) compiled by the Ministry of Economy, Trade and Industry (METI) of Japan. These are indices of industrial production, producer's shipments, and producer's inventory of finished goods. Details of IIP are available on the website (METI, Japan, 2015). Here, we simply note that the indices are compiled based on nearly 500 items (496 for production and shipment indices; 358 for inventory). We use this data for 240 months, from January 1988 to December 2007 (the data itself is available from January 1977, but some items are missing before January 1988). Two classification schemes of the IIP are available: indices classified *by industry* and indices classified *by use of goods*. We adopt the latter classification scheme, which is listed in Table 6.1. Its concept, featuring input–output interindustrial relations, is illustrated in Fig. 6.2. We emphasize that the inner loop of production existing in the economic system may give rise to a nonlinear feedback mechanism to complicate the dynamics of the system; outputs are reused by the system as inputs for its production activities.

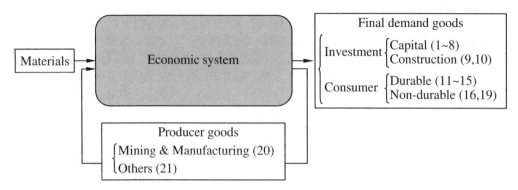

Fig. 6.2 Input–output relationship in industrial activities of the economic system as measured by IIP in Japan. The numbers in the parentheses denote the classification index g in Table 6.1. (From Iyetomi et al. (2011a))

Table 6.1 Classifications of 21 goods in IIP. The center column is the identification number, g. (From Iyetomi et al. (2011b))

Final Demand Goods		
Investment Goods		
Capital Goods	1	Manufacturing Equipment
	2	Electricity
	3	Communication and Broadcasting
	4	Agriculture
	5	Construction
	6	Transport
	7	Offices
	8	Other Capital Goods
Construction Goods	9	Construction
	10	Engineering
Consumer Goods		
Durable Consumer Goods	11	Household Use
	12	Heating/Cooling Equipment
	13	Furniture & Furnishings
	14	Education & Amusement
	15	Motor Vehicles
Non-durable Consumer Goods	16	Household Use
	17	Education & Amusement
	18	Clothing & Footwear
	19	Food & Beverage
Producer Goods		
	20	Mining & Manufacturing
	21	Others

We denote the IIP by $S_{\alpha,g}(t_j)$, where $\alpha = 1, 2$, and 3 denotes production (value added), shipments, and inventory, respectively. Moreover, $g = 1, 2, \ldots, 21$ denotes the 21 categories of goods shown in Table 6.1, and $t_j = j\Delta t$ with $\Delta t = 1$ month, and $j = 1, 2, \ldots, N$ for $N (= 240)$ months starting in January, 1988 and ending in December, 2007. Figure 6.3 shows the temporal change of the averaged IIP data for production, shipments, and inventory during the period January 1988 to December 2009. The ongoing global recession is traced back to the subprime mortgage crisis in the U.S., which became apparent in 2007. The economic shock has affected Japan without exception, leading to a dramatic drop in the production activities of the country, as shown in the figure. The logarithmic growth rate $r_{\alpha,g}(t_j)$ is then defined as

$$r_{\alpha,g}(t_j) := \log_{10}\left[\frac{S_{\alpha,g}(t_{j+1})}{S_{\alpha,g}(t_j)}\right], \tag{6.2}$$

where j runs from 1 to $N' := N - 1 (= 239)$. And $w_{\alpha,g}$ denotes its normalized growth rate:

$$w_{\alpha,g}(t_j) := \frac{r_{\alpha,g}(t_j) - \langle r_{\alpha,g} \rangle}{\sigma_{\alpha,g}}, \tag{6.3}$$

where $\sigma_{\alpha,g}$ is the standard deviation of $r_{\alpha,g}$ over time.

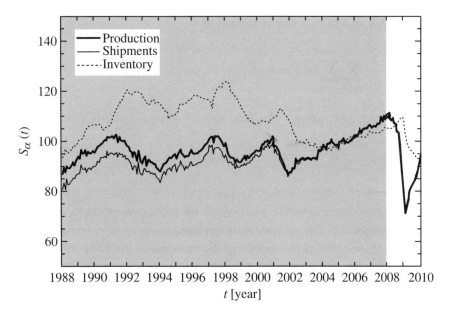

Fig. 6.3 Averaged IIP data S_α for production (thick solid line), shipments (thin solid line), and inventory (dotted line) as a function of time t. The correlation matrix is calculated using the data in the gray shaded area, from January 1988 through December 2007. (Adapted from Iyetomi et al. (2011a))

6.1.2 Power spectrum

The study of business cycles presents two basic problems. The first has to do with identifying the periodicity and the second, with identifying the common factors that cause sectoral comovements. In this section, we deal with the periodicity issue. Some economists are skeptical of the stable periodicity of business cycles (Diebold and Rudebusch, 1990). It is certainly true that when we examine the *sample path property* of a stochastic process, we are faced with difficulties in discovering, or even clearly defining, the periodicity. However, this does not necessarily mean that we cannot usefully explore the periodicity of business cycles. We resort to standard spectral analysis (see Granger (1966), for example).

The Fourier decomposition of the normalized growth rate $w_{\alpha,g}(t_j)$ is carried out in the same way as in Eq. (5.29):

$$w_{\alpha,g}(t_j) = \frac{1}{\sqrt{N'}} \sum_{k=0}^{N'-1} \widetilde{w}_{\alpha,g}(\omega_k) e^{-i\omega_k t_j}, \tag{6.4}$$

where the Fourier frequency ω_k is given by

$$\omega_k = 2\pi k/(N'\Delta t). \tag{6.5}$$

The averaged power spectrum $p(\omega_k)$ is defined as

$$p(\omega_k) = \frac{1}{M} \sum_{\alpha=1}^{3} \sum_{g=1}^{21} |\widetilde{w}_{\alpha,g}(\omega_k)|^2, \tag{6.6}$$

where $M = 63$, the total number of the time series, and $p(\omega_k)$ satisfies the sum rule:

$$\sum_{k=0}^{N'-1} p(\omega_k) = N'. \tag{6.7}$$

Figure 6.4 compares two results for $p(\omega)$ based on the seasonally-adjusted data and the non-adjusted original data. The original data has high peaks at $T =$ one year, six months, and three months showing the presence of significant seasonal fluctuations. We observe that they are well eliminated in the adjusted data. We will use the seasonally-adjusted data for the following analysis.

We observe the significant power at low frequencies, $T = 60$ months (five years) and $T = 40$ months (three years and six months). One might be tempted to argue that strengths of the cycles $T = 120$ and $T = 239$ are comparable to them. However, for the 239 months that our data covers, the $T = 120$ oscillation occurs only twice, and the $T = 239$, only once. Therefore, we disregard them as "one-time" events. In addition, one might argue that $T = 24$ appears to be a possible cycle. But so are many peaks for $T < 24$. In general, peaks for shorter periods tend to be the results of noise, and one needs to discard them. Furthermore, we confirmed robustness of the $T = 60$ and $T = 40$ cycles by performing two calculations: discrete Fourier transformation for chopped data and continuous Fourier transformation which is free from the discretized condition, Eq. (6.5).

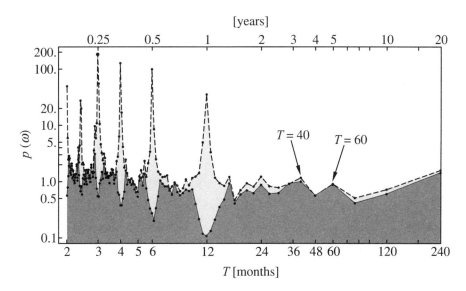

Fig. 6.4 Averaged power spectrum $p(\omega)$ of the normalized growth rate. The horizontal axes is chosen to be the period, $T := 2\pi/\omega$, which is shown in units of months at the bottom and in units of years at the top. The points connected by solid lines (with darker shading) are the spectrum for the seasonally-adjusted data, while the points connected with dashed lines (with lighter shading) are for the non-adjusted original data, multiplied by a factor of seven to make the comparison easier. (From Iyetomi et al. (2011a))

The business cycles with $T = 40$ and $T = 60$ months turned out to be significant and robust. The peaks and troughs of Japan's business cycles are officially determined by the Cabinet Office (prior to 2000, the Economic Planning Agency). In the post-war period (October 1951 – January 2002), the Japanese economy experienced 12 cycles. The average length of the twelve cycles is 50.3 months with maximum 83 months and minimum 31 months. The periodicity we have identified, $T = 40$ and $T = 60$ months is, therefore, quite reasonable.

6.1.3 Identification of dominant factors

Next, we must identify the significant common factors underlying sectoral comovements. By common factors, we mean the small number of factors that explain much of the correlation of the original data. We resort to the PCA described in Section 5.1.

The distribution of the eigenvalues $\lambda^{(n)}$ is plotted in Fig. 6.5. We see that the largest and the second largest eigenvalues, designated as $\lambda^{(1)}(\simeq 9.95)$ and $\lambda^{(2)}(\simeq 3.83)$, are well separated from the eigenvalue distribution predicted by the RMT, whereas the third largest eigenvalue $\lambda^{(3)} (\simeq 2.77)$ is adjacent to the continuum. Therefore, only 2 eigenvalues out of a total of 63 are of statistical significance according to the RMT. The RRS simulation shows that an appreciable degree of autocorrelation is involved in the IIP data. The third largest eigenvalue $\lambda^{(3)} \simeq 2.77$ becomes very close to the upper limit, $\lambda'_+ = 2.47 \pm 0.20$, of the eigenvalues obtained by the RRS, where the error is estimated at 95% confidence level. This result reinforces the neglect of $\lambda^{(3)}$.

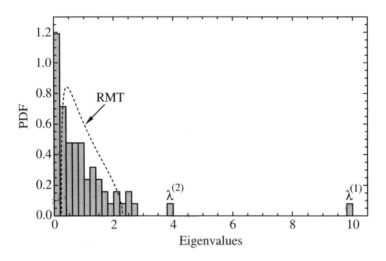

Fig. 6.5 PDF $\rho(\lambda)$ of the eigenvalues (bars) compared with the RMT prediction (dashed curve). (From Iyetomi et al. (2011a))

The eigenvectors $V^{(1)}$ and $V^{(2)}$, associated with $\lambda^{(1)}$ and $\lambda^{(2)}$, are displayed in Fig. 6.6; those vectors are called "the first eigenvector" and "the second eigenvector", respectively, hereafter. These two eigenvectors have characteristic features that distinguish them from each other. The eigenvector $V^{(1)}$ represents an economic mode in which production and shipments of all goods expand (shrink) synchronously with decreasing (increasing) inventory of produced goods. This corresponds to the market mode obtained for the largest eigenvalue in the stock market analyses (Laloux et al., 1999a; Plerou et al., 1999a), and may be referred to as the "aggregate demand" mode according to Keynes' principle of effective demand: both shipments and production in all the sectors are moved jointly by aggregate demand (Keynes, 1936). On the other hand, the eigenvector $V^{(2)}$ is a mode that apparently represents dynamics of inventory, i.e., accumulation or clearance of inventory, for most goods, including produced goods. We further find positive correlation between production enhancement and inventory accumulation for most goods. This finding indicates that production has a kind of inertia in its response to change of demands.

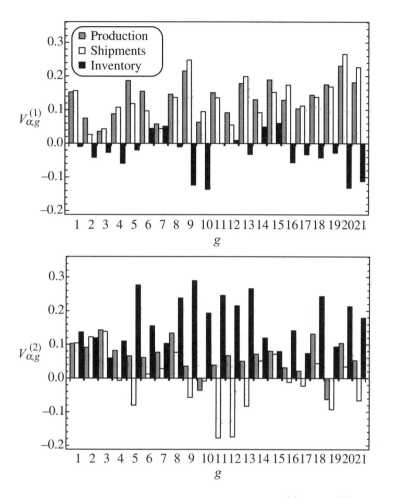

Fig. 6.6 Components of the two dominant eigenvectors, $\boldsymbol{V}^{(1)}$ and $\boldsymbol{V}^{(2)}$. The horizontal axis shows the 21 goods ($g = 1, 2, \ldots, 21$). The gray bars represent production ($\alpha = 1$); the white bars, shipments ($\alpha = 2$); and the black bars, inventory ($\alpha = 3$). (From Iyetomi et al. (2011a))

Readers may be curious about the present construction of a correlation matrix by mixing up data for production, shipments, and inventory, because these are very different species of data at first glance. Thus far, physicists have applied the RMT mainly to analyses of stock data having similar characteristics. In this sense, our approach is quite radical. However, production, shipments, and inventory form a trinity in the economic theory for business cycles, so that these variables should be treated on an equal footing. Using the two dominant eigenmodes, in fact, we are successful in proving the existence of intrinsic business cycles (Iyetomi et al., 2011b,a).

6.1.4 Dominant eigenmodes and the business cycles

We have found that (1) the $T = 60$ (months) and the $T = 40$ fluctuations are significant, and also that (2) the two largest eigenvalues, $\lambda^{(1,2)}$ of the correlation matrix, are significant based on the RMT. These two findings suggest that the contribution of the two largest eigenvalue components at $T = 60$ and $T = 40$ periods can be properly defined as the "business cycles". In this section, we measure the contribution of this "business cycle" factor in economic fluctuations.

To this end, we express the normalized growth rate in terms of the mode signals $a^{(n)}(t_j)$ associated with the eigenvectors $\boldsymbol{V}^{(n)}$:

$$w_{\alpha,g}(t_j) = \sum_{n=1}^{M} a^{(n)}(t_j) V_{\alpha,g}^{(n)}. \tag{6.8}$$

The Fourier transform of $a^{(n)}(t_j)$ is given by

$$a^{(n)}(t_j) = \frac{1}{\sqrt{N'}} \sum_{k=0}^{N'-1} \widetilde{a}^{(n)}(\omega_k) e^{-i\omega_k t_j}, \tag{6.9}$$

similar to Eq. (6.4). We then obtain the following decomposition of the power spectrum:

$$p(\omega_k) = \frac{1}{M} \sum_{n=1}^{M} \lambda^{(n)}(\omega_k), \tag{6.10}$$

where

$$\lambda^{(n)}(\omega_k) = \left|\widetilde{a}^{(n)}(\omega_k)\right|^2, \tag{6.11}$$

with

$$\lambda^{(n)} = \frac{1}{N'} \sum_{k=0}^{N'-1} \lambda^{(n)}(\omega_k). \tag{6.12}$$

Summing both sides of Eq. (6.10) over k recovers Eq. (5.12):

$$\frac{1}{N'} \sum_{k=1}^{N'-1} p(\omega_k) = \frac{1}{M} \sum_{n=1}^{M} \lambda^{(n)} = 1. \tag{6.13}$$

Equation (6.10) shows that the eigenvectors corresponding to the larger eigenvalues have larger contribution to the power spectrum *on average*. For instance, the relative contribution of the first eigenvector is 16% as $\lambda^{(1)}/M \approx 0.16$ ($M = 63$). If the contributions of all the eigenvectors were the same, it would be only 1.6% ($1/M \approx 0.016$). The sum of the relative contribution of the first and the second eigenvector components amounts to 21%. The dominance of two large eigenvalues is clear. Also Fig. 6.7 elucidates what part of the power spectrum the large eigenvalues have significant contributions. We observe that the eigenmodes of the larger eigenvalues tends to have greater contributions as T gets larger.

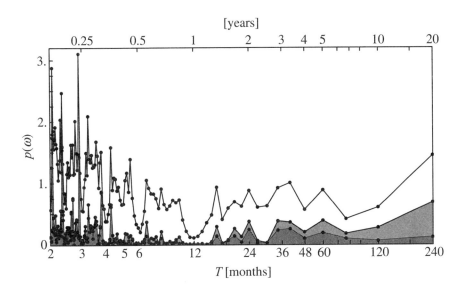

Fig. 6.7 Contributions to the total power spectrum of the two dominant eigenmodes. The solid line at the top represents the total power spectrum $p(\omega_k)$. The darker shading represents the partial power spectrum $\lambda^{(1)}(\omega_k)/M$ of the first eigenmode, while the lighter shading, that of the second eigenmode. (From Iyetomi et al. (2011a))

We have identified the two dominant factors $\boldsymbol{V}^{(1)}$ and $\boldsymbol{V}^{(2)}$ as demand factors, not productivity shocks. It is important to recognize that just because the relative contribution of $\boldsymbol{V}^{(1)}$ and $\boldsymbol{V}^{(2)}$ is roughly 40%, it does not mean that the residual 60% is explained by other systematic factors such as "productivity shocks" emphasized by the RBC theory. The residual 60% is explained by $\boldsymbol{V}^{(j)}$ ($j = 3, 4, \ldots$), but the significance of $\boldsymbol{V}^{(j)}$ ($j = 3, 4, \ldots$) is rejected by the RMT. The residual 60% is, therefore, not really a part of systematic "business cycles", but rather caused by various unidentified random factors. In fact, Slutzky (1937b) suggested that the summation of pure random shocks might generate apparently "cyclical" fluctuations; in effect, he rejected the existence of systematic "business cycles". Contrary to his assertion, our findings confirm the presence of business "cycles" which *systematically* account for 40% of economic fluctuations. Such "business cycles" defined based on $\boldsymbol{V}^{(1)}$ and $\boldsymbol{V}^{(2)}$ at $T = 40$ and 60 are basically caused by demand factors. In the next section, we will investigate the cyclical behavior of production, shipments and inventory in greater detail.

6.1.5 Production and shipments

We have already confirmed that the two eigenmodes of the correlation matrix give significant contributions to the $T = 40$ and 60 Fourier components. Let us now examine the behavior of production and shipments in these dominant modes. To do so, we first decompose the IIP data to Fourier and eigenmode components by combining Eqs. (6.8) and (6.9):

$$w_{\alpha,g}(t_j) = \frac{1}{\sqrt{N'}} \sum_{n=1}^{M} \sum_{k=0}^{N'-1} \widetilde{a}^{(n)}(\omega_k) e^{-i\omega_k t_j} V_{\alpha,g}^{(n)} \tag{6.14}$$

$$= \frac{2}{\sqrt{N'}} \sum_{n=1}^{M} \sum_{k=1}^{(N'-1)/2} \Re\left(\widetilde{a}^{(n)}(\omega_k) e^{-i\omega_k t_j}\right) V_{\alpha,g}^{(n)}. \tag{6.15}$$

Here we used the facts that $\widetilde{a}^{(n)}(0) = 0$ because of $\widetilde{w}_{\alpha,g}(0) = 0$; $\widetilde{a}^{(n)}(\omega_k) = \widetilde{a}^{(n)}(\omega_{N'-k})^*$ because of $a^{(n)}(t_j)$ being real; and that $N' - 1 = 238$ is even. The "business cycle component" is obtained by limiting the sum in Eq. (6.15) to over only $k = 4$ ($T = 60$) and $k = 6$ ($T = 40$) and $n = 1, 2$.

In Fig. 6.8, we compare the sum of the components of the first and the second eigenvectors at the $T = 40$ and $T = 60$ periods with the original data for production, shipments, and inventory. To make comparison easier, we apply a simple moving average operation to the original data over a one-year period. For the original data so smoothed out, we detect hardly any lead–lag relation between production and shipments. In contrast, the extracted "business cycles" shows that shipments clearly precede production with an average lead time of four months.

In Fig. 6.8, we also observe a close one-to-one correspondence between the original data and the extracted business cycles. The peaks and troughs of two cycles coincide very well, although their amplitudes slightly differ possibly owing to economic fluctuations with longer cycles ($T > 60$) and external shocks. A linear combination of two sinusoidal functions oscillating at different periods gives rise to an extended cyclic motion with the period of $T = 120$. Figure 6.8 thus spans two cycles of the endogenous business fluctuations. The characteristic behavior of the original data shown in the bottom panel is well captured by the "business cycles" defined here.

We find that production lags behind shipments over the extracted "business cycles". It does so because inventory adjustments tend to lag behind the initial changes in shipments. Assume shipments or final demand increased in the economy as a whole. This will induce a contemporaneous increase of production, and at the same time, depletion of inventory stock in many sectors as well as for many types of goods; this is well captured by the first eigenmode which we interpret as the "aggregate demand". Now, an increase in final demand and depletion of inventory stock make firms revise their expectations of future demand, and accordingly the level of desired inventory stock. If demand does not increase further, firms produce more for the purpose of inventory accumulation. This is captured by the second eigenmode which we interpret as the "inventory adjustments". As Metzler (1941) shows, inventory adjustments take time. This causes production to lag behind shipments. Our analysis shows that the average lag is about four months. In the case of inventory stock, an *increase* in shipments initially causes *depletion* of inventory, and, therefore, the lag becomes even longer than in the case of production. We find that the average lag of inventory behind shipments is about 11 months. Thus, we can see that shipments lead production and inventory within the framework in which changes in real demand or shipments are the major shocks. Note that productivity shocks emphasized by the RBC, by definition, instantaneously change production. There is no room for production to lag behind shipments in such a framework.

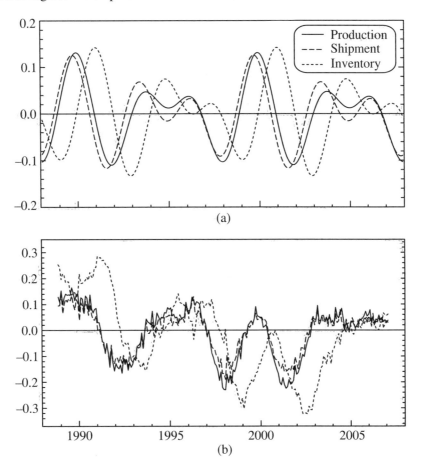

Fig. 6.8 Comparison between the extracted business cycles and the original data. Note (a) the sum of the $T = 60, 40$ components of the first and the second eigenvectors for production, shipments and inventory, each of which is averaged over 21 goods. (b) The original data are smoothed out by taking a one-year simple moving average. We notice that the systematic "business cycles" are well extracted from the original data. Since the Fourier components of $T = 120$ and 240 are *not* included in plot (a), that part of the behavior of the real data shown in (b) is not reproduced in plot (a). (From Iyetomi et al. (2011a))

6.1.6 The 2008–09 economic crisis

The 2008–09 financial crisis triggered by the bursting of the housing bubble in the U.S. immediately spread all over the world, and caused great damage to the global economy, reminiscent of the Great Depression. Based on the analyses in the previous sections, we update the IIP data to cover the period from January 2008 ($j = 241$) through June 2009 ($j = 258$), and perform an out-of-sample test.

The extended data shown in Fig. 6.3 demonstrates a dramatic drop in production in Japan. Panel (a) of Fig. 6.9 shows the volatility $P(t)$ of the IIP defined by

$$P(t) := \sum_{\alpha=1}^{3} \sum_{g=1}^{21} |w_{\alpha,g}(t)|^2. \qquad (6.16)$$

We observe the exceptionally large amplification of $P(t)$ during the 2008–09 economic crisis.

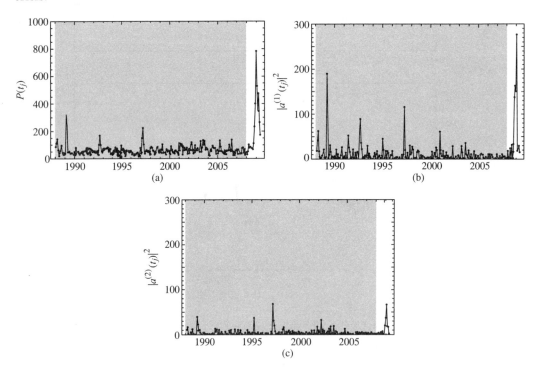

Fig. 6.9 Decomposition of volatility of the extended IIP data. (a) The total volatility. (b) The partial contribution of the first eigenmode to the total volatility. (c) The partial contribution of the second eigenmode. The gray shaded zone depicts the in-sample period used for determining the dominant modes. (From Iyetomi et al. (2011a))

This situation, therefore, provides us with a perfect opportunity for an out-of-sample test of the validity of the dominant eigenmodes that we have identified using RMT. To this end, we decompose the extended IIP data based on exactly the same eigenvectors as in Eq. (5.92). Then, $P(t)$ can be expressed in terms of the mode amplitude coefficients $a^{(n)}(t)$ as follows:

$$P(t) = \sum_{n=1}^{63} |a^{(n)}(t)|^2. \tag{6.17}$$

Panels (b) and (c) of Fig. 6.9 show the partial contributions of the largest and the second largest modes to $P(t)$, respectively. Plainly, an unprecedented rise in the volatility of the first dominant factor, $|a^{(1)}(t)|^2$, is largely responsible for a rise in $P(t)$ during the 2008–09 crisis. The contribution of the second dominant factor, $|a^{(2)}(t)|^2$, is not as significant as that of the first dominant factor.

The relative contributions of the first and second eigenmodes as defined by

$$\pi_n(t) := \frac{|a^{(n)}(t)|^2}{P(t)}. \tag{6.18}$$

are shown in Fig. 6.10. The total sum of $\pi_n(t)$ over n is normalized to unity. The first dominant eigenmode accounts for almost a half of the total power of fluctuations during the 2008–09 recession. Note that the *relative* contribution of the first eigenmode, namely about one half, is comparable to that for the in-sample period. This result demonstrates that our factor model holds for the 2008–09 crisis despite the fact that a fall in industrial production during the period as shown in Fig. 6.3 was truly unprecedented.

The first eigenmode essentially explains the fall in industrial production in 2008–09. As shown in Fig. 6.11 the production fell in response to a sudden decline in exports. We emphasize that the fall of exports was caused by a sudden shrinkage of the world trade owing to the financial crisis accompanied by the global recession, and that *it was basically exogenous to the Japanese economy*. It was certainly independent of technological shocks. By analyzing the components of eigenvectors (see Fig. 6.6), we identified the first dominant factor as the aggregate demand. To the extent that we identify a sudden fall of exports, an exogenous negative aggregate-demand shock, as the basic cause of the 2008–09 recession, the dominance of the first eigenmode provides us with another justification for our interpreting it as the aggregate demand factor. This confirms, once again, the validity of the "Old Keynesian View" (Tobin, 1993).

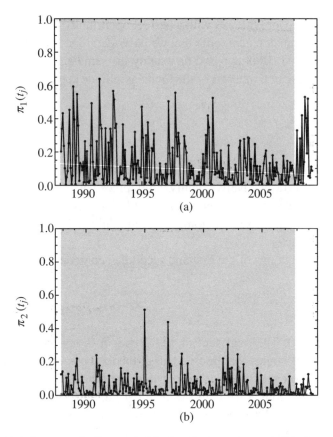

Fig. 6.10 Relative contributions of (a) the first mode and (b) the second mode, corresponding to Fig. 6.9. Panels (a) and (b) correspond to panels (b) and (c) in Fig. 6.9, respectively. (From Iyetomi et al. (2011a))

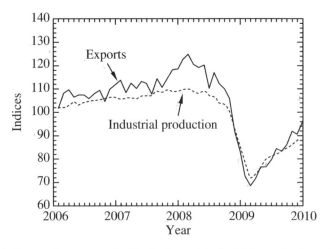

Fig. 6.11 Indices of exports (solid line) and industrial production (dashed line), normalized to 100 in 2005. (Adapted from Iyetomi et al. (2011a))

6.1.7 Summary

In his comments on Shapiro and Watson (1988) entitled "Sources of Business Cycle Fluctuations," Hall (1988) made the following remark:

> All told, I find the results of this paper harmonious with the emerging middle ground of macroeconomic thinking. There are major unobserved determinants of output and employment operating at business-cycle and lower frequencies. Some are technological, some are financial, some are monetary (Hall, 1988, [p.151]).

To discover such unobserved determinants of business cycles, the standard identifying restriction is that only technology shocks affect output permanently. As discussed in the Introduction, we maintain that this assumption is untenable. Using production, shipments, and inventory data, we analyze the causes of business cycles more directly than in the standard approach.

We have identified (i) two periods $T = 60$ and 40 months for the Japanese business cycles, and also (ii) the two significant factors causing comovements of IIP across 21 goods. Taking advantage of the information revealed by disaggregated data, we can reasonably interpret the first factor as the aggregate demand, and the second factor as intentional inventory adjustments. It is difficult to interpret the two dominant factors as productivity shocks. Within the framework of the standard RBC, productivity shocks are expected to affect production, shipments, and inventory together. The two dominant factors that we have identified do not fulfill this condition (see Fig. 6.6).

We have also found that, for the part of shipments and production explained by the two factors, shipments lead production by a few months. This lead–lag relationship between shipments and production caused by inventory adjustment is not seen clearly in the original data. However, inventory adjustments which generate this lead–lag relationship between shipments and production are an important part of "business cycles" defined by the two dominant factors.

Furthermore, an out-of-sample analysis of the 2008–09 recession demonstrates that the first dominant factor largely explains this major episode. We know that it was basically caused by a sudden fall of exports beginning the fourth quarter of the year 2008 (Figure 6.11). Therefore, to the extent that exports can be reasonably regarded as exogenous *real* aggregate demand shocks to the Japanese economy, we can identify the first dominant factor as the real aggregate demand. All these findings suggest that the major cause of business cycles is *real demand* shocks accompanied by significant inventory adjustments.

It is important to recognize that this conclusion of ours does not necessarily mean that technology shocks are unimportant. On the contrary, it is very plausible that a major driving force of fixed investment is technology shocks. Furthermore, one might argue that technology shocks are also a driving force of consumption by way of introduction of new products (Aoki and Yoshikawa, 2002). The point is that technology shocks affect the macro economy by way of changes in aggregate demand, notably investment, in the short run. For the purpose of studying business cycles, it is wrong to consider technology shocks as direct shifts of the production function.

The conclusion we drew from our analyses of the Japanese industrial production data is clear. To study whether the same results hold up for other economies is obviously an important research agenda. We believe that the methods used here can be usefully applied to the study of business cycles in other economies. However, the theory of such "cycles" of aggregate demand awaits further investigation. The multiplier–accelerator model of Samuelson (1939) and Hicks (1950b) may be a way to proceed.

6.2 Synchronization in Japanese Business Cycles

6.2.1 Background

We explore an alternative view of **business cycles** explained in Section 6.1 to develop a deeper understanding of **economic shock**s. Assume that two pendulum clocks, representing industrial sectors, with slightly different oscillation periods are hanging on a beam. The pendulums of these two clocks oscillate with the same period and the two clocks show the same time (Huygens et al. 1966). When we move one pendulum clock to another beam, the oscillation periods are not the same, and they gradually deviate. A pendulum clock is considered a limit cycle oscillator, which receives energy from the rest of the system and releases energy and entropy to the rest of the system during oscillation. The observed oscillation with the same period is induced by the interaction between the two pendulum clocks mediated by the beam. This phenomenon is called **synchronization**, or **entrainment** (Kuramoto, 2012; Ikeda et al., 2012).

If we divide the aggregate business cycles into industrial sectors, fluctuation of industrial sectors must be synchronized with the same oscillation period; otherwise aggregate business cycles will not be observed. Here, note that the division into industrial sectors is regarded as a macroscopic resolution. This work is an empirical study of synchronization in Japanese business cycles using data from the Indices of Industrial Production (IIP).

6.2.2 Time series of the production indices

We analyze data from the IIP (Seasonal Adjustment Monthly Index) for a long period of 240 months (monthly data from January 1988 to December 2007). The database includes indices of production, shipment, and inventory for 16 industrial sectors: steel (s1), non-ferrous metal (s2), metal products (s3), machinery (s4), transportation equipment (s5), precision machinery (s6), ceramic products (s7), chemicals (s8), petroleum and coal products (s9), plastic products (s10), paper and pulp (s11), textile (s12), food (s13), other industries (s14), mining (s15), and electric machinery (s16). Our analysis employs just the production indices for all 16 sectors ($N = 16$).

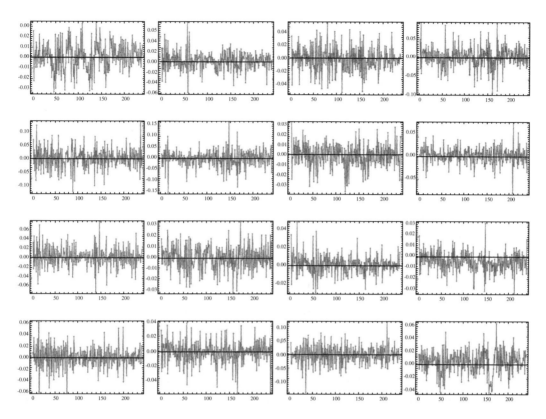

Fig. 6.12 Time series of log returns of the indices for the following 16 sectors (from the top left to the bottom right): steel (s1), non-ferrous metal (s2), metal products (s3), machinery (s4), transportation equipment (s5), precision machinery (s6), ceramic products (s7), chemicals (s8), petroleum and coal products (s9), plastic products (s10), paper and pulp (s11), textile (s12), food (s13), other industries (s14), mining (s15), and electric machinery industry (s16) (from Ikeda et al. (2013c)).

Obtaining stationary time series is an important reprocessing step of time series analysis. First, we calculate the log returns of the indices for the 16 industrial sectors. We confirm the stationary property of these time series using the unit root test explained in Section 2.4.1. The sector-wise time series of log returns of the indices are shown in Fig. 6.12. Here, the length of the time series is $L = 239$.

We then calculate the Fourier series expansion of the time series of log returns of the indices, and remove some high- and low-frequency components. The remaining Fourier components are described later.

6.2.3 Phase time series

Suppose we have two oscillators, with phases $\theta_1(t)$ and $\theta_2(t)$, respectively. We define synchronization as the phases locking $n\theta_1(t) + m\theta_2(t) = $ const., where n and m are integers, while the amplitudes can be different. In a limited case of $n = -m$ and const. $= 0$, discussion with a correlation coefficient is appropriate. However, in the case of $n = -m$ and const. $\neq 0$, where the phase difference means a delay, direct evaluation of the phase is adequate, and does not require the correlation coefficient. This is because the correlation coefficient ρ varies depending on the delay δ. For example, for the trigonometric function with a period of oscillation equal to 2π, we have $\rho = 1$ for $\delta = 0$, $\rho = 0$ for $\delta = \pi/2$, and $\rho = -1$ for $\delta = \pi$. This simple example illustrates that the correlation coefficient is not suitable for the case of phase locking where we have a phase difference or delay.

We calculate the Hilbert transform of the log return of production index $x(t)$ and name it $y(t)$. A complex time series is obtained by adopting the time series $y(t)$ as an imaginary part.

$$z(t) = x(t) + iH[x(t)] = x(t) + iy(t) = A(t)\exp[i\theta(t)]. \tag{6.19}$$

To see how the Hilbert transform actually used in this section is calculated, refer to Section 5.2.2. In Fig. 6.13, the trajectories of the time series are shown in the complex plane, where the log returns of the indices $x_s(s = 1, \cdots, L)$ and the Hilbert transform $y_s = H[x_s](s = 1, \cdots, L)$ are used as the horizontal and vertical axis, respectively, for the 16 sectors.

By applying a band-path filter, the Fourier components of the oscillation period from 24 to 80 months ($r = 3$ to 10, 230 to 237) are included in the graphs of Fig. 6.13. We notice some irregular rotational movement arising from the non-periodic nature of business cycles.

Figure 6.14 shows the time series of the phase $\theta(t)$ for the 16 sectors. These plots include Fourier components of the oscillation period from 24 to 80 months. We observe a linear trend of rotational movement with some fluctuations for the sectors. Irregular rotational movements are behind the small jumps of phases in Fig. 6.14, and notably, trajectories pass near the origin of the plane, as seen in Fig. 6.13.

Business Cycles 183

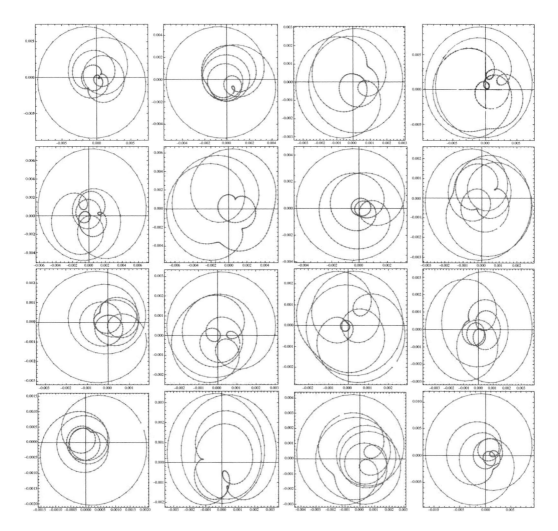

Fig. 6.13 Trajectories of the time series in the complex plane, where the horizontal axis is x and the vertical axis is $y = H[x]$, for the following 16 sectors (from the top left to the bottom right): steel (s1), non-ferrous metal (s2), metal products (s3), machinery (s4), transportation equipment (s5), precision machinery (s6), ceramic products (s7), chemicals (s8), petroleum and coal products (s9), plastic products (s10), paper and pulp (s11), textile (s12), food (s13), other industries (s14), mining (s15), and electric machinery (s16). These plots include Fourier components of the oscillation period from 24 to 80 months (from Ikeda et al. (2013c)).

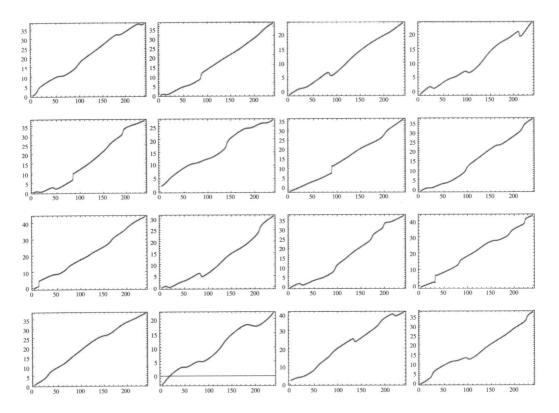

Fig. 6.14 Time series of the phase $\theta(t)$ for the following 16 sectors (From the top left to the bottom right): steel (s1), non-ferrous metal (s2), metal products (s3), machinery (s4), transportation equipment (s5), precision machinery (s6), ceramic products (s7), chemicals (s8), petroleum and coal products (s9), plastic products (s10), paper and pulp (s11), textile (s12), food (s13), other industries (s14), mining (s15), and electric machinery (s16). These plots include Fourier components of the oscillation period from 24 to 80 months (from Ikeda et al. (2013c))

6.2.4 Frequency entrainment

The presence of frequency entrainment and phase locking would be direct evidence of synchronization. The angular frequency ω_i and the intercept $\tilde{\theta}_i$ are estimated by the least squares fitting of the time series of phase $\theta_i(t)$ using Eq. (6.20),

$$\theta_i(t) = \omega_i t + \tilde{\theta}_i, \tag{6.20}$$

where i indicates the industrial sector. The intercept $\tilde{\theta}_i$ fluctuates over time because of economic shocks. In the least squares fitting, we estimate the average value of the intercept $\tilde{\theta}_i$. In Fig. 6.15, the estimated angular frequencies ω_i for all 16 sectors are plotted for the

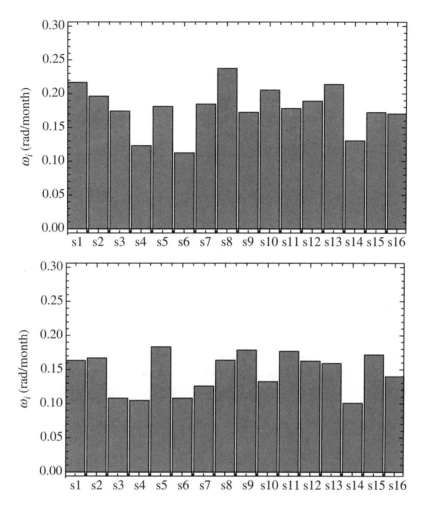

Fig. 6.15 The estimated angular frequencies ω_i for the Fourier components: 18–80 months (upper panel) and 24–80 months (lower panel). We observe that the estimated angular frequencies ω_i are entrained to be almost identical for the sectors, implying the presence of frequency entrainment (from Ikeda et al. (2013c)).

Fourier components: 18–80 months and 24–80 months. We notice that the estimated angular frequencies ω_i are entrained to be almost identical for the sectors, which means that frequency entrainment is present. When the higher frequency Fourier components are included in the analysis, the deviations of the estimated angular frequencies ω_i among the sectors are expected to gradually increase. This increase in the deviations can be explained by the presence of larger jumps of phases owing to the trajectories passing near the origin of the complex plane for the higher frequency Fourier components.

6.2.5 Phase locking

Phase locking means that phase differences for all pairs of oscillators are constant. However, this condition is rarely seen to exactly satisfy the actual time series owing to irregular fluctuation, i.e., economic shock. We introduce an indicator $\sigma(t)$ of the phase locking as

$$\sigma(t) = \left[\frac{1}{N}\sum_{i=1}^{N}\left\{\frac{d}{dt}(\theta_i(t) - \omega_i t) - \mu(t)\right\}^2\right]^{1/2}, \tag{6.21}$$

$$\mu(t) = \frac{1}{N}\sum_{i=1}^{N}\frac{d}{dt}(\theta_i(t) - \omega_i t). \tag{6.22}$$

In the case of frequency entrainment, Eq. (6.21) and Eq. (6.22) become

$$\sigma(t) = \left[\frac{1}{N}\sum_{i=1}^{N}\left\{\tilde{\theta}_i(t) - \frac{1}{N}\sum_{j=1}^{N}\tilde{\theta}_j(t)\right\}^2\right]^{1/2}. \tag{6.23}$$

Here, temporal dependence of the intercept is explicitly expressed as $\tilde{\theta}_i(t)$. The indicator $\sigma(t)$ is equal to zero when the intercepts for all oscillators are the same. On the other hand, if the indicator $\sigma(t)$ satisfies the following relationship, we call this partial phase locking,

$$\sigma(t) \ll \omega_i. \tag{6.24}$$

The estimated indicator of the phase locking $\sigma(t)$ is plotted in Fig. 6.16 for the Fourier components: 18–80 months and 24–80 months. Figure 6.16 shows that the indicator $\sigma(t)$ is much smaller than ω_i for most of the period. This implies the presence of partial phase locking. Thus, both frequency entrainment and phase locking are seen as direct evidence of synchronization.

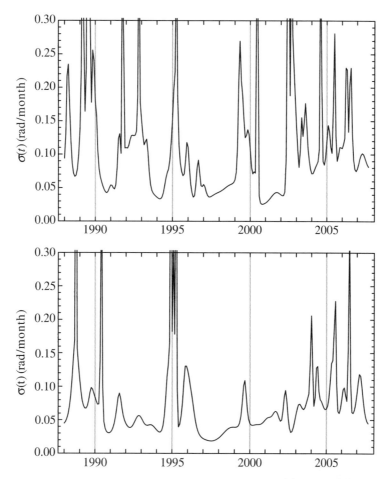

Fig. 6.16 The estimated indicators of the phase locking $\sigma(t)$ are plotted for the Fourier components: 18–80 months (upper panel) and 24–80 months (lower panel). The indicator $\sigma(t)$ is much smaller than ω_i for most of the period. This means that partial phase locking is observed (from Ikeda et al. (2013c)).

6.2.6 Common shock and individual shock

The log returns of the indices x_i are decomposed into amplitude $A_i(t)$ and phase $\theta_i(t)$ using Eq. (6.19). An interesting question to ask is: Which quantity carries the information about economic shocks, the amplitude $A_i(t)$ or the phase $\theta_i(t)$? The averages of these quantities over the industrial sectors are written as:

$$\langle A(t) \rangle = \frac{1}{N} \sum_{i=1}^{N} A_i(t) = \frac{1}{N} \sum_{i=1}^{N} \frac{x_i(t)}{\cos \theta_i(t)}, \tag{6.25}$$

$$\langle \cos \theta(t) \rangle = \frac{1}{N} \sum_{i=1}^{N} \cos \theta_i(t). \tag{6.26}$$

The average amplitudes $\langle A(t) \rangle$ and average phases $\langle \cos\theta(t) \rangle$ are shown in Fig. 6.17 and Fig. 6.18, respectively. They show that the information about economic shocks is carried by the phase $\theta_i(t)$, not by the amplitude $A_i(t)$. Japan experienced severe economic recessions in 1992, 1998, and 2001. The 1992 recession was caused by the collapse of the **bubble economy**. In 1998, there was the banking crisis, when some major banks went bankrupt. The 2001 recession was caused by the bursting of the information technology bubble. The **average amplitudes** $\langle A(t) \rangle$ in Fig. 6.17 have large values in 1992, 1998, and 2001. On the contrary, the **average phases** $\langle \cos\theta(t) \rangle$ in Fig. 6.18 show a sharp drop in all these three periods. It should be noted that the phase $\theta_i(t)$ is more sensitive to economic shock than the amplitude $A_i(t)$. Thus, it is appropriate to interpret the average phases $\langle \cos\theta(t) \rangle$ and residual $\cos\theta_i(t) - \langle \cos\theta(t) \rangle$ as common shock and individual shock, respectively.

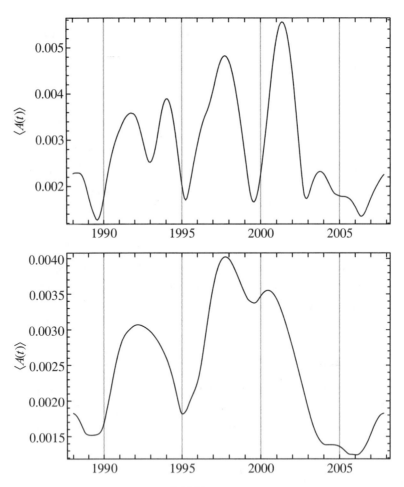

Fig. 6.17 The average amplitudes $\langle A(t) \rangle$ for the Fourier components: 18–80 months (upper panel) and 24–80 months (lower panel). These figures clearly show that the information about economic shocks is carried by the phase $\theta_i(t)$, and not by the amplitude $A_i(t)$ (from Ikeda et al. (2013c)).

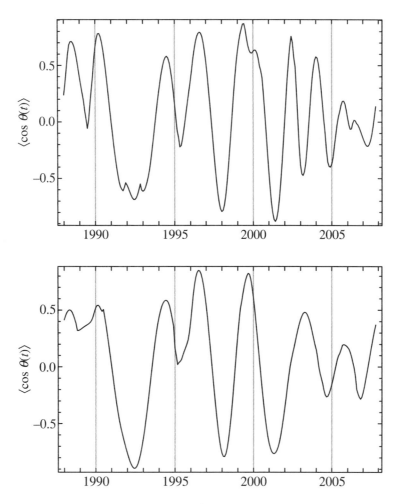

Fig. 6.18 The average phases $\langle \cos\theta(t) \rangle$ for the Fourier components: 18–80 months (upper panel) and 24–80 months (lower panel). These phases show a sharp drop in 1992, 1998, and 2001. It should be noted that the phase $\theta_i(t)$ is more sensitive to economic shock than the amplitude $A_i(t)$ (from Ikeda et al. (2013c)).

Figure 6.19 shows the **individual shock**s $\cos\theta_i(t) - \langle \cos\theta(t) \rangle$ for each industrial sector. This shows that many individual shocks occur all the time, and a number of them seem to occur randomly. More noteworthy is the contraction of the individual shocks in 1992, 1998, and 2001, when Japan was faced with severe economic recessions. All industries were exposed to the **common shock**s $\langle \cos\theta(t) \rangle$ during these periods.

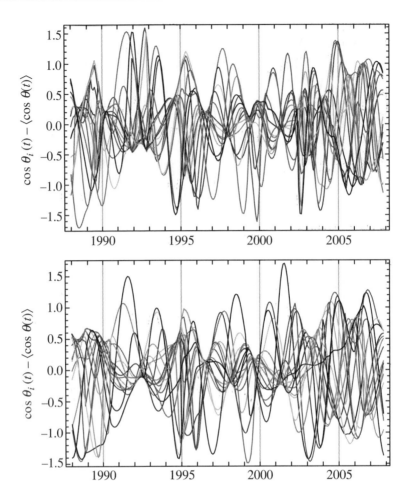

Fig. 6.19 The individual shocks are shown for the Fourier components: 18–80 months (upper panel) and 24–80 months (lower panel). This depicts that we have many individual shocks all the time, and many of them seem to occur randomly. More noteworthy is the contraction of the individual shocks in 1992, 1998, and 2001, when we had severe economic recessions (from Ikeda et al. (2013c)).

6.2.7 Summary

We analyzed the Indices of Industrial Production (Seasonal Adjustment Monthly Index) for a long period of 240 months (January 1988 to December 2007) to develop a deeper understanding of economic shocks. The angular frequencies ω_i estimated using the Hilbert transform were almost identical for the 16 sectors in the indices. Moreover, the indicator of the phase locking $\sigma(t)$ showed the presence of partial phase locking for the sectors. These provide direct evidence of synchronization in Japanese business cycles. We also showed that the information on economic shock is carried by the phase time series. We separated common shocks from individual shocks using phase time series $\theta_i(t)$ and interpreted the average phases $\langle \cos \theta(t) \rangle$ and residual $\cos \theta_i(t) - \langle \cos \theta(t) \rangle$ as the common shock and individual shock, respectively. The former dominated the economic shocks in 1992, 1998, and 2001. During the same periods, the individual shocks were fewer. Japan sees many individual shocks all the time, and these seem to occur randomly. The obtained results suggest that business cycles may be described as dynamics of the coupled limit-cycle oscillators exposed to random individual shocks.

6.3 Synchronization in International Business Cycles

6.3.1 Background

Business cycles have a long history of being subjected to theoretical studies (Haberler, 1946; Burns and Mitchell, 1946; Hatanaka, 1964). Synchronization (Huygens et al., 1966) in international business cycles, in particular, have attracted economists and physicists as an example of self-organization in the time domain (Krugman, 1996). Synchronization of business cycles across countries has been discussed using correlation coefficients between GDP time series (Stock and Watson, 2005). However, this method remains a preliminary first step, and a more definitive analysis using a suitable quantity to describe the business cycles is needed.

We analyze quarterly GDP time series for Australia, Canada, France, Italy, the United Kingdom, and the United States. The purpose of studying international business cycles is to answer the following questions:

(i) Can one obtain direct evidence of synchronization in international business cycles?

(ii) If so, what is the mechanism of such synchronization?

(iii) In relation to question (ii), what types of economic shocks play an important role in business cycles?

(iv) What is the economic origin of the synchronization?

In analyzing business cycles, an important issue is the significance of individual (micro) versus common (macro) shocks. Foerster et al. (2008), using factor analysis, showed that the volatility of the United States industrial production could be largely explained by common shocks, and partly by cross-sectoral correlation owing to individual shocks transferred through trade linkage. In this section, we take a different approach to analyzing the shocks in order to explain synchronization in international business cycles.

6.3.2 GDP time series

We analyze the quarterly GDP time series (OECD Quarterly National Accounts, QNA) for Australia, Canada, France, Italy, the United Kingdom, and the United States from Q2 1960 to Q1 2010 to study the synchronization in international business cycles. Extracting a trend component is an important pre-processing step in time series analysis. First, the growth rate of GDP $x_i(t)$ defined as $x_i(t) = (\text{GDP}_i(t) - \text{GDP}_i(t-1))/\text{GDP}_i(t-1)$ $(i = 1, \cdots, N)$ were calculated for the six countries ($N = 6$). The time series $x_i(t)$ for the six countries are shown in Fig. 6.20. We then calculate the Fourier series expansion of the time series $x_i(t)$. Given the identified business cycle periods of the analyzed countries, the high and low frequency Fourier components were removed, leaving the Fourier components from two to 10 years. The band-pass filtering of the growth rate of the GDP time series for the six countries is shown in Fig. 6.21.

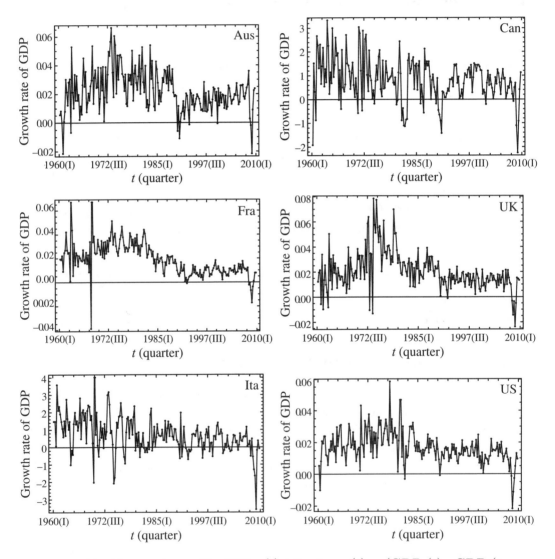

Fig. 6.20 The growth rate of the GDP $x_i(t)$ defined as $x_i(t) = (\text{GDP}_i(t) - \text{GDP}_i(t-1))/\text{GDP}_i(t-1)$ for six countries: Australia, Canada, France, the UK, Italy, and the US (from Ikeda et al. (2013a)).

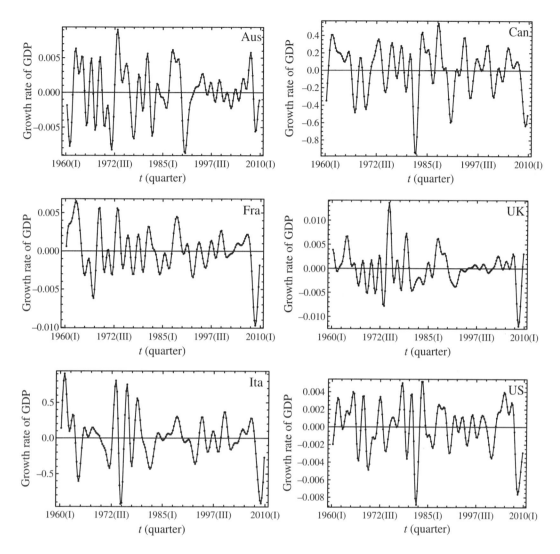

Fig. 6.21 We calculate Fourier series expansion of the time series $x_i(t)$. Given the identified business cycle periods of the analyzed countries, the high and low frequency Fourier components were removed, leaving the Fourier components from two to 10 years. The band-pass filtering of the growth rate of the GDP time series are shown for Australia, Canada, France, the UK, Italy, and the US (from Ikeda et al. (2013a)).

Business cycles with a period of four to six years are usually considered to be caused by adjustments in stock, such as inventory stock. We apply a band-pass filter to the time series of inventory changes to remove high and low frequency components, leaving components from the period of three to eight years. Frequency components were chosen for better visibility of the cycling trajectory. The obtained time series for the six countries are as shown in Fig. 6.22.

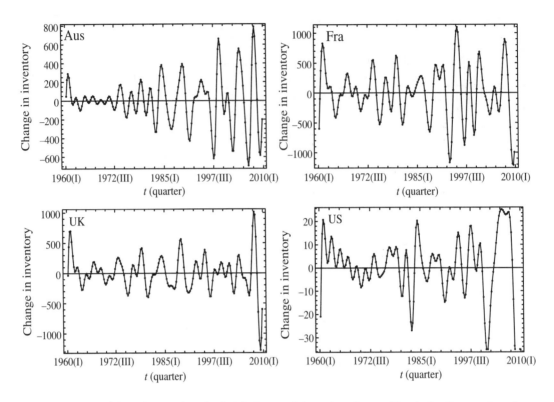

Fig. 6.22 The time series obtained after applying a band-pass filter to the time series of inventory changes. Frequency components were chosen for better visibility of the cycling trajectory. The obtained time series are shown for the six countries Australia, France, the UK, and the US (from Ikeda et al. (2013a)).

Figure 6.23 depicts trajectories in the two-dimensional plane of the GDP growth rate and the changes in inventory. These commonly used figures suggest the existence of a **limit-cycle** in business cycles.

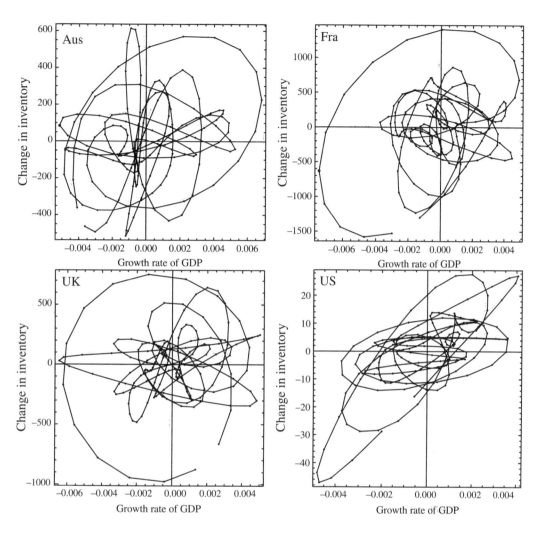

Fig. 6.23 The trajectories in the two-dimensional plane of the GDP growth rate and the changes in inventory for Australia, France, the UK, and the US. These commonly used figures suggest the existence of a limit-cycle in business cycles (from Ikeda et al. (2013a)).

6.3.3 Phase time series

The **Hilbert transform** is a method for analyzing the correlation of two time series with a lead–lag time relationship. To see how the Hilbert transform actually used in this section is calculated, refer to Section 5.2.2.

Figure 6.24 depicts the obtained trajectories in the complex plane. Fourier components of oscillation for the period from two to 10 years were included in the graphs of Fig. 6.24. Some irregular rotational movement was observed owing to the non-periodic nature of business cycles.

The time series of phase $\theta_i(t)$ for the six countries studied are depicted in Fig. 6.25. Fourier components of oscillation for the period from two to 10 years were included in these plots. We observe linear trend of the phase development with some fluctuations. The small jumps in phases in Fig. 6.25 were caused by irregular rotational movement, especially the trajectories that passed near the origin of the plane, as observed in Fig. 6.24.

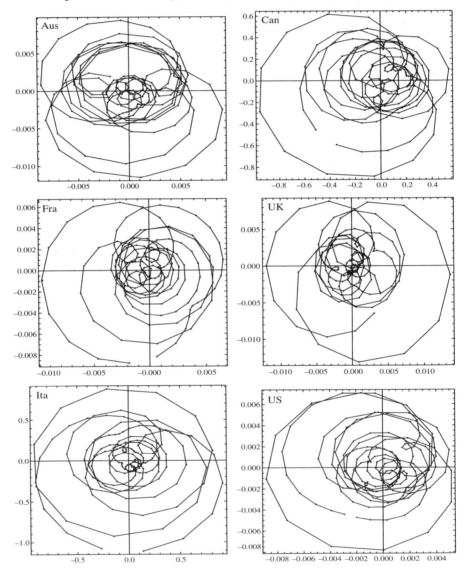

Fig. 6.24 The trajectories of the time series of growth rate of GDP in the complex plane for Australia, Canada, France, the UK, Italy, and the US. Fourier components of oscillation for the period from two to 10 years were included. Some irregular rotational movement was observed owing to the non-periodic nature of the business cycles (from Ikeda et al. (2013a)).

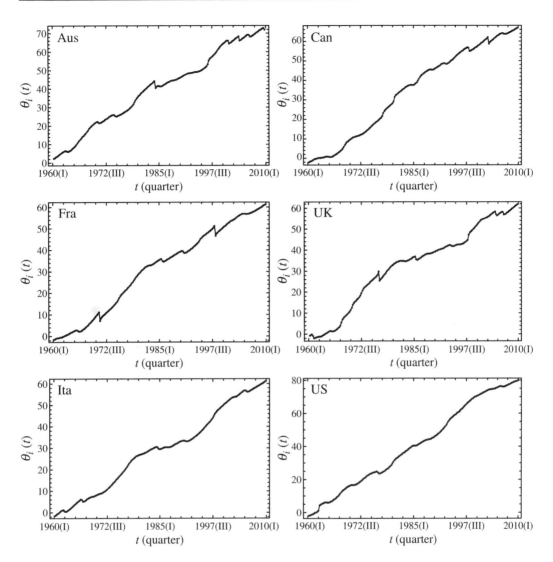

Fig. 6.25 The time series of phase obtained using Hilbert transform are shown for the six countries: Australia, Canada, France, the UK, Italy, and the US. Fourier components of oscillation for the period from two to 10 years were included in these plots. We observed the linear trend of the phase development with some fluctuations for the six countries (from Ikeda et al. (2013a)).

6.3.4 Frequency entrainment

We expect to see frequency entrainment and phase locking as direct evidence of the synchronization. Angular frequency ω_i and intercept $\tilde{\theta}_i$ are estimated by fitting the time series of the phase $\theta_i(t)$ using the relationship,

$$\theta_i(t) = \omega_i t + \tilde{\theta}_i, \tag{6.27}$$

where i indicates a country. The estimated angular frequencies ω_i for all the six countries are plotted in 6.26. We observe that the estimated angular frequencies ω_i are almost identical for the six countries. This implies the presence of frequency entrainment.

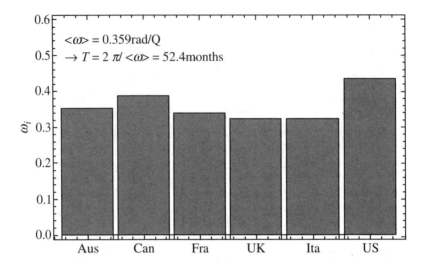

Fig. 6.26 The estimated angular frequencies ω_i for all the six countries: Australia, Canada, France, the UK, Italy, and the US. We observe that the estimated angular frequencies ω_i are almost identical for the six countries. This implies the presence of the frequency entrainment (from Ikeda et al. (2013a)).

6.3.5 Phase locking

Phase locking is the condition in which phase differences for all pairs of oscillators are constant. However, this is rarely seen to satisfy precisely the actual time series owing to irregular fluctuations, i.e., economic shocks. Therefore, we use the indicator $\sigma(t)$ of the phase locking defined by Eq. (6.21) and Eq. (6.22). Indicator $\sigma(t)$ is equal to zero when the phase differences for all pairs of oscillators are constant. On the other hand, if indicator $\sigma(t)$ satisfies the following relationship, it is known as partial phase locking.

$$\sigma(t) \ll \omega_i, \tag{6.28}$$

The estimated indicator of phase locking $\sigma(t)$ is plotted in Fig. 6.27, which shows that indicator $\sigma(t)$ is much smaller than ω_i for most of the period. This means that there was partial phase locking. As a result, both frequency entrainment and phase locking are obtained as direct evidence of the synchronization.

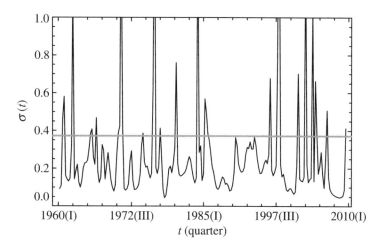

Fig. 6.27 The estimated indicator of phase locking $\sigma(t)$. The indicator $\sigma(t)$ is much smaller than ω_i for most of the period. This means that partial phase locking is present. As a result, both frequency entrainment and phase locking are obtained as direct evidence of the synchronization (from Ikeda et al. (2013a)).

6.3.6 Common shocks versus individual shocks

time series $x_i(t)$ is decomposed into amplitude $A_i(t)$ and phase $\theta_i(t)$ using Eq. (6.19). The average amplitudes $\langle A(t) \rangle$ calculated using Eq. (6.25) and the average phases $\langle \cos \theta(t) \rangle$ calculated using Eq. (6.26) are shown in Fig. 6.28.

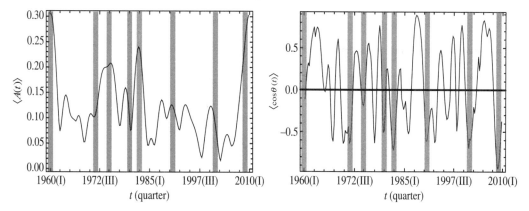

Fig. 6.28 The average amplitudes $\langle A(t) \rangle$ and the average phases $\langle \cos \theta(t) \rangle$. The United States experienced eight recessions after 1960: Q1 1961, Q4 1970, Q1 1975, Q3 1980, Q4 1982, Q1 1991, Q4 2001, and Q2 2009. The recessions in 2001 and 2009 were because of the bursting of the information technology bubble and the collapse of Lehman Brothers, respectively. The key to understanding business cycles is phase $\theta_i(t)$, not amplitude $A_i(t)$ (from Ikeda et al. (2013a)).

The United States experienced eight recessions after 1960: Q1 1961, Q4 1970, Q1 1975, Q3 1980, Q4 1982, Q1 1991, Q4 2001, and Q2 2009. The recessions in 2001 and 2009 were because of the bursting of the information technology bubble and the collapse of Lehman Brothers, respectively. The value of the average amplitudes $\langle A(t) \rangle$ in Fig. 6.28 are large in 1961, 1975, 1982, and 2009. On the contrary, the average phases $\langle \cos \theta(t) \rangle$ in Fig. 6.28 show a sharp drop in all of the eight recessions described earlier. Therefore, we conclude that the key to understanding business cycles is phase $\theta_i(t)$, not amplitude $A_i(t)$.

We focus on phase $\theta_i(t)$ in order to extract the common shocks (comovement or synchronization of shocks) of the business cycles for the six countries. For this, we analyzed time series $z_i(t) = \cos \theta_i(t)$ using the principal component analysis and the random matrix theory explained in Sections 5.1 and 5.3.1 (Mehta, 2004; Laloux et al, 1999b; Plerou et al, 1999b, 2002a). We consider the eigenvalue problem

$$\boldsymbol{C} \boldsymbol{V}^{(n)} = \lambda^{(n)} \boldsymbol{V}^{(n)} \tag{6.29}$$

where $\lambda^{(n)}$ and $\boldsymbol{V}^{(n)}$ are the eigenvalue and the corresponding eigenvector, respectively, for the correlation matrix $\boldsymbol{C}$, whose element is the correlation coefficient between countries i and j and is calculated by

$$c_{ij} = \frac{\langle (z_i(t) - \langle z_i \rangle)(z_j(t) - \langle z_j \rangle) \rangle}{\sqrt{(\langle z_i^2 \rangle - \langle z_i \rangle^2)(\langle z_j^2 \rangle - \langle z_j \rangle^2)}}, \tag{6.30}$$

where $\langle \cdot \rangle$ indicates the time average for the time series. The obtained eigenvalues and eigenvectors are shown in Table 6.2. We assume that the eigenvalues are arranged in decreasing order $(n = 1, \cdots, N)$.

Table 6.2 Eigenvalues and eigenvectors

Parameter	$n=1$	$n=2$	$n=3$	$n=4$	$n=5$	$n=6$
$\lambda^{(n)}$	2.767	1.033	0.793	0.613	0.444	0.346
$\boldsymbol{V}^{(n)}(1)$	-0.416	0.110	-0.271	0.849	-0.058	0.128
$\boldsymbol{V}^{(n)}(2)$	-0.472	-0.096	0.401	-0.051	-0.578	-0.517
$\boldsymbol{V}^{(n)}(3)$	-0.447	0.412	0.001	-0.166	0.667	-0.395
$\boldsymbol{V}^{(n)}(4)$	-0.256	-0.657	-0.643	-0.211	0.066	-0.196
$\boldsymbol{V}^{(n)}(5)$	-0.432	0.387	-0.271	-0.451	-0.329	0.526
$\boldsymbol{V}^{(n)}(6)$	-0.387	-0.475	0.526	-0.012	0.321	0.493

Business cycles fall into comovements and individual shocks. This classification is made using the random matrix theory. $\lambda_+ = 1.37743$ was estimated using Eq. (5.104) for $N = 6$ and $T = 199$. The results indicate that only the largest eigen mode ($n = 1$) is meaningful and other eigen modes are regarded as random noise. The comovement is reconstructed as a mode signal using Eq. (5.15) with $N = 1$. The individual shock is reconstructed with the remaining eigen modes. It should be noted that each element of eigenvector $\boldsymbol{V}^{(1)}$ has the same sign, which means that all the countries experience the same change in GDP. The resulting time series of economic shocks are shown in Fig. 6.29. We always observe significant individual shocks, which seem to occur randomly.

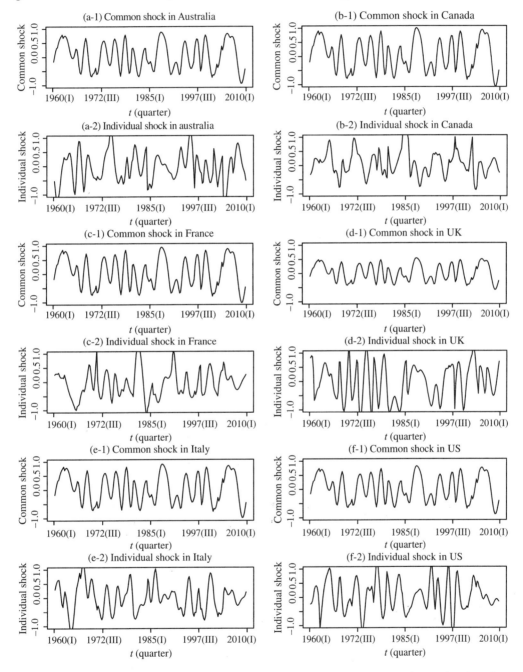

Fig. 6.29 The time series of common and individual shocks are shown for the six countries (from Ikeda et al. (2013a)).

A natural interpretation of individual shocks is that they are "technological shocks". Our analysis demonstrates that fluctuations of average phases offer a good explanation of business cycles, particularly recessions. As it is highly unlikely that all of the countries are subject to common negative technological shocks, the results suggest that pure "technological shocks" cannot explain business cycles ((Iyetomi et al. (2011b)).

6.3.7 Summary

We analyzed the quarterly GDP time series for Australia, Canada, France, Italy, the United Kingdom, and the United States from Q2 1960 to Q1 2010 to study synchronization in international business cycles, and obtained the followings results:

(i) The angular frequencies ω_i estimated using the Hilbert transform are almost identical for the six countries. This indicates the presence of frequency entrainment. Moreover, the indicator of phase locking $\sigma(t)$ shows that partial phase locking is present for the analyzed countries. This is direct evidence of synchronization in international business cycles.

(ii) Furthermore, we also showed that information from economic shocks is carried by phase time series $\theta_i(t)$. We separated the comovement and individual shocks using the random matrix theory. A natural interpretation of individual shocks is that they are "technological shocks". This analysis demonstrates that fluctuations of average phases offer a good explanation of business cycles, particularly recessions. As it is highly unlikely that all countries are subject to common negative technological shocks, the results suggest that pure "technological shocks" cannot explain business cycles.

6.4 Coupled Limit-Cycle Oscillator Model

In this section, we discuss the mechanism of synchronization. The existence of a limit-cycle in business cycles has been suggested as shown in Fig. 6.23. Based on this result, we develop a model for international business cycles based on the coupled limit-cycle oscillator model (Kuramoto, 2012).

According to Ikeda et al. (2012), since changes in "kinetic energy" are equal to summed "power," one can obtain the following power balance equation,

$$\frac{d}{dt}\left[\frac{1}{2}I_i\dot{\theta}_i^2\right] = R_i - L_i - K_d\dot{\theta}_i^2 + \sum_{j=1}^{N} k_{ji} \sin \Delta\theta_{ji}. \tag{6.31}$$

This model has the trade linkage structure depicted in Fig. 6.30. If the power is balanced, the oscillator rotates with constant speed $\dot{\theta}_i$.

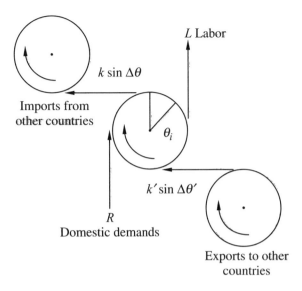

Fig. 6.30 We develop a model for international business cycles based on the coupled limit-cycle oscillator model. When the inertia term is small enough compared with the dissipation term ($\ddot{\theta}_i \ll \alpha_i \dot{\theta}_i$), the power balance equation leads us to obtain the Kuramoto oscillator, i.e., the coupled limit-cycle oscillator model. This model has the trade linkage structure as shown in the figure (from Ikeda et al. (2013a)).

When the inertia term is small compared with the dissipation term ($\ddot{\theta}_i \ll \alpha_i \dot{\theta}_i$), the power balance equation leads to the Kuramoto oscillator, i.e., the coupled limit-cycle oscillator model (Ikeda et al., 2013c),

$$K_d \dot{\theta}_i = R_i - L_i + \sum_{j=1}^{N} k_{ji} \sin \Delta\theta_{ji}. \tag{6.32}$$

Without loss of generality, Eq. (6.32) can be rewritten as,

$$\dot{\theta}_i = Q_i + \sum_{j=1}^{N} \kappa_{ji} \sin \Delta\theta_{ji}. \tag{6.33}$$

A theoretical study of this model shows that synchronization of oscillators is observed when the interaction parameters κ_{ji} are above a certain threshold.

The parameter estimation is explained as follows. Using a discretized form of the model:

$$\theta_{i,t+1} = \beta_i \theta_{i,t} + Q_i + \sum_{j=1}^{N} \kappa_{ji} \sin \Delta\theta_{ji}. \tag{6.34}$$

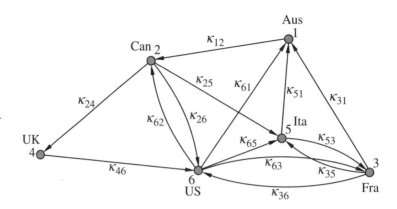

Fig. 6.31 We conduct the regression analysis to estimate the model parameters. The validity of the model implies that the origin of the synchronization is the interaction arising from international trade. The network structure of the model is shown in the figure. Here the edge between i and j is visible, if the confidence interval of the interaction parameter κ_{ij} does not cross zero (from Ikeda et al. (2013a)).

We conduct a **regression analysis** to estimate the model parameters. The results are summarized in Table 6.3, Table 6.4, Table 6.5, Table 6.6, Table 6.7, and Table 6.8. They show that the coupled limit-cycle oscillator model fits the phase time series of the GDP growth rate very well. The validity of the model implies that the origin of the synchronization is the interaction arising from international trade. The network structure of the model is shown in Fig. 6.31. Here the edge between i and j is visible, if the confidence interval of the interaction parameter κ_{ij} does not cross zero.

Table 6.3 Estimation of Model Parameters for Australia

Parameter	Estimation	Std. Error	t value	$\Pr(>|t|)$
β_1	0.996	0.001	590.090	$< 2 \times 10^{-16}$
Q_1	0.412	0.076	5.427	2×10^{-7}
κ_{21}	-0.030	0.070	-0.433	0.665
κ_{31}	-0.174	0.063	-2.766	0.006
κ_{41}	0.051	0.047	1.080	0.281
κ_{51}	0.221	0.061	3.624	3×10^{-4}
κ_{61}	0.156	0.057	2.742	0.006

Multiple R-squared: 0.999, Adjusted R-squared: 0.999
F-statistic: 6.311×10^4 on 6 and 191 DF, p-value: $< 2.2 \times 10^{-16}$

Table 6.4   Estimation of Model Parameters for Canada

Parameter	Estimation	Std. Error	t value	$\Pr(>|t|)$
β_2	0.996	0.001	814.971	$< 2 \times 10^{-16}$
Q_2	0.404	0.050	8.023	1×10^{-13}
κ_{12}	0.100	0.044	2.266	0.024
κ_{32}	-0.025	0.052	-0.478	0.633
κ_{42}	0.033	0.038	0.854	0.394
κ_{52}	0.014	0.046	0.309	0.757
κ_{62}	0.300	0.045	6.549	5×10^{-10}

Multiple R-squared: 0.999, Adjusted R-squared: 0.999
F-statistic: 1.265×10^5 on 6 and 191 DF, p-value: $< 2.2 \times 10^{-16}$

Table 6.5   Estimation of Model Parameters for France

Parameter	Estimation	Std. Error	t value	$\Pr(>|t|)$
β_3	0.995	0.002	445.277	$< 2 \times 10^{-16}$
Q_3	0.397	0.076	5.206	5×10^{-7}
κ_{13}	0.007	0.059	0.126	0.900
κ_{23}	0.073	0.073	0.990	0.323
κ_{43}	0.041	0.059	0.690	0.491
κ_{53}	-0.145	0.072	-1.996	0.047
κ_{63}	0.154	0.064	2.394	0.017

Multiple R-squared: 0.999, Adjusted R-squared: 0.999
F-statistic: 4.851×10^4 on 6 and 191 DF, p-value: $< 2.2 \times 10^{-16}$

Table 6.6   Estimation of Model Parameters for UK

Parameter	Estimation	Std. Error	t value	$\Pr(>|t|)$
β_4	0.996	0.002	459.586	$< 2 \times 10^{-16}$
Q_4	0.427	0.077	5.545	9×10^{-8}
κ_{14}	-0.041	0.060	-0.683	0.495
κ_{24}	-0.188	0.073	-2.550	0.011
κ_{34}	-0.053	0.066	-0.811	0.418
κ_{54}	0.086	0.073	1.168	0.244
κ_{64}	0.114	0.067	1.681	0.094

Multiple R-squared: 0.999, Adjusted R-squared: 0.999
F-statistic: 4.494×10^4 on 6 and 191 DF, p-value: $< 2.2 \times 10^{-16}$

Table 6.7   Estimation of Model Parameters for Italy

Parameter	Estimation	Std. Error	t value	Pr($> |t|$)
β_5	0.999	0.001	997.140	$< 2 \times 10^{-16}$
Q_5	0.330	0.034	9.658	$< 2 \times 10^{-16}$
κ_{15}	0.011	0.026	0.429	0.668
κ_{25}	-0.095	0.032	-2.960	0.003
κ_{35}	0.112	0.035	3.221	0.001
κ_{45}	-0.034	0.027	-1.288	0.199
κ_{65}	0.094	0.028	3.268	0.001

Multiple R-squared: 0.999, Adjusted R-squared: 0.999
F-statistic: 2.222×10^5 on 6 and 191 DF, p-value: $< 2.2 \times 10^{-16}$

Table 6.8   Estimation of Model Parameters for USA

Parameter	Estimation	Std. Error	t value	Pr($> |t|$)
β_6	0.998	$8e-01$	1236.896	$< 2 \times 10^{-16}$
Q_6	0.450	0.039	11.511	$< 2 \times 10^{-16}$
κ_{16}	-0.002	0.033	-0.077	0.938
κ_{26}	-0.159	0.036	-4.323	2×10^{-5}
κ_{36}	-0.104	0.035	-2.983	0.003
κ_{46}	-0.105	0.032	-3.256	0.001
κ_{56}	0.056	0.032	1.715	0.087

Multiple R-squared: 0.999, Adjusted R-squared: 0.999
F-statistic: 2.718×10^5 on 6 and 191 DF, p-value: $< 2.2 \times 10^{-16}$

Furthermore, the mechanism of synchronization in international business cycles can be confirmed using simulations of the model as follows. In these simulations, a set of simultaneous differential equations,

$$\dot{\theta}_i = Q_i + \sum_{j=1}^{N} \kappa \sin \Delta \theta_{ji}, \tag{6.35}$$

is solved numerically with the assumed parameters, where the average and standard deviation are chosen to be the same as the regression estimations. For instance, parameter Q_i is a uniform random variable over the interval $(0.35, 0.45)$, and the initial values $\theta_i(0)$ are uniform random variables over the interval $(-\frac{\pi}{2}, \frac{\pi}{2})$. The interaction strength parameter κ is chosen to be in the range 0.0 to 0.008. The results of the simulation for different values of κ are shown in Fig. 6.32. In the case of $\kappa = 0.008$, synchronization is clearly reproduced. The simulations show that the threshold of the strength parameter κ lies between 0.007 and 0.008. These results suggest that business cycles may be understood as dynamics of **comovements** described by the coupled limit-cycle oscillators being exposed to random individual shocks.

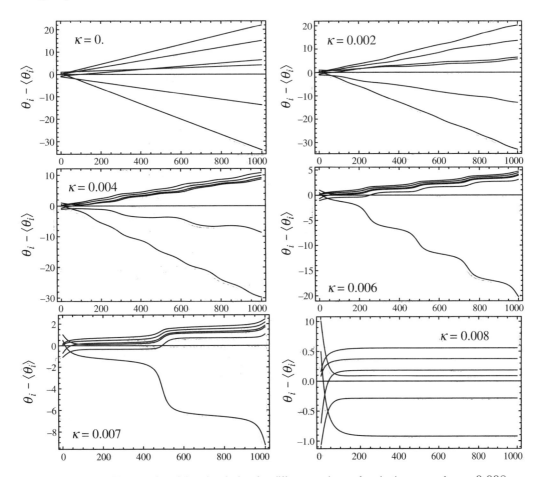

Fig. 6.32 The results of the simulation for different values of κ. In the case of $\kappa = 0.008$, synchronization is clearly reproduced. The simulations show that the threshold of the strength parameter κ is between 0.007 and 0.008. These results suggest that business cycles may be understood as dynamics of comovements described by the coupled limit-cycle oscillators being exposed to random individual shocks (from Ikeda et al. (2013a)).

Finally, we analyze the relation between the size of trade and interaction strength. Exports and imports relative to GDP are shown for Australia, France, the United Kingdom, and the United States in Fig. 6.33. The ratios have increased for the past 20 years for all four countries. Trade data show that the imports (exports) relative to GDP is high except for the United States. These figures show that the importance of international trade has increased and therefore the interaction between countries is expected to have been strong. To clarify the relation between the size of trade and interaction strength, we divide the past 40 years into four periods, i.e. period 1 (1961–1980), period 2 (1971–1990), period 3 (1981–2000), and period 4 (1991–2010), and estimate the parameter estimations using regression analysis.

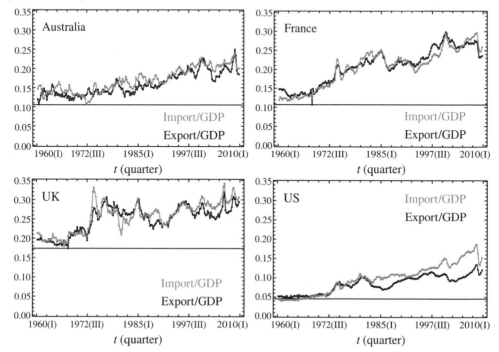

Fig. 6.33 Exports and imports relative to GDP for four countries: Australia, France, the United Kingdom and the United States. The ratios have increased for the past 20 years for all four countries. Trade data show that the imports (exports) relative to GDP is high except for the United States. These figures show that the importance of international trade has increased and therefore the interaction between countries is expected to have been strong (from Ikeda et al. (2013a)) [see Color Plate at the end of book].

The overall strength indicators $S_i (i = 1, \cdots, N)$, defined by

$$S_i = \frac{1}{N} \sum_{j=1}^{N} \kappa_{ji}^2, \tag{6.36}$$

are shown in Fig. 6.34. The statistical error ϵ_i, defined by

$$\epsilon_i = \sqrt{\sum_{j=1}^{N} \left(\frac{2\kappa_{ji}}{N}\right)^2 \sigma_{ji}^2}, \tag{6.37}$$

is shown with indicator S_i. Here σ_{ji} is the standard error of κ_{ji}. The temporal change of the interaction strengths is shown for the six counties in Fig. 6.32, which depicts that the interaction strength indicators have increased over the past 40 years.

In summary, the results suggest that business cycles may be understood as dynamics of comovements described by the coupled limit-cycle oscillators exposed to random individual shocks. The interaction strength in the model increases with an increase in the size of exports and imports relative to GDP. Therefore, a significant part of the comovements comes from international trade.

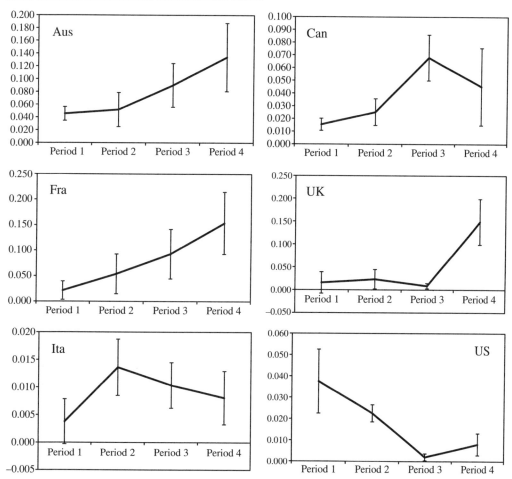

Fig. 6.34 The temporal change of the interaction strengths for Australia, Canada, France, the UK, Italy, and the US. These figures depict that the interaction strength indicators have increased for the past 40 years. The results clearly show that the interaction strength indicator became large in parallel with the increase in the size of exports and imports relative to GDP. Therefore, we conclude a significant part of the comovement came from international trade (from Ikeda et al. (2013a)).

7

Price Dynamics and Inflation/Deflation

> My conjecture is economists let small accidents of intellectual history matter too much. If we had behaved like scientists, things could have turned out very differently.
>
> Paul Romer

Deflation is a threat to the macro economy. Japan had suffered from deflation for more than a decade, and now, Europe is facing it. To combat deflation under the zero interest bound, the Bank of Japan and the European Central Bank have resorted to quantitative easing, or increasing the money supply.

Inflation/deflation is defined in terms of aggregate prices (weighted averages) such as the consumer price index (CPI). Despite its importance, dynamics of the aggregate price index is yet to be fully understood. Recent empirical works have uncovered the little known dynamics of micro prices (see Klenow and Malin (2011)). To understand inflation/deflation, one must first understand what the comovements in the set of individual prices are and how they move. This is what we set out to do in this chapter. We shall also study correlations between the extracted comovements and macroeconomic variables (indices) (Yoshikawa et al., 2015).

7.1 Individual Price Data

Japanese monthly data are available on line; for our analysis here, we examine 830 goods and services for 402 months from January 1980 to June 2013 in the following three categories:

Price Dynamics and Inflation/Deflation 211

IPI: Import Price Index, compiled by the Bank of Japan (BoJ) for 75 import goods (Bank of Japan, 2014).

DCGPI: Domestic Corporate Goods Price Index, compiled by the BoJ, for 420 goods traded between companies (Bank of Japan, 2014). This was later called Producer Price Index (PPI).

CPI: Consumer Price Index, compiled by the Statistics Bureau of the Ministry of Internal Affairs and Communications, covering 335 goods and services (Statistics Bureau, 2014).

The time series we analyze are logarithmic monthly growth rates defined by,

$$r_\alpha(t) := \log_{10}\left[\frac{p_\alpha(t+1)}{p_\alpha(t)}\right], \tag{7.1}$$

where $p_\alpha(t)$ denotes the price of the good/service α at time t. This is a common practice, as use of the logarithm ensures huge changes do not dominate the analysis as in the case of a linear growth rate. A bird's eye view of the normalized time series $\hat{r}_\alpha(t)$ is shown in Fig. 7.1.

This plot shows that the notion of 'time invariance' in a firm's price change practice (or the hazard rate of price) is irrelevant. The frequency of individual price changes and synchronization are not constant but time-varying; however, the existing literature routinely assumes otherwise. Moreover, they change in clusters, and not simultaneously in the economy as a whole. In this respect, there is a significant gap between observed facts and theory because, in standard theory, changes in money, supposedly the most important macro disturbance, more or less uniformly affect all prices. We now have to identify the comoving pattern in this data set and see which mode is dominant when and how.

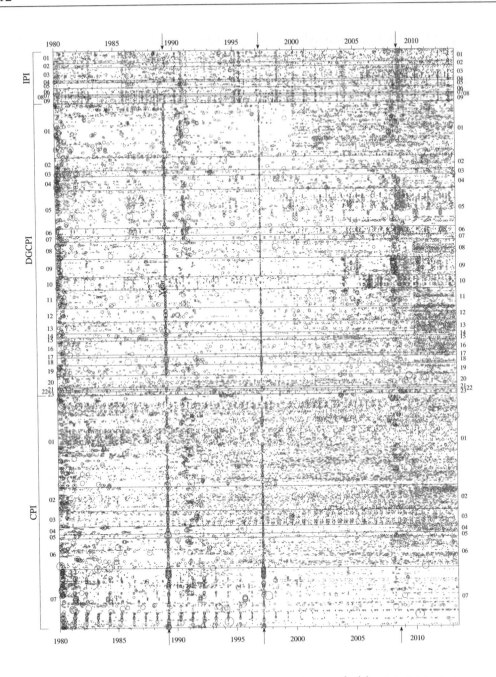

Fig. 7.1 Standardized logarithmic monthly growth rate $\widehat{r}_\alpha(t)$ of individual changes for $|\widehat{r}_\alpha(t)| > 1.0$. Three arrows correspond to the month when Japanese consumption tax (VAT) of 3% was introduced, the month it was raised to 5%, and the time when the Lehman Brothers declared bankruptcy during the subprime mortgage crisis, from left to right (from Yoshikawa et al. (2015)) [see Color Plate at the end of book].

7.1.1 CHPCA+RRS results

Figure 7.2 displays the results of our analysis. It shows that there are 26 significant eigenmodes (among 200 non-trivial eigenmodes)[1].

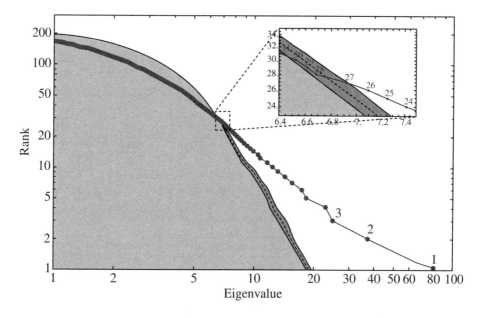

Fig. 7.2 Eigenvalues obtained by the CHPCA and the RRS results: From the inset, we see that the top 26 modes are significant (from Yoshikawa et al. (2015)).

The top solid curve in Fig. 7.3 shows the cumulative value of the eigenvalues

$$S_n := \sum_{k=1}^{n} \lambda^{(k)} \tag{7.2}$$

on the ordinate and n on the abscissa. The black vertical line is at $n = 26$, where $S_{26} = 397.45$, which shows that the 26 significant modes cover $S_{26}/830 = 0.48$, or 48% of all time series behavior. Again, the dotted curve in this plot is the corresponding result for PCA, showing that CHPCA eigenmodes correspond to stronger correlation than that of PCA consistently through all the modes.

---

[1] Since the time segment $T = 401$ is shorter than the number of time series $N = 830$, there are $N-(T-1)/2 = 630$ zero eigenvalues in this system.

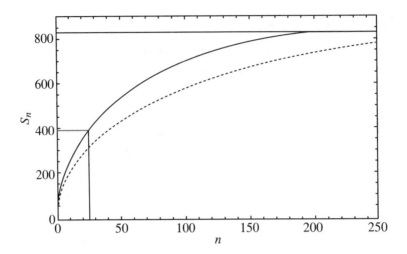

Fig. 7.3 The cumulative eigenvalues S_n defined by Eq. (7.2) (from Yoshikawa et al. (2015)).

7.1.2 A case study

We may use the significant eigenvectors to "cleanse" the data of noises, In this subsection, we take the case of "The Great Recession" 2008–2009.

Figure 7.4 shows (a) the original time series and (b) the one projected to the 26 significant eigenmodes, with the latter being clearer than the former. Before March 2008 and after May 2009, prices were rather quiet; all the action – "coordinated action" occurred in the in-between periods, when CPI was on the rise in September 2008 and then when the DCGPI and CPI went down after the sub-prime mortgage crisis, signified by the Lehman Brothers collapse in September 2008[2]. This shows yet another advantage of our method. Using only the significant comovements, we can cleanse the data to see what is actually happening, without it becoming buried in a sea of noises or the individual stochastic movements.

Furthermore, the aggregate CPI can be constructed using only these cleansed data. Central banks, in fact, use "True core" CPI, where individually fluctuating items such as food (as they are affected mainly by temperature and sunlight, not the macroeconomic situation) are removed, in order to measure the true nature of economic situations. Furthermore, some countries such as the U.S. remove the energy sector. This variation in the definition of the aggregate CPI is not desirable when it comes to a comparison of the macroeconomic state of various countries. Our method, based solely on CHPCA+RRS solves this problem. Running the whole analysis yields the definition of the real true core CPI from the toolbox without any subjective preoccupation, allowing an objective study of the macroeconomy.

---

[2]In Japan, "Lehman Crisis" is the term used to refer to this crisis.

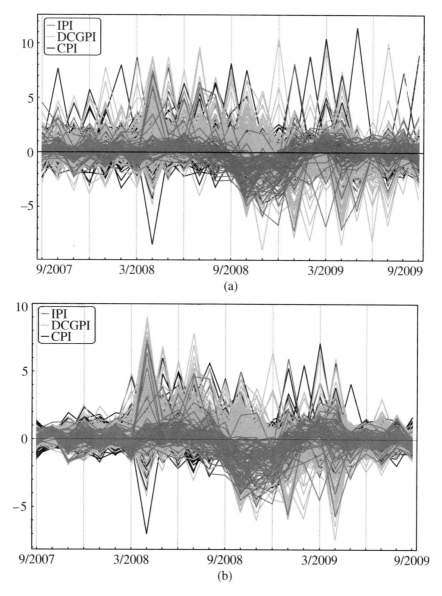

Fig. 7.4 The real part of the time series in and out of the Great Recession period. Plot (a) shows the original time series, and (b) the cleansed time series (from Yoshikawa et al.(2015)) [see Color Plate at the end of book].

The components of the first eigenvector are shown in 7.5 with overlays given for the subcategories "oil-related goods", "services", and "construction-related goods", as shown in the legend.

This clearly shows what is happening in the Japanese economy. The import of raw materials (IPI) moves fast; they are turned into goods and sold to other companies downstream, causing the change in the DCGPI (PPI), and then at the end of the stream, becomes the goods/services meant for the consumer. For detailed numerical analysis of the properties of the eigenmodes including the lower modes, see Yoshikawa et al. (2015).

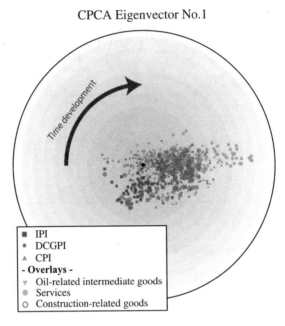

Fig. 7.5 The first eigenvector components in the complex plane (from Yoshikawa et al. (2015)) [see Color Plate at the end of book].

7.2 Correlation between Mode-signals and Macroeconomic Indices

Within the context of macroeconomics and monetary policy, we are primarily interested in the behavior of an aggregate price index such as the CPI. The standard method of regressing changes in the aggregate price index on various macro-variables is aimed at answering this question, of course. However, the traditional analysis of the NKPC based on macro data has its clear limitations (Mavroeidis et al., 2014). Meanwhile, the empirical studies on micro prices have amply demonstrated that the aggregate price index, which is nothing but the weighted average of individual prices,

$$P(t) = \sum_{\alpha} g_\alpha p_\alpha(t), \qquad (7.3)$$

where g_α is a weight, contains much micro noises. For the purpose of extracting "systemic" movements of the aggregate price index from individual prices, we construct eigenmodes based on the correlation matrix of the complexified rates of change of individual prices. In this section, we explore which macro variables these significant eigenmodes correlate with. The macroeconomic indices used in our study are the following:

1. **Wage index**: This is a seasonally adjusted wage index based on contractual cash earnings for establishments with 30 employees or more (source: Monthly Labor Survey; Ministry of Health, Labor and Welfare)
2. **Overtime hours worked**: These include morning work, overtime work, or work on a day off (source: Monthly Labor Survey; Ministry of Health, Labor and Welfare)
3. **Unemployment rate**: This refers to the seasonally adjusted X-12-ARIMA (source: Labor Force Survey; Ministry of Internal Affairs and Communications)
4. **Building starts**: This refers to all dwellings with the total floor area seasonally adjusted by Decomp (The Institute of Statistical Mathematics, 2014) using the period 12. (source: Ministry of Land, Infrastructure, Transport and Tourism)
5. **Monetary base (base money)**: This comprises the sum of banknotes in circulation, coins in circulations, current account deposits held by financial institutions at the Bank of Japan; seasonally adjusted X-12-ARIMA, and average amounts outstanding (source: Bank of Japan)
6. **Money stock M2**: This is the quantity of money held by money holders (corporations, households, and local governments including municipal enterprises). It refers to the sum of currency in circulation and deposits. The money issuers are the Bank of Japan, domestically licensed banks (excluding the Japan Post Bank), foreign banks in Japan, Shinkin Central Bank, Shinkin banks, the Norinchukin Bank, and the Shoko Chukin Bank (source: Bank of Japan)
7. **Exchange rate (Yen/US Dollar)**: This is the monthly average of the spot rate at 17:00 in JST, Tokyo market (source: Bank of Japan)
8. **Crude oil (petroleum) price index**: This refers to the simple average of three spot prices (Dated Brent, West Texas Intermediate, and the Dubai Fateh, USD, 2005=100) (source: IMF Primary Commodity Prices)

The time period is exactly as the data of micro prices. All the macro indices except building starts, money stock M2, exchange rate, and crude oil price index are seasonally adjusted. We use the logarithmic monthly rate of change (see Eq. (6.2)) of these indices, except for the money stock M2, for which only year-to-year changes are available. All the macro variables except for money stock M2 are found to be stationary by the Dicky–Fuller test and the Phillip–Perron tests. For this reason, we also study the logarithmic monthly rate of change of the (year-to-year change of) money stock M2, which we have found to be stationary, and we give this variable a number **6a**.

Figure. 7.6 is the plot of all these variables, where all but money stock are the original value (not the rate of change). Money stock is available only as the year-to-year ratio and is plotted here as the logarithmic rate of change. In examining the correlation of these macro variables with mode signals, we explicitly take into account leads and lags. For this purpose, we complexify these time series variables by using the method explained in Section 5.2.2 and denote its standardized (with mean $= 0$ and standard deviation $= 1$) time series by $\widetilde{M}_j(t)$.

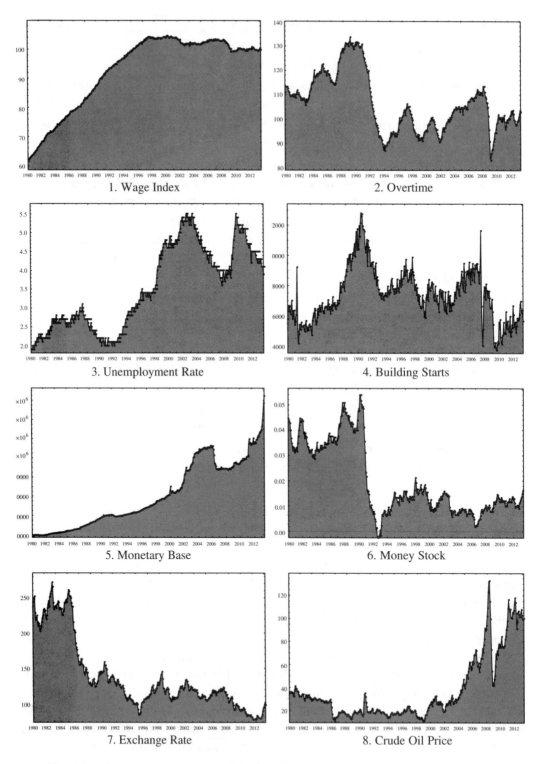

Fig. 7.6 The eight macroeconomic indices that we study (from Yoshikawa et al. (2015)).

To investigate the correlation between these macroeconomic variables and the factors which drive systemic changes of individual prices, namely, mode-signals $a^{(n)}(t)$, we calculate the following correlation coefficient:

$$\mathcal{A}_{j,n} := \frac{1}{T} \sum_{t=1}^{T} \widetilde{M}_j^*(t) a^{(n)}(t), \tag{7.4}$$

where the index j runs from 1 to 8 for the eight macro variables and the index n runs from 1 to 26 for the 26 mode-signals $a^{(n)}(t)$ (defined in Eq. (5.90)) that represent systemic comovements of the individual prices of goods and services. Note that because the mode-signals satisfy Eq. (5.16) we have normalized it by dividing by $\sqrt{\lambda^{(n)}}$ so that $|\mathcal{A}_{j,n}| \in [0, 2\sqrt{\lambda^{(n)}}]$.

To determine whether the resulting value of $\mathcal{A}_{j,n}$ implies significant correlation between the macro index j and the mode-signal n, we use the RRS method reviewed in Section 5.3.3. Specifically, we calculate the distribution of the time-shifted correlation

$$\mathcal{A}_{j,n}^{(\text{RRS})}(\tau) := \frac{1}{T\sqrt{\lambda^{(n)}}} \sum_{t=1}^{T} \widetilde{M}_j^*(t) a^{(n)}(\text{Mod}(t+\tau, T) + 1), \tag{7.5}$$

for $\tau = 1, 2, \cdots, T$ and compare the distribution of the strength of correlations, namely their absolute values, to the absolute value of $\mathcal{A}_{j,n}$.

The results are shown in Fig. 7.7. In the figure, the absolute values $|\mathcal{A}_{j,1}|$ of the first mode-signal are shown by thick bars for seven macro variables from top to bottom. The black dot shows the median of the distribution of $|\mathcal{A}_{j,1}^{(\text{RRS})}(\tau)|$, the dashed bars, the "$1\sigma$ range", which is the range where 68% of the RRS results are contained, the solid bars, the "2σ range", where 95 % are contained.[3]

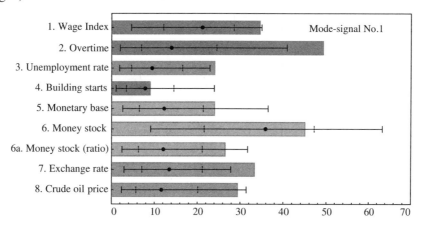

Fig. 7.7 The absolute values of the correlation coefficient $\mathcal{A}_{j,1}$ with their RRS ranges (from Yoshikawa et al. (2015)).

---

[3]Note that the RRS result does not obey a normal distribution. These ranges are obtained by excluding 16% of the largest and 16% of the smallest values obtained by random rotation for the "1σ" range and similarly for the "2σ" range.

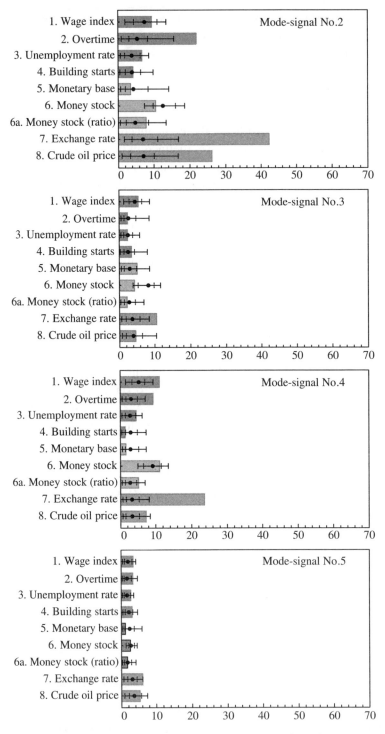

Fig. 7.8 The absolute values of the correlation coefficient $|\mathcal{A}_{j,n}|$ for $n = 2$–5, along with their RRS ranges (from Yoshikawa et al. (2015)).

From Figs. 7.7 and 7.8, we draw the following conclusions.

1. The first eigenmode shows significant correlations with overtime hours worked, the unemployment rate and the exchange rate.
2. The exchange rate is *very* significantly correlated with the second and the forth mode-signals.
3. Neither monetary base nor money stock shows a significant correlation with any of the significant modes.

Further examination by Yoshikawa et al. (2015) shows that the absolute value of the first eigenvector for the CPI and services is much larger than that for prices of imported and oil-related goods. The result suggests that the first eigenmode represents the factors which drive domestic prices. It is consistent with the finding in the current analysis that the first eigenmode has significant correlations with overtime hours worked and the unemployment rate.

This analysis demonstrates that systemic movements of micro prices are not correlated with money. In standard macroeconomics models, money is the most important macro variable which affects prices by way of expectations. Our results cast a serious doubt on this standard framework.

Further work has been done not by using individual prices, but by using semi-aggregated prices, which revealed yet another surprising feature of prices. We now proceed to examine this.

7.3 Collective Motion of Prices and Business Cycles

Which came first, the long-term deflation or the depression called the Lost Decade in Japan? To address this basic issue in economics, we apply the CHPCA to a set of Japanese economic data collected over the past 30 years, comprising individual price indices of the middle classification level (imported goods, producer goods, consumption goods and services), indices of business conditions (leading, coincident, lagging), and the yen–U.S. dollar exchange rate (Kichikawa et al., 2016). Such a combinatorial CHPCA enables us to delve into the dynamic linkage of price movements with economic conditions and foreign exchange rates. Adopting the RRS method as a null hypothesis, we extract statistically meaningful correlations out of the noisy data; the RRS destroys only cross correlations, preserving autocorrelations involved in multivariate time series. The statistical test identifies two significant eigenmodes with the largest and second-largest eigenvalues.

Figures 7.9 and 7.10 depict the two eigenvectors on a complex plane. The lead–lag relations among domestic prices in the two modes are quite similar. This fact infers that there exists a collective motion of individual prices arising from the mutual interactions between them. However, careful observation of the behavior of the eigenvectors tells us that the collective motion is driven quite differently in the two modes. Its driver is the exchange rate at the upper stream side in the first mode. However, its impact on the whole economy is restrictive. The second mode, on the other hand, indicates that the collective motion of prices is excited by total demand at the lower stream side and its impact is of much more influence on the whole economy.

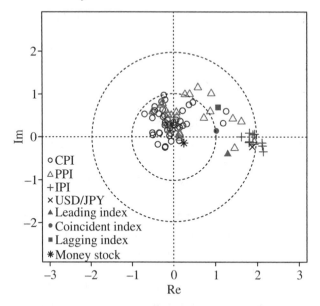

Fig. 7.9 The eigenvector components for the first mode in the CHPCA on the complex plane, showing the collective motion of the individual prices together with the macroeconomic indicators. (From Kichikawa et al. (2016))

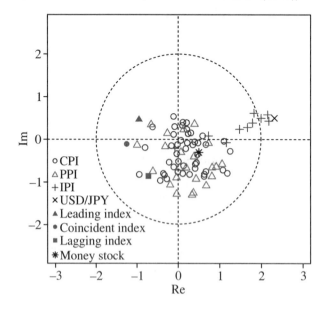

Fig. 7.10 The same as Fig. 7.9, but for the second mode. (From Kichikawa et al. (2016))

Figure 7.11 shows the contributions of the first and second eigenmodes to the coincident index of business conditions. The business cycles are thus decomposable into two components in terms of the collective motion of prices. The second-mode component should be focused on more seriously because it gives information on the condition of the whole economy, i.e., on whether current economic growth is driven by an increase in the total demand or not. From Fig. 7.11, we see that the economic upturn toward the crash of the bubble economy late in 1990 was basically led by the first-mode component. A similar situation is observable immediately after the launch of Abe's second Cabinet, in December 2012. In corporation with the Cabinet, the Bank of Japan has set an inflation target of 2% to promote recovery of the Japanese economy from its long-standing depression. It is clear that the policy, Abenomics, was initially successful. However, its work on the economy is so weak that it cannot excite the second mode of the collective motion of prices; even the second-mode component began to decline early in 2014. This should continue to be monitored.

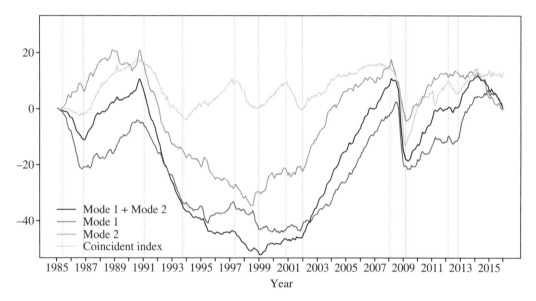

Fig. 7.11 Contributions of the first and second eigenmodes to the coincident index of business conditions compared with its standardized data, where all the results are temporally accumulated with the same initial value to facilitate comparison. The red and blue vertical lines indicate troughs and peaks in the Japanese business cycles, respectively. (From Kichikawa et al. (2016)) [see Color Plate at the end of book].

7.4 Summary

Deflation and inflation are macroeconomic phenomena. However, we cannot fully understand them by only exploring macro data because the behavior of aggregate prices such as the consumer price index depends crucially on the interactions of micro prices. On the one hand, our results confirm that aggregate prices change significantly, moving up or down, as the level of real output changes.

The correlation between aggregate prices and money, on the other hand, is not significant. The major factors affecting aggregate prices other than the level of real economic activity are the exchange rate and the prices of raw materials represented by the price of oil. Japan suffered a deflation for more than a decade beginning at the end of the last century. More recently, Europe faces a threat of deflation. Our analysis suggests that it is difficult to combat deflation only by expanding the money supply.

In addition, we have established that the collective motion of prices exhibits concrete lead/lag relations irrespective of their drivers, the foreign exchange rate and the total demand. Furthermore, the macro behavior of prices has definite dynamic linkages with the indices of business conditions; the real economic condition leads the price changes. These empirical facts support the old Phillips curve (Phillips, 1958; Tobin, 1972).

8

Complex Networks, Community Analysis, Visualization

> I read somewhere that everybody on this planet is separated by only six other people. Six degrees of separation. Between us and everybody else on this planet.
> **In a play written by John Guare**

As pointed out in the Introduction, a macro economy consists of many micro agents. These agents are not independent of each other, but form networks, i.e., they have relationships among each other. An understanding of their interactions is crucial for an understanding of a macro economy.

This chapter explains the basic concepts and techniques for understanding the structure of economic networks, and then applies them to economic networks. Economic networks are not random nor regular, but exhibit fat-tailed distributions for the numbers and weights of relationships. They also form densely knit groups that are only loosely connected with each other, called clusters or communities.

Basic tools cover

- fundamental graph search algorithms that are applied to identify upstream and downstream in networks and also to model financial distress propagation
- a class of random networks used as a null model, namely a configuration model
- the concept of modularity derived from assortative mixing and the configuration model
- detection of communities by modularity and its extension
- visualization of large-scale networks by physical simulations

in addition to other graph-theoretical tools. Applications include

- a nationwide production network comprising of a million firms and millions of supplier–customer relationships
- a community structure in correlation with the stock market found by the RMT and its extension
- a globally-coupled network of equities and currencies in the world found by the CHPCA method
- a world input–output database for countries and industrial sectors
- global production data for G7 countries

We will show that one can retrieve interesting information and implications by applying network analyses to these applications.

8.1 Basic Tools

8.1.1 Graph search and simple applications

Computational algorithms for simulation on a network are often based on graph-search algorithms and their modifications. Graph search refers to a way of visiting nodes and/or edges exhaustively by putting "marks" on them in a systematic manner. It is analogous to the well-known Greek tale about the hero, Theseus, who successfully escaped from a labyrinth of Minotaur by following Ariadne's thread, but is far more elaborate.

We shall explain elementary graph-search algorithms in order to use them in simple applications, which are potentially useful to economic networks, and not fully explained elsewhere. Specifically, the simple applications include

- Enumerating the neighbours of neighbours with no duplication. More generally, for counting the number of nth neighbours at n links away from a node or a set of nodes. This will be useful to evaluate the impact of a shock occuring at a given point or portion in an economic network.
- Identifying upstream and downstream in economic networks. Especially, for directed graph, this identification problem is what they call the "bow-tie" structure of the largest strongly connected component, its in-component, and out-component.
- Applying a basic graph-search algorithm to the so-called DebtRank algorithm recently developed to quantify systemic risks owing to financial stress in financial networks.

For more elaborate algorithms, readers are highly recommended to refer to excellent textbooks and guiding books in computer science such as Cormen et al. (2001); Mehlhorn and Näher (1999); Siek et al. (2002), as well as the concise exposition given by Newman (2010), Part III.

Representation of graphs

Let us summarize the fundamental definitions of graphs, namely, a mathematical representation of a network, in a glossary.

We assume that the readers are familiar with the following concepts:

Nodes Basic constituents of the network. A node is a terminology in computer science, also called a vertex, a site (percolation theory in physics), an actor (sociology), or an agent (economics).

Links Relationship connecting two nodes. A link is a terminology in computer science, often also called an edge, a bond (physics), or a tie (sociology).

Graph A graph G is a pair of a set of nodes, V, and a set of links, E, that is, $G = (V, E)$. This is a mathematical representation of a network. A directed graph consists of directed links, i.e., each link goes from one node to another in a one-way direction. An undirected graph is one where each link can be regarded as a two-way directed link from a node to another and vice versa.

A **directed graph** or **digraph** has each link $e \in E$ as an ordered pair of nodes, $e = (u, v)$ ($u, v \in V$). An **undirected graph** has each link $e \in E$ as a set of nodes $e = \{u, v\}$ ($u, v \in V$ and $u \neq v$). In this section, we denote a link of an undirected graph by (u, v) (assuming $(u, v) = (v, u)$).

If $e = (u, v) \in E$, v is said to be adjacent to u. In the case of an undirected graph, if v is adjacent to u, u is adjacent to v, trivially. In the case of a directed graph, a link $e = (u, v)$ starts from the node u and ends at v. e is said to be incident from u and to v in this case.

A self-loop is a directed link starting from and ending at the same node. It is usually convenient to consider an undirected link as a pair of two oppositely directed links as we shall see later. Multiple links (edges) or parallel links (edges) means that there is more than one link between the same pair of nodes. Often a link has a characteristic of *weight*, which is usually a positive number attributed to the link. Multiple links between a pair of nodes can be interpreted as links with an integer-valued weight.

The number of links connected to a node is called the node's degree. For a directed graph, there is both in-degree and out-degree, the numbers of in-coming links and out-going links, respectively. The empirical distribution for degrees is called a degree distribution.

If the set of nodes can be grouped into two groups such that there are no links within each group, and all the links connect between one group and the other, the graph is called a bipartite graph. Bipartite graphs appear in many economic networks as we shall see.

For computational purpose, it is convenient to represent graphs as shown in Fig. 8.1 and Fig. 8.2.

Adjacency list is a list of nodes incident to a node u, namely

$$\text{Adj}[u] := \{v \in V | (u, v) \in E\}. \tag{8.1}$$

In Figs. 8.1 and 8.2, the adjacency lists are represented by a data structure of *list* in computer science, where a pointer that points to the next element is represented by the asterisk * and the terminating symbol by slash. The order of nodes put in the list is arbitrary. It should be kept in mind that graph-search results may depend on the adjacency list's order of nodes.

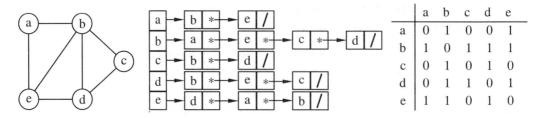

Fig. 8.1 Undirected graph, adjacency list and matrix

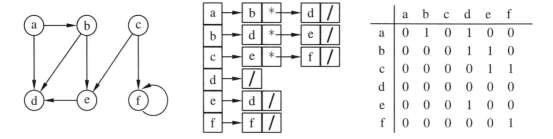

Fig. 8.2 Directed graph, adjacency list and matrix

Adjacency matrix is an array of elements:

$$A_{ij} = \begin{cases} 1 & ((i,j) \in E), \\ 0 & ((i,j) \notin E), \end{cases} \qquad (8.2)$$

as shown also in Figs. 8.1 and 8.2.

Let the numbers of nodes and links be denoted by $|V|$ and $|E|$ respectively. Many economic networks are *sparse* in the sense that $|E| \ll |V|^2$. However, we shall consider *dense* networks such as a correlation matrix and its corresponding network. The adjacency list is often useful for large but sparse graphs. The adjacency matrix for a large graph is usually used to analyze mathematical properties including eigenvalues and eigenvectors for the matrix.

Depth-first search

The graph-search algorithm, called **depth-first search** (DFS) is an algorithm to visit all the nodes of a graph in such a recursive way as to follow one of the most recently visited node v's unexplored links, (v, w), and to mark w by a "color" as visited if it is not yet visited. When there are no more unexplored links, a "backtrack" brings one to the node that had brought one to the node v. If all the nodes that can be backtracked are colored, one starts from an uncolored node and the algorithm is run again.

Suppose that the color WHITE is put to the node "not yet visited"; GRAY means "visited for the first time"; BLACK is put to the node, if all the links which are adjacent to it were explored and done. We show the algorithm in Algorithm 8.1, and its core in Algorithm 8.2.

Algorithm 8.1: Depth-first search

```
1   // Initialize
2   for each node v in V[G]
3       color[v] ← WHITE
4       p[v] ← NIL
5   time ← 0   // global time
6   for each node v in V[G]
7       if color[v] is WHITE
8           DFSvisit(G,v)
```

Algorithm 8.2: DFSVISIT(G, v)

```
1   color[v] ← GRAY
2   time ← time + 1
3   discover_time[v] ← time
4   for each node w in Adj[v]
5       if color[w] is WHITE
6           // (v,w) is a tree edge
7           p[w] ← v
8           DFSvisit(G,w)
9       else if color[w] is GRAY
10          // (v,w) is a back edge
11      else if color[w] is BLACK
12          // (v,w) is a cross edge or a forward edge
13  color[v] ← BLACK
14  time ← time + 1
15  finish_time[v] ← time
```

Line 3 of Algorithm 8.2 records the time of first visit, and line 15, the time of finished exploration for the node.

Note that the line 7 of Algorithm 8.2 visits the node w for the first time after following the unexplored link (v, w) incident to the most recently colored v, and determines the "parent" of w, denoted by $p[w]$, as v. This parent–child relationship determines a tree structure in the graph, which is called a DFS tree. After exiting from a recursion, one does a "backtrack" in this algorithm.

It is not easy to prove why this algorithm works. Instead, it is recommended that the reader apply it to examples, including Fig. 8.2 as shown in Fig. 8.3. Note that each diagram in Fig. 8.3 is the state right after line 3 or line Fig. 8.3 in Algorithm 8.2, when the computational time is updated.

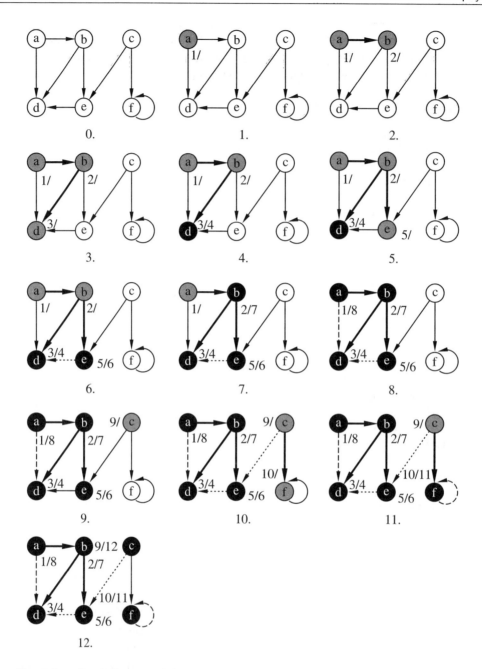

Fig. 8.3 Depth-first search for a directed graph

Examining the example, one can understand that in general we can obtain one or more DFS trees after completing the search. The links in the tree are $(p[v], v)$ $(p[v] \neq \text{NIL})$. The tree is a rooted tree the root of whose *discover_time* is its minimum. If in a DFS tree v is reachable from u, u is called an ancestor of v, and v is a descendant of u. There may be pairs of nodes that are not an ancestor or a descendant of each other. The root of the DFS tree is the ancestor of all the nodes in it. For a given adjacency list, one has a unique set of DFS trees. After completion, any link in the graph can be categorized into either of tree edge that forms the tree, back edge (u, v) from a node u to its ancestor v, forward edge (u, v) from a node u to its descendant v, or cross edge that connects a pair of nodes that are not an ancestor or a descendant of each other.

When a link (u, v) is explored for the first time with the color of u as GRAY, the link is a tree edge if v is WHITE, a back edge if it is GRAY, or either a forward edge or a cross edge if it is BLACK. In the last case, if $discover\_time[u] < discover\_time[v]$, the link (u, v) is a forward edge, and if $discover\_time[u] > discover\_time[v]$, the link is a cross edge.

In Fig. 8.3, a tree edge is represented by a bold line, a back edge by a dashed line, a forward edge by a broken line, and a cross edge by a dotted line. We show the obtained DFS trees and the classification of links in Fig. 8.4.

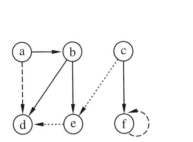
(a) Example of completion of DFS search

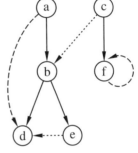
(b) Example of DFS trees and classication of links

Fig. 8.4 DFS trees [see Color Plate at the end of book].

In the case of an undirected graph, when v is BLACK, and the same link (v, u) was explored as before in the opposite direction, it turned out that all the links are either tree edges or backward edges.

Strongly connected component

In the case of a directed graph, if there is a path from a node u to another node v, v is said to be reachable from u. When u and v are mutually reachable from each other, we define a binary relation denoted by $u \sim v$. Because the relation defines an equivalence, the nodes can be categorized into equivalence classes as V_i ($i = 1, \ldots, p$). Denoting the set of links by E_i, with either end of each link falling into the same equivalence class V_i, one has a unique set of subgraphs $G_i = (V_i, E_i)$. Each G_i is called a **strongly connected component**.

One can modify the DFS algorithm to find all the strongly connected components, which is given in Algorithm 8.3 and Algorithm 8.4. (See Cormen et al. (2001) for a proof).

Algorithm 8.3: Strongly connected components

```
1   // Initialize
2   for each node v in V[G]
3       color[v] ← WHITE
4       p[v] ← NIL
5       scc_id[v] ← NIL   // v belongs to which SCC
6   time ← 0   // global time
7   num_scc ← 0   // number of SCCs
8   for each node v in V[G]
9       if color[v] is WHITE
10          SCCvisit(G,v)
```

Algorithm 8.4: SCCvisit (G, v)

```
1   color[v] ← GRAY
2   time ← time + 1
3   discover_time[v] ← time
4   root[v] ← v   // static hash
5   Push(S,v)   // static stack
6   for each node w in Adj[v]
7       if color[w] is WHITE
8           p[w] ← v
9           SCCvisit(G,w)
10  color[v] ← BLACK
11  time ← time + 1
12  finish_time[v] ← time
13
14  for each node w in Adj[v]
15      if scc_id[w] is NIL
16          if discover_time[root[w]] < discover_time[root[v]]
17              root[v] ← root[w]
18  if root[v] is v
19      do
20          w ← Pop(S)
21          scc_id[w] ← num_scc
22      while w is not v
23      num_scc ← num_scc + 1
```

$root[v]$ is the node of first visit time being minimum in the strongly connected component that includes the node v (see the example in Fig. 8.5). Stack means a data structure of a "stack" of documents on a desk, last-in first-out.

An example is shown in Fig 8.5. Each resulting strongly connected component, $G_i = (V_i, E_i)$, is represented by an area of gray. If the links connecting the nodes of different components of V_i and V_j ($i \neq j$) are represented as a single link connecting V_i and V_j, one can obtain a reduced or coarse-grained graph. In such a graph, from the definition of the components, there are no cycles (a path starting from one component and ending at the same component) so that one can see the "upstream" and "downstream" in the reduced graph.

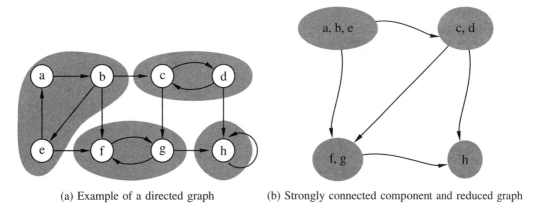

(a) Example of a directed graph   (b) Strongly connected component and reduced graph

Fig. 8.5 Decomposition into strongly connected component

Breadth-first search

The graph-search algorithm, called **breadth-first search** (BFS) is an algorithm to visit all the nodes of a graph in such a systematic way as to follow one of the most *early* visited node v's unexplored links, (v, w), and to mark w by a "color" as visited if it is not yet visited.

Color each node by WHITE for unvisited, GRAY if it is visited for the first time, and BLACK if all the links emanating from it were explored. We show the process in Algorithm 8.5. Often, one starts the BFS from a particular node, which is denoted by s.

Algorithm 8.5: Breadth-first search BFS(G, s)

```
1   // Initialize
2   for each node v in V
3       color[v] ← WHITE
4       d[v] ← INFINITY
5       p[v] ← NIL
6   color[s] ← GRAY
7   d[s] ← 0
8
9   Q ← empty queue
10  Enqueue(Q,s)
11  while Q ≠ empty
12      v ← Dequeue(Q)
13      for each node w in Adj[v]
14          if color[w] is WHITE
15              color[w] ← GRAY
16              d[w] ← d[v] + 1
```

```
17            p[w] ← v
18            Enqueue (Q,w)
19       color[v] ← BLACK
```

Queue is a data structure similar to a "queue" of waiting people, first-in first-out. It stores a list of nodes to be examined for emanating links from them from old to new ones. ENQUEUE in line 10 and line 18 of Algorithm 8.5, represents adding an element to the end of the queue; DEQUEUE in line 12 represents picking the element at the head of the queue.

In following the link (v, w) from the earliest visited node v, if the node w is not visited, w's parent $p[w]$ is defined by v in line 17. This relationship of parent and child determines a tree in the graph, which is called a *BFS tree*.

For all the paths from v to w, the one with the minimum length (number of links) is called **shortest path**, and its length is **distance**, $d(v, w)$. The distance is defined for a pair of nodes that are reachable from one to the other. Once the parent–child relationship is determined in the BFS tree, one can calculate the distance $d(s, w)$ from that staring node s to all the nodes w that are reachable from s in line 16 (verify the example shown in Fig. 8.6).

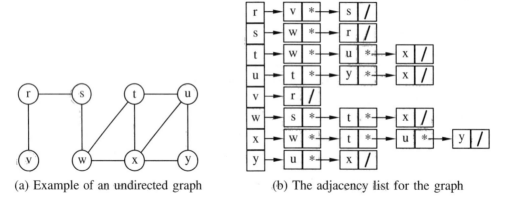

(a) Example of an undirected graph   (b) The adjacency list for the graph

Fig. 8.6   An undirected graph for BFS

An example of an undirected graph is given in Fig 8.6. The result is shown in Fig 8.7, each diagram of which corresponds to the state right after the initialization of line 10 or after line 19. The queue at each step is also shown in each diagram.

Examining the example, one can understand the following. The BFS tree consists of the links $(p[v], v)$ $(p[v] \neq \text{NIL})$. The root of the tree is the starting node s. For the distance from s, it holds that $d[p[v]] = d[v] + 1$. For other links, except the tree edges, $(u, v) \in E$, it holds that $d[v] \leq d[u] + 1$. In the case of an undirected graph, either of the relations, $d[v] = d[u]$ or $d[v] = d[u] + 1$, holds.

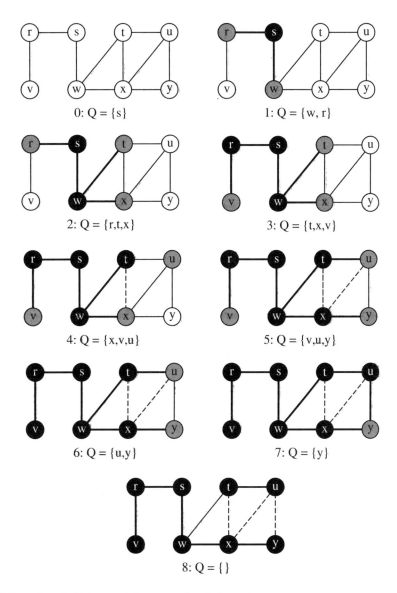

Fig. 8.7 Breath-first search for an undirected graph

In Fig 8.7, a tree edge is represented by a bold line, and others by dashed lines. The resulting BFS tree is shown Fig 8.8.

By using the BFS for each node as a starting point, one can calculate the distances for all pairs of nodes. **Average path length** is the average of the distances, and **diameter** refers to the maximum distance in them.

Note that for a given pair of nodes, more than one shortest path can exist. It is not difficult to modify the BFS algorithm to count all possible shortest paths for all the pairs of nodes. Such a calculation is used for **betweenness centrality**. Interested readers can refer to an efficient algorithm by Brandes (2001); Newman and Girvan (2004a).

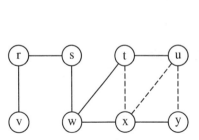 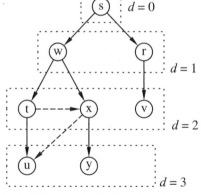

(a) Example of completion of BFS (b) A BFS tree (solid lines) and distanced

Fig. 8.8 A BFS tree

Number of (friends')^n friends

Suppose we want to calculate the number $T_n(s)$, which is the number of nodes at distance n from the node s. $T_2(s)$ is the number of friends' friends, $T_3(s)$ is the number of friends' friends' friends. $T_n(s)$ can be uniquely determined for each node s. The diameter is the maximum for all the nodes v such that $T_n(v) > 0$.

Since BFS can determine the distance of a node from the starting point, when the node was visited for the first time, one can record the list of nodes and their distances. After completing the search, it is easy to calculate $T_n(s)$ from the node s. Interested readers are encouraged to write an algorithm for this by modifying the BFS algorithm.

8.1.2 Random graphs

There exists an infinite class of random graphs for the obvious reason that what is defined by "random" depends on the assumption of a particular model. Here, we pick up just two classes of models, the so-called Poisson random graph and a null hypothesis of randomizing a graph while preserving the degree sequence.

Poisson random graph

The simplest model for random graph is to assume that the probability of a link between any pair of nodes is independent and equal. As we shall see shortly, in a limit of infinite number of nodes, the degree distribution coincides with a Poisson distribution. Thus, this class is called a **Poisson random graph**. (This is one of the classes that P. Erdős and A. Rényi studied extensively in the 1950s and 60s).

Assume that the probability of having a link between any pair of nodes is given by a constant, p. Then the degree distribution follows a binomial distribution:

$$f(k) = \binom{N-1}{k} p^k (1-p)^{n-1-k}. \tag{8.3}$$

The average and variance of the distribution are respectively

$$\langle k \rangle = (N-1)p \simeq Np, \tag{8.4}$$

$$\langle k^2 \rangle - \langle k \rangle^2 = (N-1)p(1-p) \simeq Np(1-p). \tag{8.5}$$

Define a generating function for the probability distribution, Eq. (8.3), and calculate it as

$$\begin{aligned} g(x) &:= \sum_{k=0}^{\infty} f(k) x^k \\ &= \sum_{k=0}^{N-1} \binom{N-1}{k} p^k (1-p)^{N-1-k} x^k \\ &= (1 + px - p)^{N-1}. \end{aligned} \tag{8.6}$$

In the limit of $N \to \infty$ and $Np \to \text{const.} := z$, we have

$$g(x) \to e^{z(x-1)}. \tag{8.7}$$

Solving Eq. (8.6) for p_k, we obtain

$$\begin{aligned} f(k) &= \frac{1}{k!} \left. \frac{d^k g(x)}{dx^k} \right|_{x=0} \tag{8.8} \\ &= \frac{1}{k!} z^k e^{-z}, \tag{8.9} \end{aligned}$$

a Poisson distribution. The average and variance are respectively

$$\langle k \rangle = z = Np, \tag{8.10}$$

$$\langle k^2 \rangle - \langle k \rangle^2 = z, \tag{8.11}$$

which coincides with the limits of Eq. (8.4) and Eq. (8.5).

Note that for a Poisson random graph, one has the relation, $\langle k^2 \rangle \sim \langle k \rangle^2$. In many economic networks, it is not adequate to use this class of random networks because such a relationship does not hold empirically as we shall see.

Configuration model

The so-called **configuration model** is a class of random graphs with a given degree *sequence*, rather than degree distribution. Each node's degree is fixed, rather than merely preserving the entire degree distribution, but the links are randomized in the following way.

Consider an undirected graph with N nodes and M links. Cut each link in the middle to have $2M$ "stubs" (as follows).

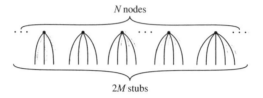

Then, connect those stubs randomly, uniformly random over the stubs, like this drawing:

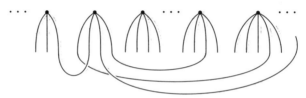

It is obvious that after this operation, each node's degree is exactly the same as before. Note that any topological structure that may have been present for the original pairs of nodes and any group of nodes are totally destroyed. For example, the information of degree correlation is lost after this randomization.

For any given degree sequence, this randomizing operation generates all possible graphs with exactly the same degree sequence. This class of random networks is the configuration model. A recent mathematical development in random graphs is the study of the statistical properties of this class of models. See Molloy and Reed (1998); Newman et al. (2001); Newman (2010), for examples. For economic networks, it is important to use the configuration model for the empirical data as a null hypothesis, by which we can identify significant statistical properties in the real data and compare them with this class of random graphs.

Excess degree

Choose a node randomly, in a random graph of the configuration model, and follow a link randomly from the links incident to the node to reach a node at the other end of the link. Consider the probability that the degree of the reached node is exactly equal to k, including the followed link. Another way to reach a node is to first choose a link randomly from all the links, and select a node randomly from the two ends of the link. Consider the probability that the degree of the selected node is exactly equal to k including the link chosen first.

For a random graph of the configuration model, these two ways yield the same probability. (The reader is encouraged to consider the reason here). Note that the probability is *not* equal to $p(k)$, but is proportional to $k \times p(k)$, as easily understood from the following:

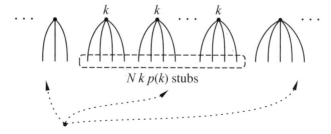

In either case, we are interested in how many links the reached node has, except the link that leads to it, namely the remaining links. Let us call the number of remaining links as **excess degree**. Trivially, excess degree plus one is equal to degree.

Denoting the probability that the node reached by either of the ways has excess degree ℓ by $q(\ell)$, the earlier argument shows that $q(\ell) \propto (\ell+1)\,p(\ell+1)$. Normalizing the probability, we have

$$q(\ell) = \frac{(\ell+1)\,p(\ell+1)}{\sum_{k=0}^{\infty} k\,p_k} = \frac{1}{\langle k \rangle}\,(\ell+1)\,p(\ell+1). \tag{8.12}$$

The fact that $q(k)$ is proportional to $(k+1)p(k+1)$, not $p(k+1)$ is a simple fact but gives us important insights into some problems. Suppose that a firm in an economic network is chosen at random, and then one of the firms connecting to the firm chosen first is selected. This process of selecting a firm is not a random sampling from the network, because a firm with larger degrees is more probable to be selected than those with smaller degrees. In other words, big firms and small firms are selected in a biased way with different probabilities.

This fact is also important to consider while examining the problem of propagation of financial stress (see the next chapter). Another interesting consequence is the evaluation of the clustering coefficient.

8.1.3 Basic properties of network structures

Let us briefly summarize the basic properties of network structures which we shall use in this and the next chapter.

- Unary property: degree
- Binary properties: assortative mixing, degree correlation
- Ternary property: clustering coefficient
- Higher-order structure: betweenness

In particular, we shall focus on the important concept of assortative mixing to explain, in later sections of this chapter, modularity, and see an application of a configuration model to evaluate the statistical significance of a clustering coefficient, as well as a graph-search algorithm applied to efficient calculation of betweenness centrality.

Degree distribution

Economic networks quite often exhibit fat-tailed degree distributions, namely the property of scale-free network. Specifically, in the region of a large degree k, the PDF asymptotically obeys

$$f(k) \approx \left(\frac{k}{k_0}\right)^{-\mu-1}, \tag{8.13}$$

where k_0 is a certain scale of degree, or equivalently, the cumulative distribution has the asymptotic form:

$$\bar{F}(k) := \sum_{k'=k}^{\infty} f(k') \approx \left(\frac{k}{k_0}\right)^{-\mu}. \tag{8.14}$$

See Chapter 3 for fat-tailed distributions and related topics, and Section 2.1 for stochastic processes to generate power-law distributions. To estimate μ and k_0, one can use the so-called Hill's estimation and its extension (Newman, 2005; Clauset et al., 2009).

Assortative mixing

The most important binary structure of a network is the nodes' characteristics appearing for each pair of nodes, because it is related to how links are formed. Note that any pair of nodes will be statistically independent in their characteristics in the configuration model, so one can use the model as a null hypothesis.

Let us take an example of Table 8.1. given in Newman (2003a) for sexual partnerships in a survey in San Francisco, California, 1992, studied in epidemiology (see the references therein for original works). Obviously partners are formed within the same race, which is called **assortative mixing**.

Table 8.1 Assortative mixing in sexual partnerships, a survey in epidemiology.

	categories	women				total
		black	hispanic	white	other	
men	black	506	32	69	26	633
	hispanic	23	308	114	38	483
	white	26	46	599	68	739
	other	10	14	47	32	103
	total	565	400	829	164	1958

Assume a node can have characteristics that are categorized into K categories. For M links in the whole network, examine all the pairs of categories at each two ends of nodes for each link. For a directed graph, denote by n_{ij} the number of directed links such that the start and end at nodes have characteristics i and j ($i,j = 1,\ldots,K$). For an undirected graph, let us use the convention that each link consists of two opposite links, and do the same calculation[1]. In either case, we have

$$\sum_{i,j} n_{ij} = M. \tag{8.15}$$

We have the following contingency table:

	1	...	j	...	K	total
1	n_{11}	...	n_{1j}	...	n_{1K}	$n_{1\bullet}$
$\vdots$	$\vdots$		$\vdots$		$\vdots$	$\vdots$
i	n_{i1}	...	n_{ij}	...	n_{iK}	$n_{i\bullet}$
$\vdots$	$\vdots$		$\vdots$		$\vdots$	$\vdots$
K	n_{n1}	...	n_{nj}	...	n_{nK}	$n_{K\bullet}$
total	$n_{\bullet 1}$	...	$n_{\bullet j}$	...	$n_{\bullet K}$	M

where

$$n_{i\bullet} := \sum_j n_{ij}, \tag{8.16}$$

$$n_{\bullet j} := \sum_i n_{ij}. \tag{8.17}$$

Define $e_{ij} := n_{ij}/M$. From Eq. (8.15), we have

$$\sum_{i,j} e_{ij} = 1. \tag{8.18}$$

And let us define

$$a_i := \sum_j e_{ij}, \tag{8.19}$$

and

$$b_j := \sum_i e_{ij}. \tag{8.20}$$

In terms of e_{ij}, a_i, and b_j, the contingency table reads

---
[1] Equivalently, count a link with (i,i) as 1, and (i,j) ($i \neq j$) as $1/2$ in the calculation.

	1	...	j	...	K	total
1	e_{11}	...	e_{1j}	...	e_{1K}	a_1
⋮	⋮		⋮		⋮	⋮
i	e_{i1}	...	e_{ij}	...	e_{iK}	a_i
⋮	⋮		⋮		⋮	⋮
K	e_{n1}	...	e_{nj}	...	e_{nK}	a_K
total	b_1	...	b_j	...	b_K	1

Assortativity coefficient (Newman, 2003a) is defined as

$$r = \frac{\sum_{m=1}^{K} e_{mm} - \sum_{m=1}^{K} a_m b_m}{1 - \sum_{n=1}^{K} a_n b_n}. \tag{8.21}$$

The first term of the numerator is the fraction of the links connecting nodes with the same characteristic. Note that in the configuration model for this network, one has

$$e_{ij} = a_i b_j, \tag{8.22}$$

because the probability that a randomly chosen link has a node with characteristic i at either end of the link is a_i, and the probability that the other end of the link has characteristic j is b_j. The second term of the numerator, therefore, is such a factor that under the null hypothesis of a random graph in the configuration model, the assortative coefficient is exactly zero. The denominator is a normalization factor so that in the case of perfect assortative mixing, $e_{ij} = 0$ for all $i, j = 1, \ldots, K$ $(i \neq j)$, the coefficient becomes unity.

Now one can easily think of a variety of characteristics including

- Attributes of nodes, not related to network structure itself, such as firm-size (revenue and sales, number of employees, etc.), industrial sectors and geographical locations in economic networks; age and personal income in social networks, and so forth.
- Attributes of nodes derived from network structure, such as degree correlation.
- Categories obtained by some methodology, typically by a clustering algorithm. The frequently used quantity of **modularity** or Q value is given by

$$Q = \sum_p e_{pp} - \sum_p a_p b_p. \tag{8.23}$$

This can measure if the categorization of nodes into different clusters or communities labeled by p is good, because if the clustering is good, one should have a large value of Q.

To evaluate the statistical error by using the class of random graphs in the configuration model, one can use either a Jackknife re-sampling of bootstrap method or the so-called Delta method of large sample theory. See Newman (2003a).

According to the Jackknife re-sampling, the statistical error for the assortative coefficient r can be calculated by

$$\sigma_r^2 = \sum_{i=1}^{M} (r_i - r)^2, \tag{8.24}$$

where r_i is the value of Eq. (8.21) calculated for a network without the link i. Similarly, we can calculate for the error of the modularity or Q value, σ_Q^2.

By the Delta method, one has for Eq. (8.21),

$$\sigma_r^2 := \mathrm{Var}[r] = \frac{1}{M} \frac{\sum_i a_i b_i + \left(\sum_i a_i b_i\right)^2 - \sum_i a_i b_i (a_i + b_i)}{\left(1 - \sum_i a_i b_i\right)^2}, \tag{8.25}$$

under the null hypothesis of Eq. (8.22)[2]. Similarly for Eq. (8.23) under the null hypothesis of Eq. (8.22), one has

$$\sigma_Q^2 := \mathrm{Var}[Q] = \frac{1}{M} \left[\sum_i a_i b_i + \left(\sum_i a_i b_i\right)^2 - \sum_i a_i b_i (a_i + b_i) \right]. \tag{8.26}$$

As an exercise, in the example of Fig. 8.1, one has $M = 1958$, $K = 4$ and we obtain

$$r = 0.6214, \tag{8.27}$$

for which Eq. (8.25) yields under the null hypothesis $r = 0$

$$\sigma_r = 0.0144, \tag{8.28}$$

so that we observe that $r \sim 40\sigma_r$. Similar values can be obtain by the Jackknife re-sampling method.

Similarly, one can calculate $Q = 0.4301$ and $\sigma_Q = 0.0100$ under the null hypothesis of $Q = 0$.

Degree correlation

One can understand assortative mixing by using the simplest attributes of nodes derived from network structure, namely degree. This is equivalent to considering the correlation

---

[2] The corresponding formula in Newman (2003a) has a typo or error.

between degrees for pairs of nodes, with each pair corresponding to a link. For simplicity, we consider an undirected graph, but this can easily be extended to the directed case.

Since the degrees at either end of a link is trivially more than one, what should be considered is the excess degrees, explained in Section 8.1.2. Moreover, note that a degree is not a nominal variable that tells if a link is present or not, but an ordinal variable. Hence, let us consider instead of Eq. (8.21),

$$\langle i\,j \rangle - \langle i \rangle \langle j \rangle := \sum_{ij} ij\,(e_{ij} - q_i\,q_j), \tag{8.29}$$

where e_{ij} represents the ratio of the links with the pair of excess degrees (i, j). e_{ij} is symmetric and satisfies Eq. (8.18).

Define

$$q_i = \sum_j e_{ij} = \sum_j e_{ji}, \tag{8.30}$$

and maximize Eq. (8.29) under the constraints of Eq. (8.18) and Eq. (8.30) to see that $e_{ij} = q_i \delta_{ij}$, where δ_{ij} is the Kronecker symbol. To normalize Eq. (8.29), therefore, define **degree correlation** by

$$r := \frac{1}{\sigma_q^2} \sum_{ij} ij\,(e_{ij} - q_i\,q_j), \tag{8.31}$$

where

$$\sigma_q^2 := \sum_k k^2 q_k - \left(\sum_k k q_k \right)^2. \tag{8.32}$$

Note that Eq. (8.30) is equal to the distribution, $q(i)$, in the excess degree distribution of Eq. (8.12). σ_q^2 is equal to the variance for the distribution.

It is easy to show that by denoting the excess degrees at either end of each link a as (j_a, k_a), Eq. (8.31) can be rewritten as

$$r = \frac{\frac{1}{M} \sum_a j_a k_b - \left[\frac{1}{M} \sum_a \frac{1}{2}(i_a + j_a) \right]^2}{\frac{1}{M} \sum_a \frac{1}{2}(i_a^2 + j_b^2) - \left[\frac{1}{M} \sum_a \frac{1}{2}(i_a + j_a) \right]^2}. \tag{8.33}$$

Here the summation $\sum_a$ is over all the links.

Since the excess degree is simply equal to degree *minus* 1, one can have exactly the same formula interpreting j_a and k_a simply as corresponding degrees. For each link, consider two copies of pairs of degrees, namely (i,j) and (j,i) (in the undirected case) and a scatter plot. The Pearson correlation coefficient for this is exactly equal to Eq. (8.33).

One can use the Jackknife re-sampling method to evaluate the statistical error for the degree correlation. Delete a link in turn to make M copies of modified graphs, and calculate the degree correlation for each copy as r_i. The statistical error for r can then be estimated by $\sigma_r^2 = \sum_{i=1}^{M}(r_i - r)^2$.

Clustering coefficient

A triplet of nodes can form a clique, mutually connected three nodes or a triangle. Social networks of friendship and acquaintance, especially, have such ternary relationships. A **clustering coefficient** measures precisely the abundance of such triangles in the network. The global definition of clustering coefficient is given by

$$C := \frac{3 \times (\text{number of triangles})}{\text{number of connected triples}}. \tag{8.34}$$

Here a connected triple is a single node with links going to an unordered pair of others. The number of triangles is calculated in the entire network. Since a triangle has exactly three connected triples, the factor of three in the numerator is a normalization so that $0 \leq C \leq 1$. C represents the probability that a connected triple is closed as a triangle, namely that one's two "friends" are actually friends of each other too.

As easily understood, Eq. (8.34) can be written as

$$C = \frac{6 \times (\text{number of triangles})}{\text{number of paths of length 2}}. \tag{8.35}$$

Here a path of length 2 means a directed path starting from a specific node. The factor of 6 in the numerator follows from the fact that a triangle has exactly 6 paths of length 2. This expression of C can be interpreted as the probability that one's friend's friend is actually one's friend.

On the other hand, the local clustering coefficient is defined for each node i as

$$C_i := \frac{\text{number of triangles connected to node } i}{\text{number of triples centered on node } i}. \tag{8.36}$$

The denominator is the number of all pairs of neighbouring nodes of i, and the numerator is the number of triangles present in the pairs of neighbours and the node i. Let the degree of i be k_i, the denominator is equal to $k_i(k_i-1)/2$. It is conventionally defined that $C_i = 0$ for the cases of $k_i = 0, 1$. The average clustering coefficient is then defined by

$$C_{\text{local}} := \frac{1}{N} \sum_i C_i \tag{8.37}$$

The values of C and C_i may differ depending on the network under study. Note that the two definitions reverse the order of two operations; taking the ratio of triangles to triples and averaging over all nodes. C_i tends to weight for low-degree nodes more heavily than high-degree ones, because of the denominator in the definition of Eq. (8.36).

C_i is easy to calculate by the definition. The global C can be calculated by the following algorithm of the local graph-search. Here, it is assumed that nodes are ordered by labels of integers or numbers so that one can compare nodes numerically by its labels.

Algorithm 8.6: Clustering global (G)

```
1   num_triangles ← 0   // 3* (number of triangles)
2   num_triples ← 0   // number of connected triples
3   for each node u of V[G]
4      k ← Degree (u,G)
5      num_triples ← num_triples + k*(k−1)/2
6      for each node v in Adj[u]   // Adj: adjacent list
7         for each node w in Adj[u]
8            if v < w and v is adjacent to w
9               num_triangles ← num_triangles + 1
10  C ← num_triangles / num_triples
11  return C
```

For the random graph of the configuration model, let us calculate the evaluation of the global clustering coefficient as follows. Choose two nodes i and j from the adjacent list of nodes for a node v. Let the excess degrees of i and j be k_i and k_j respectively. Consider a connected triple of v, i, j. The probability that one of the excess links of k_i goes to j is $k_j/(2M)$, where M is the total number of links. It follows that the probability that i and j is connected is given by the ensemble average of the quantity, $k_i k_j/(2M)$.

By using the fact that $2M = N \langle k \rangle$ (N is the total number of nodes), we have

$$C = \frac{\langle k_i k_j \rangle_q}{N \langle k \rangle}, \tag{8.38}$$

where the ensemble average in the numerator is taken for the distribution of excess degree, Eq. (8.12). Using the calculation:

$$\sum_k k q_k = \frac{1}{\langle k \rangle} \sum_{k=0}^{\infty} k(k+1) p(k+1) = \frac{1}{\langle k \rangle} \sum_{k=1}^{\infty} (k-1) k \, p(k) = \frac{\langle k^2 \rangle - \langle k \rangle}{\langle k \rangle}, \tag{8.39}$$

it follows that

$$C = \frac{\langle k \rangle}{N} \left[\frac{\langle k^2 \rangle - \langle k \rangle}{\langle k \rangle^2} \right]^2. \tag{8.40}$$

This expression should be used for economic networks with fat-tailed degree distributions.

It should be noted that in the case of a Poisson random graph, by inserting Eq. (8.10) and Eq. (8.11) into Eq. (8.40), we have

$$C = \frac{\langle k \rangle}{N}. \tag{8.41}$$

The probability for a link being present for any pair of nodes is independently given by $\langle k \rangle / N$, which is simply equal to the clustering coefficient. For large N, $C \sim O(N^{-1})$.

Note that in the case of fat-tailed degree distribution, the behavior is completely different. Since the variance is much larger than the square of average degree, C is much larger than that given by the class of Poisson random graph.

Suppose that the degree distribution follows a power-law for $k \gg 1$. Note that the moments of degree depend on the system size N as

$$\langle k^m \rangle \propto \begin{cases} k_{\max}^{m-(\alpha-1)} & m > \alpha - 1, \\ O(1) & m < \alpha - 1 \end{cases} \tag{8.42}$$

as easily shown from the definition of moments. Here $k_{\max}$ means the maximum degree in the network. From the definition of cumulative distributions, we have

$$\bar{F}(k_{\max}) \sim \frac{1}{N}, \tag{8.43}$$

which leads to the dependency, $k_{\max} \sim N^{1/(\alpha-1)}$. By inserting this into Eq. (8.42) and using the moments for $m = 1, 2$ in Eq. (8.40), it follows that

$$C \propto \frac{1}{N} k_{\max}^{6-2\alpha} \propto N^{-\frac{3\alpha-7}{\alpha-1}}. \tag{8.44}$$

From this evaluation, for large N, C becomes asymptotically zero for $\alpha > 7/3$. For $\alpha < 7/3$, C becomes unbounded, which means that $C \sim O(1)$ is the breakdown of the argument based on random graph.

8.2 How to Identify Communities

Community structure provides us with an alternative point of view of networks (Girvan and Newman, 2002; Fortunato, 2010; Newman, 2012). Real networks are often divided into densely connected subgroups and sparse parts bridging them. It is well expected that such communities will possess different statistics from the averaged properties of the networks. Extracting communities thus provides one with a coarse-grained picture of large-scale networks and also serves as a data mining tool for valuable information hidden in a huge amount of data.

A wide variety of methods have been proposed to identify the community structure of a network computationally (Fortunato, 2010; Newman, 2012). One way is to introduce a quality function to evaluate the density of connections of each group for a given partition. The modularity Q defined by Eq. (8.23) is the most common choice among researchers. It measures the fraction of links within given communities of a network with reference to the expected fraction of the intralinks for the corresponding network randomized with the degree of each node preserved. For the modularity of weighted networks, Eq. (8.23) may be generalized by replacing the unit weight of each link by its actual weight (Newman, 2004a); for instance, the degree of each node is interpreted as the total sum of the weights associated with its connecting links.

Once the modularity has been accepted for community detection, the next thinking step is how to find the optimum decomposition of nodes by maximizing the modularity. However, we immediately encounter a serious computational problem. The number of ways to divide a network into pieces is given by the sum of the Stirling number of the second kind, so that the computational complexity dramatically increases with the network size. The true optimization of Q over all possible divisions would be actually intractable even for such a small-size network composed of several tens of nodes.

We, therefore, have to adopt approximate heuristic methods. Various optimization algorithms have been proposed so far. The readers can consult Fortunato (2010); Newman (2012) for details of each method together with its pros and cons. It should be noted that modularity has a number of local maxima as illustrated in Fig. 8.9. One of the important methodological issues is thus how to get rid of such a pitfall in carrying out the optimization. Computational efficiency is another important issue especially when dealing with large-scale networks.

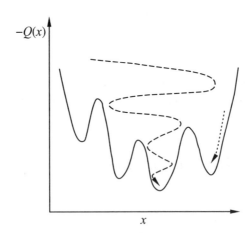

Fig. 8.9 Schematic representation of the modularity landscape, where x represents all possible partitions of nodes in a network. The dashed line depicts a tracking path of the modularity optimization based on simulated annealing; the dotted line, a steepest descent path.

Here, we present our practical experiences as shown in Table 8.2, where the greedy (Newman, 2004b; Clauset et al., 2004), the bisectional (Iino et al., 2010), and the fast unfolding (Blondel et al., 2008) methods are applied to an interfirm network in Japan. The greedy algorithm assumes that every node has its own community at the outset. Communities are then constructed gradually from bottom to top by merging a pair of communities with the largest increment of modularity at every step. Although the greedy calculation has modest speed among the three methods, it gives the worst result. This is because the greedy algorithm shares the same characteristics as the steepest descent method (see Fig. 8.9), which has no means to escape from a local maximum/minimum once it is trapped. Alternatively, the bisection method partitions nodes into two groups in a stepwise fashion from top to bottom. Simulated annealing is also built in to reinforce the optimum search in modularity at each step. In return, it requires significantly more computational time than the greedy algorithm. The fast unfolding method is a variant of the greedy algorithm. Implementation of a hierarchical scheme into the original method dramatically speeds up its computational speed and provides much better results. However, since it is stochastic in nature through sequential selection of nodes for possible community formation, the computation should be repeated using different series of random numbers to reach the optimum community structure as closely as possible.

Table 8.2 Performance of typical modularity optimization methods when applied to an interfirm network consisting of 773,670 nodes and 3,192,582 links. The computational time was measured on a personal computer with Core 2 Quad 3.0 GHz. The fast unfolding shows the best result over 1000 samples; their average and standard deviation is 0.640 and 0.008, respectively.

	Optimization Method		
	Greedy	Bisection	Fast Unfolding
Q (Number of communities)	0.552 (6543)	0.651 (120)	0.661 (165)
Computational time	467 min	1664 min	2 min

Finally, we demonstrate the power of community analysis. Figure 8.10 depicts the result of community detection for the karate club network (Zachary, 1977) at $Q = 0.42$, obtained by the fast unfolding method. The club eventually split into two groups after a dispute between the club president and the lead instructor. The division of the club is already indicated by the community structure of the network; the real splitting is exactly the same as that predicted by the community analysis. In this case, the optimization of modularity is crucial. If the greedy method was adopted for community detection, e.g., calculation by Mathematica as shown in Appendix A.4, the splitting of the club would not be predicted correctly; in fact, the optimized Q value is 0.38 with three communities instead of four.

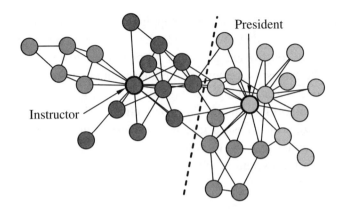

Fig. 8.10 Community structure of the karate club network (Zachary, 1977). The club eventually split into two groups after a dispute between the club president and the lead instructor, as shown by the dashed line.

8.3 Visualization

It is very difficult to analyze all connections in real networks because of their multiplicity and complexity. Visualization is another useful tool available to illuminate the structural properties of networks. Of course, visualization is not good at everything, but it greatly helps to explore the complicated structures of networks by stimulating one's intuition.

Various algorithms have been developed to visualize networks (Battista et al., 1998) in 2D or 3D physical space. Here we sketch the visualization method based on a spring–charge model. In this model, pairs of nodes with direct links are connected with springs and any pair of nodes repels each other through a repulsive Coulomb force (Hu, 2006). The total potential of such a virtual physical system is given as

$$U = \sum_{<i,j>} \left(\frac{1}{2} k_{ij} r_{ij}^2 + \frac{q_i q_j}{r_{ij}} \right), \quad (8.45)$$

where the sum is taken over all pairs of nodes, r_{ij} is the physical distance between nodes i and j, q_i is a Coulomb charge of node i, and k_{ij} is a spring constant between nodes i and j which are directly connected ($k_{ij} = 0$ for unlinked pairs). The attractive force owing to the spring keeps tightly connected nodes close in a space. On the other hand, the repulsive Coulomb force tends to distribute nodes uniformly over the available space and prevent entanglement of the network. If we assume that q_i and k_{ij} take identical values for every node and every pair, q and k, respectively, Eq. (8.45) can be written in a scaled form as

$$U = \left(k q^4 \right)^{1/3} \sum_{<i,j>} \left(\frac{1}{2} R_{ij}^2 + \frac{1}{R_{ij}} \right), \quad (8.46)$$

with

$$R_{ij} = \left(\frac{k}{q^2}\right)^{1/3} r_{ij}. \tag{8.47}$$

In this case, one is required to change the functional form of the potential if one likes to have an essentially different layout of a network.

Experimentally, a physical system has a well-ordered structure in the lowest energy state, as demonstrated by the crystals of materials. We thereby expect that the ground state of the spring–charge system would be a leading candidate for the optimized configuration of nodes in a given network. The molecular dynamics (MD) method (Allen and Tildesley, 1987; Frenkel and Smit, 2002) works well to reproduce crystallization of materials. One can achieve the minimization of the total energy by gradually cooling a system with any initial configuration; precisely speaking, the final result is nearly the best solution, a crystalline state with some defects, in general. Figure 8.12 shows the optimized layout of the network thus obtained in Fig. 8.11.

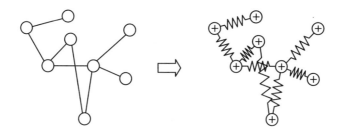

Fig. 8.11 Spring–charge model for network visualization.

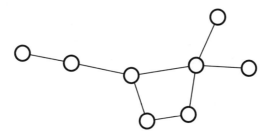

Fig. 8.12 Optimized layout of the network in Fig. 8.11.

The MD simulation of a system consisting of charged particles of the order of a million or more is not a simple calculation, because of the long-range nature of the Coulomb force. The computational task is of the order of N^2 without any skillful algorithms. To speed up the computation, one may implement the hierarchical tree algorithm (Barnes and Hut, 1986b; Pfalzner, 1996) in one's MD code. The whole space is divided into cells at hierarchical levels as shown in Fig. 8.13. The interactions of a given particle with other remote particles are replaced by the interactions with the cells to which those particles belong. As the distance between particles is larger, cells at lower levels are used. On the other hand, the force on the particle is calculated directly for its neighbors residing in the same cell or in adjacent cells at the highest level. This trick reduces the computational task to $O(N \log N)$.

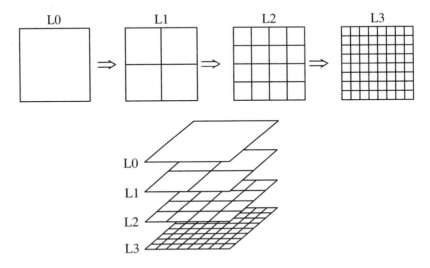

Fig. 8.13 Hierarchical domain decomposition. In this example, the whole space (L0 level) is decomposed into three hierarchical levels (L1, L2, L3).

In addition, we can use the hierarchical domain decomposition for coarse-graining of a network. This is because the optimized layout embedded in physical space should reflect the connectivity of a network; the closer the nodes are, the more they are related to each other. The coarse-graining process is illustrated in Fig. 8.14. We applied the process to the power grid network to illuminate backbone structures of the network at different resolution scales. The results are shown in Fig. 8.15 together with the optimized layout of the original network.

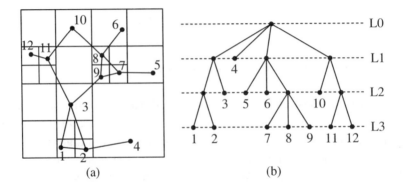

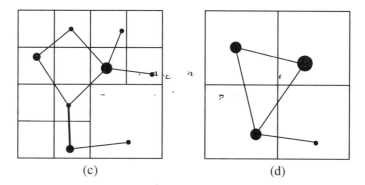

Fig. 8.14 Illustrative example of the coarse-graining process of a network based on the hierarchical domain decomposition in Fig. 8.13. Panel (a) is an original network with four-level domain decomposition (L0, L1, L2, L3); (b), a tree diagram corresponding to (a); (c), the coarse-grained network at the L2 level; and (d), the coarse-grained network at the L1 level.

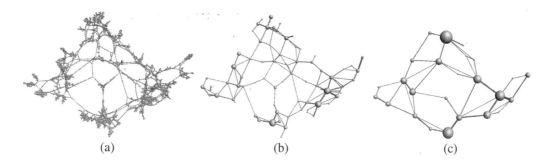

Fig. 8.15 Application of the coarse-graining procedure, illustrated in Fig. 8.14, to the power grid network. Panel (a) shows the original network in 3D space; (b), the coarse-grained network at $8 \times 8 \times 8$ decomposition level (L3) of the whole space (L0); and (c), the coarse-grained network at $4 \times 4 \times 4$ decomposition level (L2).

8.4 Nationwide Production Network

Engine of a real economy is the entire line of economic activities of firms; firms purchase intermediate goods and services from "upstream" firms, value add them, and sell the goods and services to "downstream" firms. The net sum of value added in the whole network is basically the net total production in a nation, that is, gross domestic product (GDP). The entire line of these processes of putting value added in turn forms a giant web of production ranging from upstream to downstream, down to consumers. A real economy has its engine in such a network of production as well as fuels of labor and financing. Let us call such a network the production network or supplier–customer network.

It has been a formidable task to study the structure of the production network owing to the lack of empirical data. Recently, however, such data on a nationwide scale become available in Japan thanks to credit research agencies who have been collecting and accumulating relevant information including who are suppliers and customers of *individual* firms so as to measure the potential risk to a firm's sales, profit, and production.

A supplier–customer link $i \to j$ is defined, if firm i is a supplier to another firm j, or equivalently, j is a customer of i. While it is difficult to record every transaction of supply and purchase among firms, it is also pointless to have a record that a firm buys a pencil from another. Necessary for a study of the structure of production network are data of links such that the relationship $i \to j$ is crucial for the business activities of either end or both of i and j.

Empirical data have been gathered and compiled on this idea; if a firm nominates a certain number of its suppliers and customers as crucial ones, then those links are recorded in a database. It is assumed that the links playing important roles in the production network are recorded at either end of each link as we describe earlier, while we should understand that it is possible to drop relatively unimportant links from the data. Although amounts of transactions provide information of weights on links, such data are only partially available at the moment, so let us simply ignore the weights in the sequel.

Our data for a production network is based on a survey done by Tokyo Shoko Research (TSR), one of the leading credit research agencies in Tokyo, with the Research Institute of Economy, Trade and Industry (RIETI). This is, to our knowledge, the largest data available for an empirical study of production network. In fact, the directed network constructed from the database comprises a million firms and millions of supplier–customer relationships among them. The data were compiled in July 2011.

It turns out that the directed network has a largest or giant strongly connected component, explained in Section 8.1.1, including most of the firms contained in the data set (see Section 9.1.2 for the details). The component can be identified by an application of depth-first search, given in Algorithm 8.1. Denoting the number of nodes and links by N and M respectively, we have

$$N = 986,185, \qquad (8.48)$$

$$M = 4,402,270. \qquad (8.49)$$

In this section, we shall apply to this network the tools explained in Section 8.1.

8.4.1 Large-scale structure of the production network

Diameter and average distance

Application of the breadth-first search, the algorithm which was given in Algorithm 8.5, shows that the diameter of a graph is the maximum length for all ordered pairs (i, j) of the shortest path from i to j. The average distance is the average length for all those pairs (i,j). We found that

$$\text{Diameter} = 22, \tag{8.50}$$

$$\text{Average distance} = 4.6, \tag{8.51}$$

which indicates the "small-world" property of the production network, as often encountered in many social networks.

Degree distributions

A firm has suppliers for and customers of it, whose numbers are in-degree and out-degree, respectively. The empirical data show that the distributions for in-degree and out-degree are fat-tailed. See Figure 1 in Fujiwara and Aoyama (2010).

Both for in-degree and out-degrees of a firm, the distribution has a heavy tail that can be characterized by a power-law, given by Eq. (8.14). We estimated the exponent μ by maximum likelihood (MLE), i.e., the Hill's estimate, in a tail region $k > k_0$. In Figure 1 in Fujiwara and Aoyama (2010), the estimates are shown for $k_0 = 40$, namely $\mu = 1.35 \pm 0.02$ for in-degree and $\mu = 1.26 \pm 0.02$ for out-degree, by solid lines. Here the errors correspond to 1.96σ (99% significance level) of the estimated standard errors σ.

The first two moments of in- and out-degree are

$$\langle k_{\text{in}} \rangle \equiv \langle k_{\text{out}} \rangle = 4.003, \tag{8.52}$$

$$\langle k_{\text{in}}^2 \rangle = 1.041 \times 10^3, \quad \langle k_{\text{out}}^2 \rangle = 1.036 \times 10^3. \tag{8.53}$$

When viewed as an undirected graph, we have

$$\langle k \rangle = 2M/N = 8.006, \tag{8.54}$$

$$\langle k^2 \rangle = 3.070 \times 10^3. \tag{8.55}$$

Firms with largest in-degrees belong to the sectors of manufacturing and construction among others, including heavy industry, electrical machinery (e.g., Hitachi, Mitsubishi, Panasonic, Toshiba), automobiles (Toyota, Nissan, Honda), metal production, and so on. Large construction companies are also included. Firms with the largest out-degrees are worldwide traders, distributors of construction-related materials, metals, petroleum, mechanical and electrical instruments, and general wholesale companies, as well as the manufacturing firms mentioned for in-degrees.

Correlation with firm-size

A large firm is likely to possess a large number of suppliers to purchase various intermediate goods and services; similarly, it often has many customers to sell its products. Firm-size can be measured in many ways; by stock variables such as total–assets, number of employees, and so forth, or by flow variables of sales, profits, and so on.

The firm-size, however measured, obeys a power-law, being well known as a Zipf's law. For the nodes in the network, we examined financial data. The cumulative distribution for the sales of those nodes (0.73 million) is shown in Figure 2 of Fujiwara and Aoyama (2010). Zipf's law, $\bar{F}(x) \propto x^{-\alpha}$, is obvious for sales x. The exponent is close to unity, $\alpha = 0.96 \pm 0.02$ by MLE estimated for $x > 10^4$ million yen.

Figure 2 in Fujiwara and Aoyama (2010) depicts a scatter-plot for total degree and sales. Correlation between the firm's degree and size is positive. The statistical significance can be quantified by non-parametric statistics, such as Kendall's rank correlation, τ. For the data, $\tau = 0.391$ (p-value $< 10^{-7}$), which shows a significant positive correlation between the degree and firm-size. We used different quantities for firm-size, such as profits and the number of employees, and obtained very similar results. In addition, when considering either in- or out-degree, we can observe that each has a positive correlation with firm-size.

Clustering coefficients

Unlike many social networks, the supplier of a firm's supplier is not likely also to be the firm's supplier, and similarly for customers, because such a process of production is redundant for most cases. Let us regard the network as an undirected graph, and apply Eqs. 8.34 and 8.37, and Algorithm 8.6.

The global clustering coefficient, defined by Eq. (8.34) or Eq. (8.35), is the mean probability that two firms who have a common supplier/customer are also suppliers/customers of each other. The undirected graph of our data set yields

$$C = 1.87 \times 10^{-3} = 0.187\%. \tag{8.56}$$

To compare this value with that for a class of random graphs having the same degree sequence but randomly rewired links, we use the expected value of global clustering coefficient given by Eq. (8.40).

Putting the values Eq. (8.54), Eq. (8.55) and N into Eq. (8.40), we have $C = 1.81 \times 10^{-2}$. The observed value Eq. (8.56) is, therefore, merely 10% of Eq. (8.40), and shows weaker transitivity than what is expected by chance. This is reasonable because triangular relations, during the selection of suppliers and customers, are suppressed in the formation of them.

The average of the local clustering coefficient is, on the other hand, calculated as

$$C_{\text{local}} = 4.58\%. \tag{8.57}$$

for the same data set by using Eq. (8.37) and Algorithm 8.6.

Degree correlation

For each node, the in-degree and out-degree are highly correlated. This is consistent with what we saw earlier that each quantity has positive correlation with firm-size.

For each link, to see the assortative mixing with respect to degrees (k_1, k_2) at both end of each link, or degree correlation, let us examine the joint distribution for (k_1, k_2). Here we ignore the direction of links, but even when taking the possible four combinations of in/out at a directed link, we obtain similar results. To test for the assortativity, we calculate the frequency $F(k_1, k_2)$ that the pair of k_1 and k_2 appears at either end of a link in the network. Then compare it with a same quantity $F_r(k_1, k_2)$ that is obtained in a randomized network with the same degree sequence. We generated 1,000 randomized networks, and quantify as the ratio F/F_r, where F_r is the average for the randomizations.

The result is shown in Figure 3 in Fujiwara and Aoyama (2010). One can observe that large-degree nodes, large firms, are connected with small-degree nodes, small firms. Hubs have a large number of suppliers and customers, but is similar to firms with intermediate size, displaying disassortativity. This can be quantified by the Pearson correlation coefficient r for (k_1, k_2). For the data, we have

$$r = -0.0747 \pm 0.0002, \tag{8.58}$$

where the error is calculated by the method given in Section 8.1.3. This indicates that r is negative with a statistical significance.

Community structure

The global connectivity, which will be examined in Section 9.1.2, shows that basically all industries are highly entangled with each other within the weakly or strongly connected component. Yet connectivity alone does not tell how dense or sparse the stream of production is distributed depending on industrial or geographical groups. Detection of community structure is to find how nodes cluster into tightly-knit groups with high density in intra-groups and lower connectivity in inter-groups.

To illustrate such a community structure, let us focus on the manufacturing sector with 0.12 million firms, in order to understand the sector's modular structure by excluding other dominant sectors including wholesale and retail trade, which obviously have a different role in the stream of production from the core of the manufacturing sector.

We use the method of maximizing modularity, explained in Eq. (8.23) and Section 8.2 and implemented in a study on a large-scale SNS (Yuta et al., 2007). While considerable studies have been conducted to develop various methods for community extraction (see Fortunato (2010) for a review), we use the modularity optimization for its clear interpretation in terms of statistical hypothesis.

As shown in Fortunato and Barthelemy (2007), however, the method can give undesired grouping, depending on the density of connections and the network size. Especially, large communities can potentially contain sub-communities. Currently, without an established method to avoid this problem of resolution limit (see Fortunato (2010) and references therein), we check the structure of detected communities by constraining the modularity optimization on each single community, especially for those with relatively large community size.

We apply the method of community extraction to the undirected subgraph whose nodes consist of only firms in the manufacturing sector. The resulting modularity is

$$Q = 0.566 \pm 0.001, \tag{8.59}$$

which indicates strong community structure (the error calculated by the Jackknife method is given by Eq. (8.24)).

The number of extracted communities exceeds a thousand, whose sizes range from a few to more than 10,000. From the database of the information on the firms, we found that many of those small communities are each located in the same geographical areas forming specialized production flows. An example is a small group of flour-maker, noodle-foods producers, bakeries, and packing/labeling companies in a rural area.

On the other hand, five large communities exceed 10,000 each in size, being possibly subject to the problem of resolution. After checking the sub-communities in the aforemntioned fashion, we obtained the communities in a hierarchy as shown in Table III in Fujiwara and Aoyama (2010).

The basic reason for the presence of such a community structure can be considered as the following, for the case of the manufacturing sector.

It turns out that large firms in a same community do not form a set of nodes that are mutually linked in nearly all possible ways, or a quasi-clique. Rather, with their suppliers and customers, they form a quasi-clique in a corresponding bipartite graph as follows. A supplier–customer link $i \to j$ for a set of nodes V ($i, j \in V$) can be considered as an edge in a bipartite graph that has exactly two copies of V as V_1 and V_2 ($i \in V_1$ and $j \in V_2$).

Large and competing firms quite often share a set of suppliers to some extent, depending on the industrial sectors, geographical locations and so on. For example, Honda (j_1), Nissan (j_2) and Toyota (j_3) possibly have a number of suppliers i of mechanical parts, electronic devices, chassis and assembling machines, etc., in common. Then the links form a clique or a quasi-clique in the bipartite graph, where most possible links from i to j_1, j_2, j_3, ... are present. This forms a portion in the original graph with a higher density than other portions. By enumerating cliques in the bipartite graph and examining them, we can see that this is actually the case for the communities extracted in the manufacuturing sector. See Fujiwara and Aoyama (2010) for more details.

8.4.2 Visualization of directed graph of production network

As explained in Section 8.3, one can use a physical simulation such as a force-directed graph drawing method to visualize large-scale networks. In addition to the molecular dynamics, one could employ an N-body simulation (Makino and Taiji, 1998; Kawai and Fukushige, 2006) with the forces of anti-gravity or Coulomb interactions among the nodes and springs of the links to visualize the connectivity, and the Barnes–Hut algorithm (Barnes and Hut. 1986a) with computational cost of $O(N \log N)$, N being the number of nodes (see (Fujiwara, 2009); (Fujita et al., 2016) as well as the closely related but different method explained by Iino and Iyetomi (2015)). Initial configuration of the nodes is

Complex Networks, Community Analysis, Visualization

determined by the Brandes–Pich's method (Brandes and Pich, 2007) of multi-dimensional scaling Borg and Groenen (2005).

In particular, Fujita et al. (2016) captures the upstream and downstream parts of the production network by its method of utilizing the direction of each link in the graph drawing.

The results are depicted in Fig. 8.16 and Fig. 8.17.

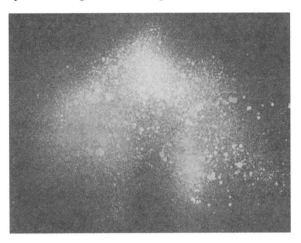

Fig. 8.16 Visualization of directed network of the production network in Japan, comprising a million firms as points. Colors of the points indicate industrial sectors of the firms. Note that upstream and downstream are depicted in the bottom and top sides respectively (from Fujita et al. (2016)) [see Color Plate at the end of book].

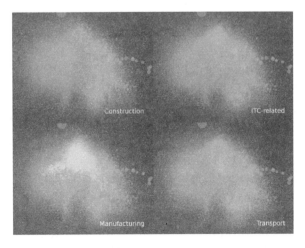

Fig. 8.17 Industrial sectors are identified for construction, ICT-related, manufacturing, and transport of automobiles (from Fujita et al. (2016)) [see Color Plate at the end of book].

8.5 Stock Correlation Network

In this section, we present a new way to delve into correlations in a stock market by casting them onto a network (Yoshikawa et al., 2011, 2012a,b, 2013). Such a network theoretic approach can be traced back to the seminal work of Mantegna (1999).

8.5.1 Group correlations

Here we revisit the TSE data as used in drawing Fig. 5.6. The correlation matrix C may be decomposed (Plerou et al., 2002b; Utsugi et al., 2004; Kim and Jeong, 2005) into three parts:

$$C = C_{\text{market}} + C_{\text{group}} + C_{\text{random}}$$

$$= \lambda^{(1)} \mathbf{V}^{(1)} \mathbf{V}^{(1)\text{t}} + \sum_{n=2}^{N_c} \lambda^{(n)} \mathbf{V}^{(n)} \mathbf{V}^{(n)\text{t}} + \sum_{n=N_c+1}^{N} \lambda^{(n)} \mathbf{V}^{(n)} \mathbf{V}^{(n)\text{t}}, \tag{8.60}$$

Table 8.3 Polarization of stocks at sector classification level in $\mathbf{V}^{(2)}$, $\mathbf{V}^{(3)}$ and $\mathbf{V}^{(4)}$. The figures in parentheses against each sector represents the number of stocks belonging to it. Values of the polarization satisfying $|P_s| \geq 2/3$ are highlighted in **boldface** just for guidance (Sectors with less than 5 stocks are excluded). (From Yoshikawa et al. (2012b))

Sector	Mode 2	Mode 3	Mode 4
Electric Appliances (72)	**−0.97**	−0.21	**0.92**
Chemicals (60)	0.10	0.19	0.04
Machinery (54)	−0.34	0.61	0.47
Foods (32)	**1.00**	−0.46	−0.12
Construction (31)	**0.71**	**0.82**	**−0.80**
Transportation Equipment (31)	−0.14	−0.29	−0.39
Wholesale Trade (21)	0.25	**0.71**	0.13
Banks (20)	−0.24	**−1.00**	**−1.00**
Iron & Steel (20)	0.45	**0.99**	0.35
Retail Trade (17)	−0.47	0.18	**−0.73**
Textiles & Apparels (17)	**0.94**	0.27	−0.27
Land Transportation (17)	**0.81**	**−0.97**	**−0.82**
Glass & Ceramics Products (15)	**0.68**	−0.36	**0.98**
Other Products (14)	−0.51	0.00	−0.47
Nonferrous Metals (14)	−0.04	0.22	**0.98**
Pharmaceutical (13)	0.62	−0.61	−0.62
Precision Instruments (13)	**−0.99**	−0.08	**0.71**

Contd...

Sector	Mode 2	Mode 3	Mode 4
Electric Power & Gas (13)	**1.00**	**−1.00**	**−1.00**
Information & Communication (10)	**−0.90**	**−1.00**	0.63
Other Financing Business (9)	−0.61	−0.52	−0.56
Metal Products (8)	0.48	0.08	−0.17
Rubber Products (7)	0.47	−0.03	−0.22
Services (6)	0.04	0.04	−0.35
Securities & Commodity Futures (6)	**−1.00**	**0.95**	**1.00**
Real Estate (6)	**0.98**	0.45	0.47
Insurance (6)	0.11	**−0.97**	**−1.00**
Marine Transportation (5)	**1.00**	0.30	0.49
Warehousing & Harbor Transportation Services (5)	**1.00**	−0.37	−0.12
Pulp & Paper (5)	**1.00**	−0.37	−0.36
Fishery, Agriculture & Forestry (4)	1.00	−1.00	0.66
Oil & Coal Products (3)	1.00	−0.43	−0.31
Mining (2)	1.00	1.00	0.27
Air Transportation (1)	1.00	−1.00	−1.00

where the same notations as used in Eq. (5.13) are adopted. The first term in Eq. (8.60) singles out the contribution of the collective motion of the market (market mode). The second term collects the contributions of group correlations among stocks. The last term is regarded as random noises involved in the data according to the RMT; that is, N_c is determined by the following condition,

$$\lambda^{(N_c)} < \lambda_+ < \lambda^{(N_c+1)}, \tag{8.61}$$

where λ_+ is defined by Eq. (5.104). In the TSE data with $Q \simeq 4.86$ and hence $\lambda_+ \simeq 2.11$, $13 (= N_c)$ statistically meaningful eigenmodes are detected. The largest eigenvalue $\lambda^{(1)}$ ($\simeq 132.95$) is one order of magnitude as large as the second largest eigenvalue $\lambda^{(2)}$ ($\simeq 15.48$). However, we focus on the group correlations here. This is analogous to observing the motion of particles in the center of mass frame; the market mode thus corresponds to the center-of-mass motion. This treatment of stock data was also adopted to study correlation structures in stock markets by Borghesi et al. (2007) and Musmeci et al. (2015).

It is well established (Plerou et al., 2002b; Utsugi et al., 2004; Kim and Jeong, 2005) that stocks belonging to the same sectors tend to appear simultaneously in the eigenvectors associated with the large eigenvalues. To quantify this collective behavior of stocks, we define the polarization $P_s^{(n)}$ of stocks in sector s in the nth eigenvector $\boldsymbol{V}^{(n)}$ as

$$P_s^{(n)} = \frac{\sum_{j \in s} V_j^{(n)}}{\sum_{j \in s} \left| V_j^{(n)} \right|}, \tag{8.62}$$

where $V_j^{(n)}$ represents the jth component of the $\boldsymbol{V}^{(n)}$ and by definition, $-1 \leq P_s^{(n)} \leq 1$. As observed in Table 8.3, most of the sectors form groups of comoving stocks, indicated by $\left|P_s^{(n)}\right| \simeq 1$, in some or all of the three dominant eigenvectors, $\boldsymbol{V}^{(2)}$, $\boldsymbol{V}^{(3)}$ and $\boldsymbol{V}^{(4)}$; all the stocks form one group in the market mode, $\boldsymbol{V}^{(1)}$.

8.5.2 Community detection

We construct a stock correlation network by regarding the group correlation matrix $\boldsymbol{C}_{\text{group}}$ as an adjacency matrix $\boldsymbol{A}$ for it. We then extract communities of comoving stocks. Assume that the network is partitioned into k communities and the community to which node i belongs is designated by index σ_i ($1 < \sigma_i < k$). Community decomposition is thereby represented by a set $\{\sigma\}$ of σ_is. To find an optimum decomposition of nodes, we adopt the following objective function, called *frustration* by Traag and Bruggeman (2009),

$$F(\{\sigma\}) = -\sum_{i,j} A_{ij}\delta(\sigma_i, \sigma_j), \tag{8.63}$$

where $A_{ij} = [\boldsymbol{C}_{\text{group}}]_{ij}$ and $\delta(\sigma_i, \sigma_j)$ denotes the Kronecker delta. Minimizing the frustration $F(\{\sigma\})$ thus means maximizing the sum of positive and negative weights within each community. The frustration is peculiar to networks with both positive and negative links, so it does not have a counterpart for networks with only positive links.

We use the simulated annealing method (see Fig. 8.9) to carry out the minimization of the frustration $F(\{\sigma\})$. To elucidate the correlation structure thus obtained, we calculated the polarization degree $P(\sigma, \sigma')$ of correlation within or between communities defined by

$$P(\sigma, \sigma') = \frac{\sum_{i \in \sigma, j \in \sigma'} [\boldsymbol{C}_{\text{group}}]_{ij}}{\sum_{i \in \sigma, j \in \sigma'} \left|[\boldsymbol{C}_{\text{group}}]_{ij}\right|}, \tag{8.64}$$

where $\sum$ sums up the weights of links within communities ($\sigma = \sigma'$) or across communities ($\sigma \neq \sigma'$). The computed results show that all the communities are full of positive weights and almost perfectly exclusive of negative weights. In contrast, the communities are interconnected to a large extent with negative links.

The stock correlation network is decomposed into four communities, named Comms. I, II, III and IV in the descending order regarding their size. The correlation structure thus obtained is depicted in Fig. 8.18. We find a triangular relationship of the communities competing against each other in the stock market. The most prominent conflicting relationship is observed among Comms. I, II and IV. There are nine sectors listed in the figure; more than 2/3 of stocks of those sectors are concentrated in a single community. We excluded small sectors consisting of less than five stocks. On the other hand, the stocks in the machinery sector, one of the leading sectors in the market, are divided into rather evenly three communities, I, II, and IV, forming the most conflicting triangle. Also the stocks in the construction sector, another major sector, is distributed over the three communities.

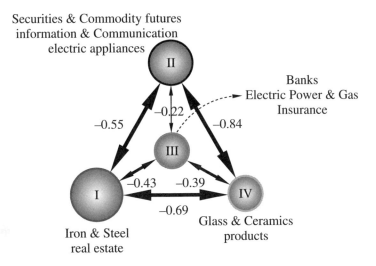

Fig. 8.18 Frustrated stock group structure in the Tokyo Stock Exchange. The communities are numbered in the order of their size; the community I is the largest one. The arrows pointing in both directions designate anti-correlation relationships between communities with the associated values of $P(\sigma, \sigma')$, Eq. (8.64), represented by their width. (From Yoshikawa et al. (2012b))

8.5.3 Frustrated correlation structure

The triangular relationship of competing communities has no stable static state, because the enemy of my enemy is my enemy, not my friend! Such a frustrated correlation structure among the stocks may give rise to complicated behavior of the market. A system of three antiferromagnetic spins on a triangle is its physical counterpart. In sociology, Heider (1944, 1946, 2013) is a pioneer who developed structural balance theory for human and country relations. Figure 8.19 shows four distinctive relations among three people. Two of them, as shown in the panels (a) and (c), are structurally stable, referred to as *balanced*. In contrast, the remaining two in the panels (b) and (d) are not stable at all, referred to as *unbalanced*. The relationship in the panel (c) is a social counterpart of the exotic structure embedded in the market. In Antal and Redner's paper (2006), the readers can find a figure depicting the evolutionary history of alliances in Europe from unbalanced to balanced structure, leading to the World War I.

To have a more comprehensive image on the frustration correlation structure among stocks, we introduce a group correlation state vector for each stock defined by

$$d_i = \begin{bmatrix} \sqrt{\lambda^{(2)}} V_i^{(2)} \\ \vdots \\ \sqrt{\lambda^{(N_c)}} V_i^{(N_c)} \end{bmatrix}. \quad (8.65)$$

Fig. 8.19 Triangular networks constructed with friendly (solid lines) or unfriendly (dashed lines) bilateral relations. Relationship among the three nodes is *balanced* in (a) and (b) and *unbalanced* in (b) and (c).

This is a vector in $(N_c - 1)$-dimensional space; note that $N_c = 13$ in the TSE. The group correlation coefficient between stocks i and j is calculated by taking the inner product of their correlation state vectors:

$$[C_{\text{group}}]_{i,j} = d_i \cdot d_j. \quad (8.66)$$

If $[C_{\text{group}}]_{i,j} < 0$, the two vectors have an obtuse angle. We cannot imagine how the correlation state vectors of stocks are arranged in a high-dimensional space, i.e., 12-dimensional space in the case of TSE. Fortunately, the stocks are clustered into four synchronizing communities and their centers of gravity is in a three-dimensional space. Figure 8.20 shows the correlation state vectors of all stocks projected onto the three-dimensional space thus determined. Certainly, it is visible that the stocks forms four communities and the centers of gravity of these communities are arranged at obtuse angles with each other.

We also applied the same analysis to S&P 500 data, collected during the period of January 2008 through December 2011. As shown in Figs. 8.21 and 8.22 corresponding to Figs. 8.18 and 8.20, respectively, the results are similar to those for the TSE data. However, sectors play a more important role in characterizing the communities in S&P 500. Also the combination of sectors forming the communities of comoving stocks is different between the two markets. For instance, the IT sector forms a rather independent community in S&P 500, while it is one of the core sectors in the second largest community in TSE. On the other hand, the financial sector occupies a relatively neutral position in TSE, while it is one of the core sectors in the largest community in S&P 500.

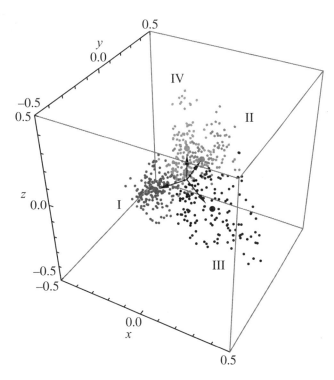

Fig. 8.20 Three-dimensional correlation state vectors of the stocks in the Tokyo Stock Exchange. The stocks in the community I are colored red, those in II are blue, those in III are green, and those in IV are brown. The arrows point to the positions of the center of gravity in individual communities [see Color Plate at the end of book].

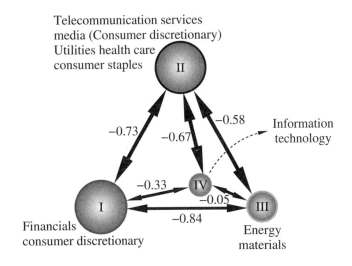

Fig. 8.21 Same as Fig. 8.18, but for the S&P 500. (From Yoshikawa et al. (2013))

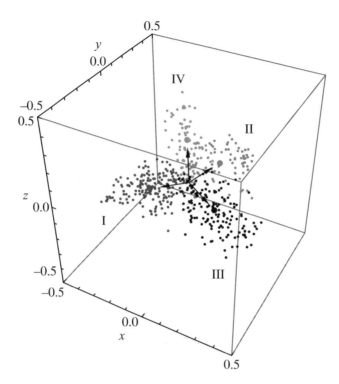

Fig. 8.22 Same as Fig. 8.20, but for the S&P 500 [see Color Plate at the end of book].

8.6 Globally-coupled Financial Networks

This section goes back to the globally-coupled financial networks which has been discussed as a case study of the CHPCA in Section 5.4. Here we extend the analysis to an identification of comoving nodes by applying the community detection technique to the networks (Vodenska et al., 2016).

We first recall the spectral decomposition of the complex correlation matrix $\tilde{C}$, Eq. (5.91). The noise elimination procedure is carried out on $\tilde{C}$ by retaining only statistically meaningful components in the decomposition; we assume there exist N_s such components. The filtered complex correlation coefficient between a pair of nodes α and β is thereby defined as

$$\tilde{C}_{\alpha\beta}^{(N_s)} = \sum_{n=1}^{N_s} \tilde{\lambda}^{(n)} \tilde{V}_\alpha^{(n)} \tilde{V}_\beta^{(n)\dagger} = w_{\alpha\beta}\, e^{i\theta_{\alpha\beta}}. \tag{8.67}$$

where $N_s = 6$ for the entire (1999–2012) and the third (2007–2012) periods, and $N_s = 5$ for the first (1999–2002) and second (2003–2006) periods. Also the magnitude $w_{\alpha\beta}$ and phase $\theta_{\alpha\beta}$ of the correlation coefficient are introduced in Eq. (8.67).

If we regard $\tilde{C}^{(N_s)}$ as an adjacency matrix, we can construct a network of the financial nodes linked to each other with the corresponding correlation coefficients as *complex* weights. The coupling strength between nodes α and β varies with their associated magnitude $w_{\alpha\beta}$ ranging from 0 to 1. In addition, the linkage has direction depending on the lead–lag relation between the two nodes. According to the current definition of Eq. (5.18), for complexification of real data, node β leads node α if $\theta_{\alpha\beta} > 0$ (the direction of this link is defined as from β to α) and vice versa if $\theta_{\alpha\beta} < 0$.

Although we are focusing on a specific problem in economy here, the network that we encounter offers a brand new general issue to be addressed by researchers in network science.

8.6.1 Lead–lag relations

We define the in-degree k_α^{in} and out-degree k_α^{out} of node α by summing all weights of the incoming links to the node or the outgoing links from it:

$$k_\alpha^{\text{in}} = \sum_{\beta(\theta_{\alpha\beta}>0)} w_{\alpha\beta}, \qquad (8.68)$$

$$k_\alpha^{\text{out}} = \sum_{\beta(\theta_{\alpha\beta}<0)} w_{\alpha\beta}. \qquad (8.69)$$

Our network is, in principle, a complete graph in which all pairs of nodes are connected, but with varied degrees of strength; all the links are not equally important.

By calculating the difference between the out-degree and in-degree of each node,

$$\Delta k_\alpha = k_\alpha^{\text{out}} - k_\alpha^{\text{in}}, \qquad (8.70)$$

we can single out four typical cases for nodes with different lead–lag relations:

(1) $k_\alpha^{\text{out}} \gg k_\alpha^{\text{in}}$

(2) $k_\alpha^{\text{out}} \ll k_\alpha^{\text{in}}$

(3) $k_\alpha^{\text{out}} \simeq k_\alpha^{\text{in}} \not\simeq 0$

(4) $k_\alpha^{\text{out}} \simeq k_\alpha^{\text{in}} \simeq 0$

The stock markets and currencies satisfying condition (1) lead the world economy. On the other hand, those satisfying condition (2) basically follow the economic trend. The nodes satisfying condition (3) are sometimes leaders and sometimes followers, and those satisfying condition (4) are for the most part isolated from the rest of the world.

Let us list typical stock markets and currencies in each of these four classifications obtained for the globally-coupled financial network averaged over the entire period. The U.S., German, U.K., and Mexican stock markets are the strongest leaders together with the Mexican Peso and the Australian Dollar. On the other hand, the stock markets of Austria, the Czech Republic, the Philippines, and Japan are the ones most lagging in the world economy, accompanied by the Slovak Koruna (replaced by the Euro in 2009), the Greek Drachma (replaced by the Euro in 2001), and the Norwegian Krone. The Italian, Dutch, and Finnish stock markets belong to the third classification. That is, they influence other markets and currencies and are also similarly influenced by others. Finally, we note that the representative financial constituents of the fourth classification include the Slovakian, Indian, and South Korean markets, which are mostly isolated from the rest of the world.

8.6.2 Synchronization networks

To shed light on the synchronization of nodes in the networks, we further impose the following condition on each pair of nodes α and β that we want to connect:

$$|\theta_{\alpha\beta}| < \theta_c, \tag{8.71}$$

where θ_c is a cutoff angle to be specified. By assuming that the weight of the link between the nodes is given by the magnitude $r_{\alpha\beta}$, we can construct a synchronization network in which nodes moving in phase are linked to each other. To detect communities of such comoving nodes, we employ the fast unfolding method (Blondel et al., 2008) for the modularity maximization. We show in the following the optimal results at the largest modularity obtained by generating 10,000 samples with $\theta_c/\pi = 0.1$.

Community structure

Maximizing the modularity, we find that the stock market and foreign exchange coupled network is mainly decomposed into four communities (C1, C2, C3, and C4) of comoving nodes in each period; the optimum modularity is 0.415, 0.516, 0.556, and 0.348 in the entire, the first, the second, and the third periods, respectively. In practice, modularity value exceeding approximately 0.3 indicates that the community decomposition is significant (Newman and Girvan, 2004b). Figure 8.23 gives a geographical idea of the community structure.

The community structure is relatively stable over both the entire period and the three sub-periods, i.e., the 1999–2002 (mild crisis), 2003–2006 (calm period), and 2007–2012 (severe crisis) periods. We can characterize the four dominant communities by looking into their financial constituents:

1. The stock market community C1, dominated by Europe, South America, U.S., and Canada as major categories that always appear in this community;
2. The currencies-only community C2, dominated by Europe and Canada;
3. The currency community C3, dominated by Russia, U.S., South America, Asia, and the Middle East; and
4. The stock market community C4, dominated by Asia (including Japan) and the Middle East.

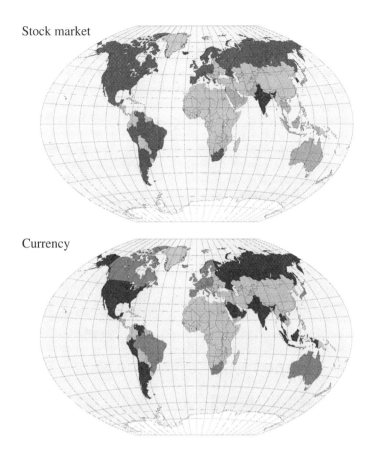

Fig. 8.23 Community structure of the financial network on a world map in the entire period, where communities dominated by stock markets are shown in red and orange, and those dominated mainly by currencies, in blue and green. (Adapted from Vodenska et al. (2016)) [see Color Plate at the end of book].

Almost all the countries of North and South America belong to one stock-market dominated community, and those of the Middle East, Asia, and Australia belong to another stock-market dominated community. In contrast, there is one currency community that encompasses the U.S. dollar, a number of Latin American currencies, the Middle East, Russia, India, and several smaller Asian countries, and another currency community that encompasses the Canadian dollar, the Brazilian Real, the Australian dollar, the Euro, and the South African Rand.

We depict the results of the community detection in the form of an adjacency matrix in Fig. 8.24, which is accompanied by Fig. 8.25 where the optimized layout of the networks in the spring–charge model are drawn with boundaries separating the communities. The components, given by $w_{\alpha\beta}$, of the adjacency matrix take values between 0 (no synchronization) and 1 (perfect synchronization). The adjacency matrix is visualized using a color code based on a temperature map scheme in which the color changes continuously from blue ($w_{\alpha\beta} = 0$) to red ($w_{\alpha\beta} = 1$) through white ($w_{\alpha\beta} = 0.5$). The community structure does not change significantly if the network is constructed with a larger or smaller cutoff for the phase differences, e.g., $\theta_c/\pi = 0.15$ or 0.05. However, the nature of the communities and the interrelationships between them display different characteristics from period to period.

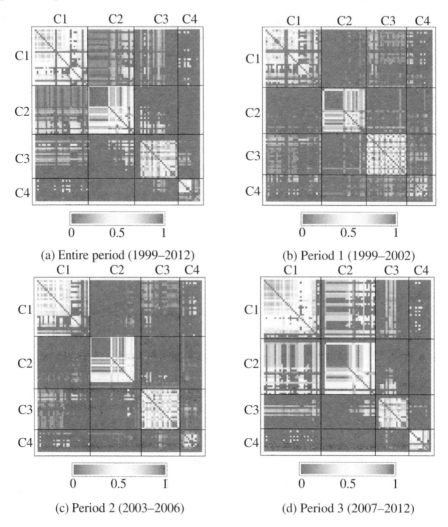

Fig. 8.24 Adjacency matrices of the financial network for the four periods, sorted according to the classification into four communities (C1, C2, C3, C4) of synchronizing nodes. (Adapted from Vodenska et al. (2016)) [see Color Plate at the end of book].

In the first period, the major equity community, C1, has the lowest strength of synchronization while one of the two currency communities, C2, is relatively isolated from the rest of the network. The other currency community, C3, is closely related to C1, and the Asian equity community C4 is not so strongly connected. In the second period, the network is decomposed into communities to the largest extent possible. We especially note that the formation of community C2 is very tight. In the third period, the community structure observed is quite similar to that in the entire period. This indicates that a global crisis has a profound influence on the global financial network. Community C1 has the highest degree of synchronization. Community C2, strongly connected to C1, suddenly changes its character while C3 is further apart from C1, as compared with the previous sub-periods. Community C4 is well established as a group of synchronizing nodes; the independence of C4 from the rest of the network increases steadily.

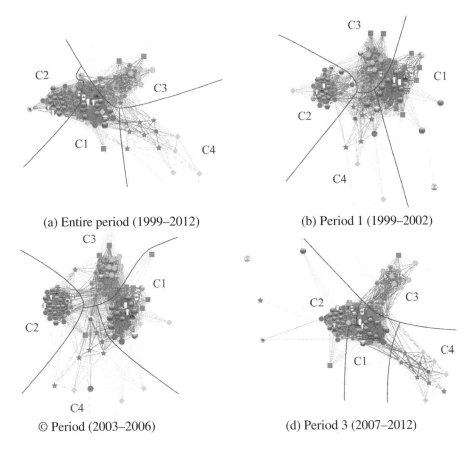

Fig. 8.25 Optimized layouts of the financial network in a spring–electrical model with boundaries separating the communities, corresponding to Fig. 8.24. (Adapted from Vodenska et al. (2016)) [see Color Plate at the end of book].

In passing, we note that the US stock market and the Yen occupy relatively peripheral positions in the financial network throughout the entire period. We also find that the Indian and Korean stock markets do not belong to C4 and remain largely isolated from the main body of the network irrespective of time period.

Temporal relationships between communities

We have already identified the four communities arising from the coherent motion of nodes in the equity–currency network. We expect that some of them will have definite lead–lag relations to each other. To quantify such a lead–lag relation between two communities, Cm and Cn, we compute the median for a weighted distribution of the phase differences between them. The distribution function $\rho_{mn}(x)$ is defined in terms of Dirac's δ function $\delta(x)$ as

$$\rho_{mn}(x) = \sum_{\alpha \in Cm,\ \beta \in Cn} w_{\alpha\beta}\, \delta(x - \theta_{\alpha\beta}) \Big/ \sum_{\alpha \in Cm,\ \beta \in Cn} w_{\alpha\beta}, \qquad (8.72)$$

where the correlation strength $w_{\alpha\beta}$ plays the role of a weight. The results are 0.053 for C1–C2, -0.010 for C1–C3, -0.205 for C1–C4, 0.276 for C2–C3, -0.318 for C2–C4, and -0.045 for C3–C4 in units of $1/\pi$. This mutual lead–lag relation between the communities indicates that C1, C2, and C4 are interrelated to each other with a solid causal relationship given by C2 $\to$ C1 $\to$ C4. In contrast, C3 has three possible positions depending on which binary correlation is emphasized, C1–C3, C2–C3, or C3–C4.

Final remark

The financial constituents are now globally coupled, so that they have rich lead–lag relations among themselves as has been just demonstrated. Suppose we carry out the same community analysis as the one in Section 8.5 by adopting the PCA instead of the CHPCA. We would still be able to detect communities of comoving nodes. However, those PCA-based communities would be simply regarded as being anti-correlated to each other. Very recently, the CHPCA has been also applied to stock prices data (Arai et al., 2015). The dominant CHPCA eigenvectors for stock markets have much simpler structures than those for the financial networks.

8.7 Community Analysis of Trade and Production Networks

8.7.1 Background

It has been known that various **collective motion**s exist in physics. A large deformation of heavy nuclei in a highly excited state, which subsequently proceeds to fission, is a typical example. This phenomenon is a collective motion due to a strong nuclear force between nucleons in a microscopic system that consists of a few hundred nucleons. Most national economies are linked by international trade and hence, economic globalization forms a complex economic network with strong links, i.e., interactions because of increasing trade. Many small, medium, and large enterprises achieve higher economic growth through free trade based on the establishment of an **Economic Partnership Agreement** (EPA), such as the **Trans-Pacific Partnership** (TPP). For this reason, we expect that various interesting collective motions owing to strong interaction will emerge in the global economy under **trade liberalization**.

The interdependent relationships of the global economy have become stronger because of an increase in international trade and investment (Tzekina et al., 2008; Barigozzi et al., 2011; He and Deem, 2010; Piccardi and Tajoli, 2012). A theoretical study using a coupled limit-cycle oscillator model suggests that the interaction terms owing to international trade can be viewed as the origin of this synchronization as described in Section 6.4 (Pikovsky et al., 2003; Ikeda et al., 2013a). This means that the international business cycle has synchronized and an **economic crisis** starting in one country now spreads across the world instantaneously. The linkages between national economies play an important role during economic crises as well as in normal times.

In this section, we analyze the industry sector-specific international trade data from 1995 to 2011 to clarify the structure and dynamics of communities consisting of industry sectors of various countries linked by international trade. Then we study the G7 Global Production Network constructed using production index time series from January 1998 to January 2015 for G7 countries. The collective motion of the G7 Global Production Network was analyzed using complex Hilbert principal component analysis, community analysis for single layer networks and multiplex networks, and structural controllability (Ikeda et al., 2014; 2016).

8.7.2 Data

World input–output database

The World Input–Output Database has been developed to analyze the effects of globalization on trade patterns across a wide set of countries (Timmer et al., 2012). This database includes annual industry sector-specific international trade data for 41 countries and 35 industry sectors for the years 1995 to 2011. Therefore the number of nodes in the international trade network is equal to 1435.

Figure 8.26 depicts the growth of international trade and business cycles. The average amount of trade per node increased monotonically from 1995 to 2011, except for the year 2008 when there was a financial crisis caused by the crash of the housing bubble in the US. As a result, the interdependent relationship of the global economy has strengthened.

The industry sector-specific international trade network is specified by nodes of industry sector α of country A and links of trade amount $w_{A\alpha,B\beta}$ between industry sector α of country A and sector β of country B. This data also includes time series of value added $V_A\alpha(t)$ for industry sector α of country A. The value added is a measure of the size of the industry sector.

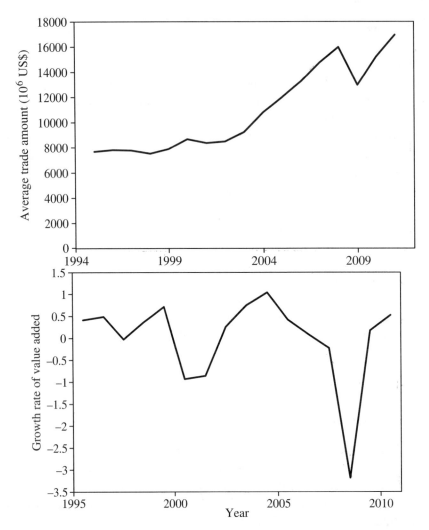

Fig. 8.26 The upper panel shows the temporal change of import in the US and the lower panel, the temporal change in the average growth rate of value added in the US. The average amount of trade per node increased monotonically from 1995 to 2011, except for the year 2008 when there was a financial crisis caused by the crash of the housing bubble in the US. As a result, the interdependent relationship of the global economy has strengthened (from Ikeda et al. (2016)).

Global production data

G7 Global Production Data is compiled using monthly time series of the production index $p(t)$, which is available for each of the G7 countries from Eurostat (n.d.); Statcan (n.d.); FRS (n.d.); METI, Japan (2015). Duration of the data is from January 1998 to January 2015 and therefore, includes the global economic crisis in 2008. The items of the G7 Global Production Data are listed in Table 8.4 and Table 8.5.

Table 8.4 G7 Global Production Data1

		Japan
1	JP01	Steel products
2	JP02	Nonferrous metal products
3	JP03	Fabricated metal products
4	JP04	Transportation equipments
5	JP05	Ceramic, stone and clay products
6	JP06	Chemical products
7	JP07	Petroleum and coal products
8	JP08	Plastic products
9	JP09	Pulp and paper products
10	JP10	Textile products
11	JP11	Food and tobacco
12	JP12	Miscellaneous
13	JP13	Mining
14	JP14	Electric appliances
15	JP15	General machinery
16	JP16	Precision machinery
		USA
17	US01	Food
18	US02	Beverage and tobacco products
19	US03	Textile mills
20	US04	Textile product mills
21	US05	Apparel
22	US06	Leather and allied products
23	US07	Wood products
24	US08	Paper
25	US09	Printing and related support activities
26	US10	Petroleum and coal products
27	US11	Chemical
28	US12	Plastics and rubber products
29	US13	Nonmetallic mineral products
30	US14	Primary metal
31	US15	Fabricated metal products
32	US16	Machinery
33	US17	Computer and electronic products
34	US18	Electrical equipment, appliance, and component
35	US19	Transportation equipment
36	US21	Furniture and related product7
37	US22	Miscellaneous

Table 8.5 G7 Global Production Data2

		Canada
38	CA01	Goods-producing industries
39	CA02	Service-producing industries
40	CA03	Industrial production
41	CA04	Non-durable manufacturing industries
42	CA05	Durable manufacturing industries
43	CA06	Energy sector
		Germany
44	DE01	Capital goods
45	DE02	Durable consumer goods
46	DE03	Intermediate goods
47	DE04	Non-durable consumer goods
		France
48	FR01	Capital goods
49	FR02	Durable consumer goods
50	FR03	Intermediate goods
51	FR04	Non-durable consumer goods
		Italy
52	IT01	Capital goods
53	IT02	Durable consumer goods
54	IT03	Intermediate goods
55	IT04	Non-durable consumer goods
		Great Britain
56	GB01	Capital goods
57	GB02	Durable consumer goods
58	GB03	Intermediate goods
59	GB04	Non-durable consumer goods

Correlation matrices are estimated for time windows of five years. Each time window is slided by one year, thus we have 13 periods for which correlation matrices are estimated. For instance, period 1 corresponds to the duration from 1998 to 2003 and is labeled as year 2001. Similarly, period 2 corresponds to the duration from 1999 to 2004 and is labeled as year 2002, and so on.

8.7.3 Community analysis

Single layer network

A community structure is detected using the maximizing modularity function (Newman, 2003b; Newman and Girvan, 2004b; Newman, 2006) for a network, as explained in Section 8.2.

$$Q_s = \frac{1}{2w} \sum_{ij} \left(w_{ij} - \frac{w_i^{\text{out}} w_j^{\text{in}}}{2w} \right) \delta(c_i, c_j), \tag{8.73}$$

$$w_i^{\text{out}} = \sum_j w_{ij}, \tag{8.74}$$

$$w_j^{\text{in}} = \sum_i w_{ij}, \tag{8.75}$$

$$2w = \sum_i w_i^{\text{out}} = \sum_j w_j^{\text{in}} = \sum_i \sum_j w_{ij}, \tag{8.76}$$

where $\delta(c_i, c_j) = 1$ if the community assignments c_i and c_j are the same, and 0 otherwise. w_{ij} are matrix elements representing the weighted adjacency matrix between node i and node j.

We identify a community structure by maximizing **modularity** using the **greedy algorithm** (Newman, 2003b; Newman and Girvan, 2004b; Newman, 2006) to each time slice of the global production network, and then identify the links between communities in adjoining years as described in the following subsection.

Jaccard index

Once the community structure is obtained for each year, the temporal evolution of the communities becomes an item of great interest. Therefore, we need to measure the similarity between communities c_i and c_j in adjoining years to obtain the linked structure of the communities. The measured similarity is the Jaccard index (Jaccard, 1912) defined as follows,

$$J(c_i, c_j) = \frac{|c_i \cap c_j|}{|c_i \cup c_j|}. \tag{8.77}$$

The range of the **Jaccard index** is defined as

$$0 \leq J(c_i, c_j) \leq 1. \tag{8.78}$$

Multiplex network

For a static network, a random network is often used as the null model. However, there are no known null models for a time-dependent network such as a global production network. Recently, **multiplex network** analysis has emerged as a methodology for detecting the community structure in a time-dependent network (Mucha et al., 2010). Most current algorithm is limited to application only to undirected network. The community structure for an undirected multiplex network is identified by maximizing the **multiple modularity**,

$$Q_m = \frac{1}{2w} \sum_{ijsr} \left(\left(w_{ijs} - \frac{w_{is}w_{js}}{2w_s} \right) \delta_{sr} + \delta_{ij} C_{jsr} \right) \delta(c_{is}, c_{jr}), \tag{8.79}$$

where the term $\delta_{ij} C_{jsr}$ is the inter-layer coupling term.

Robustness of community structures

Robustness of an identified community structure is confirmed by calculating the **variation of information** between unperturbed and perturbed networks (Karrer et al., 2008) First, we make a perturbed network by changing the links of an original unperturbed network with probability α without changing the weighted out-degree distribution. Then we compare the community division $c = \{c_1, c_2, \cdots, c_K\}$ of the unperturbed network with the community division $c' = \{c'_1, c'_2, \cdots, c'_{K'}\}$ of the perturbed network. The comparison is made using the variation of information defined by

$$V(c, c') = H(c|c') + H(c'|c) = -\sum_i^K \sum_{i'}^{K'} \frac{n_{ii'}}{N} \ln \frac{n_{ii'}}{n_{i'}} = -\sum_i^K \sum_{i'}^{K'} \frac{n_{ii'}}{N} \ln \frac{n_{ii'}}{n_i}, \tag{8.80}$$

where $n_{ii'}$, n_i, $n_{i'}$, and N are the number of nodes belonging to community c_i in the community division c and community $c'_{i'}$ in the community division c', the number of nodes belonging to community c_i in the community division c, the number of nodes belonging to community $c'_{i'}$ in the community division c', and the total number of nodes in network, respectively.

We have the larger variation of information $V(c, c')$ for the larger difference of community structure. Therefore the variation of information $V(c, c')$ is a good measure of the robustness of identified community structures.

Phase time series

The linear trend in **value added** time series $V(t)$ of the world input–output database is removed by calculating the growth rate as follows,

$$v(t) = \ln V(t) - \ln V(t-1), \tag{8.81}$$

$$x(t) = \frac{v(t) - \mathrm{E}\left[v(t)\right]}{\sqrt{\mathrm{Var}\left[v(t)\right]}}. \tag{8.82}$$

Similarly, the linear trend in production index time series $p(t)$ of the global production data is removed by calculating the growth rate as follows,

$$x(t) = \ln p(t) - \ln p(t-1). \tag{8.83}$$

This is a standard procedure to obtain stationary time series.

We calculated the **Hilbert transform** $y(t)$ (Gabor, 1946b; Rosenblum et al., 1996; Ikeda et al., 2013c) of growth rate of value added time series $x(t)$ and consequently obtained the phase time series $\theta(t)$. To see how the Hilbert transform actually used in this section is calculated, refer to Section 5.2.2.

Order parameter as a measure of synchronization

We briefly explain **synchronization** for a system consisting of two oscillators, one with a phase of $\theta_1(t)$ and one with a phase of $\theta_2(t)$. While the amplitudes can be different, synchronization is defined as the phases locking $\theta_1(t) - \theta_2(t) = \mathrm{const.}$. In general, $\mathrm{const.} \neq 0$, where the phase difference signifies a delay; therefore a direct evaluation of the phase instead of the correlation coefficient is more appropriate.

The collective rhythm produced by the whole population of oscillators is captured by a macroscopic quantity, such as the complex **order parameter** $u(t)$, defined as follows,

$$u(t) = r(t)e^{i\phi(t)} = \frac{1}{N}\sum_{j=1}^{N} e^{i\theta_j(t)}. \tag{8.84}$$

The radius $r(t)$ measures the **phase coherence**, and $\phi(t)$ represents the average phase (strogatz (2000)). The order parameter Eq. (8.84) is a good indicator of synchronization when the frequencies of oscillators are close to each other.

Construction of the global production network

Then the global production network is constructed using the global production data. The complex time series $z_\alpha(t)$ for sector α ($\alpha = 1, \cdots, N$) is normalized and $N \times N$ complex correlation matrix C is calculated. The eigenvalue $\lambda^{(n)}$ and the corresponding eigenvector $V^{(n)}$ for the **complex correlation matrix** C are calculated using the theory of CHPCA described in Section 5.2.

The filtered complex correlation matrix $C^{(N_s)}_{\alpha\beta}$ is calculated using Eq. (8.67) with the number of dominant eigen modes N_s, which is estimated using the rotational random shuffling procedure, as explained in Section 5.3.3. Amplitude $w_{\alpha\beta}$ is used as weight of the link in community analysis: Eq. (8.73), Eq. (8.74), Eq. (8.75), and Eq. (8.76). If we consider the correlation matrix $C^{(N_s)}_{\alpha\beta}$ as an adjacency matrix, we obtain a network of production nodes linked to each other with the corresponding correlation coefficients as weights. Note that the weight of the link $w_{\alpha\beta}$ between node α and node β ranges from 0 to 1, and the link has direction depending on the lead–lag relation between two nodes: β (α) leads α (β) if $\theta_{\alpha\beta}$ takes a positive (negative) value. Although the network constructed is in principle a **complete graph**, we select links with small lead–lag by setting the threshold θ_{th} on phase $\theta_{\alpha\beta}$ in the following analysis.

8.7.4 Community structure of trade network

We conduct the community analysis for each time slice of the international trade network. Figure 8.27 (a) and (b) lists examples of community structures obtained for 1997 and 2009. There were seven communities for 1997, and eight for 2009. Since we have a few small communities, the number of major communities is equal to six. The temporal change of modularity Q is shown in Fig. 8.27 (c). The value of the obtained modularity Q is about 0.2, which depends on the threshold of the weight of links $w_{A\alpha,B\beta}$. In this community analysis, we applied threshold of weight $w_{ij} > 10^7$ US $. This means that about half of the links are included in the analysis. If we increase the threshold of weight, we have larger value of modularity Q.

We then confirm the robustness of the identified community structure by using the variation of information $V(c, c')$. The results are shown in Fig. 8.28 for 1995 and 2011. From the value of modularity Q and the dependence of the variation of information on the probability of changing links α, we can say for a fact that the community structure is barely identified with the threshold of weight $w_{ij} > 10^7$ US $.

Complex Networks, Community Analysis, Visualization

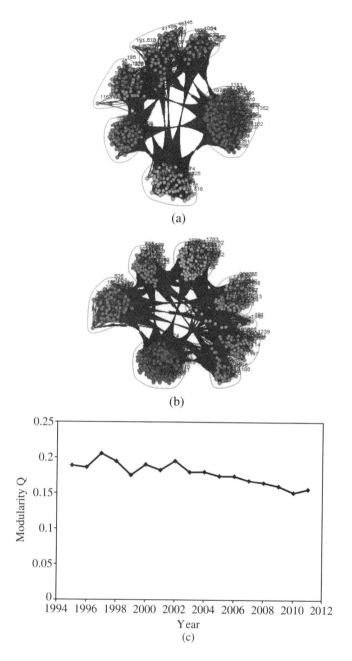

Fig. 8.27 The community structures for each time slice of the international trade network. Panels (a) and (b) lists examples of community structures for 1997 and 2009. There were seven communities for 1997, and eight for 2009. The temporal change of modularity Q is shown in panel (c). The value of the obtained modularity Q is about 2, which depends on the threshold of the weight of links $w_{A\alpha,B\beta}$ (from Ikeda et al. (2016)) [see Color Plate at the end of book].

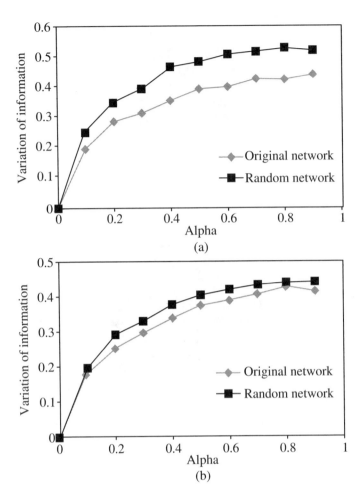

Fig. 8.28 The robustness of identified community structure was confirmed using the variation of information $V(c, c')$. From the value of modularity Q and the dependence of the variation of information on the probability of changing links α, we can say that the community structure is barely identified with the threshold of weight $w_{ij} > 10^7$ US $ (from Ikeda et al. (2016)).

8.7.5 Linked communities of trade network

The temporal evolution of communities is characterized by the link relations for communities in adjoining years. We measure the similarity of communities c_i and c_j in adjoining years by using the Jaccard index $J(c_i, c_j)$. Figure 8.29 shows the Jaccard indices between 1995 and 1996, and between 2010 and 2011. Communities are arranged in decreasing order of the number of industry sectors in each community. For instance, 95c1 (community c_1 in 1995) is the largest, and 95c2, the second largest.

Complex Networks, Community Analysis, Visualization

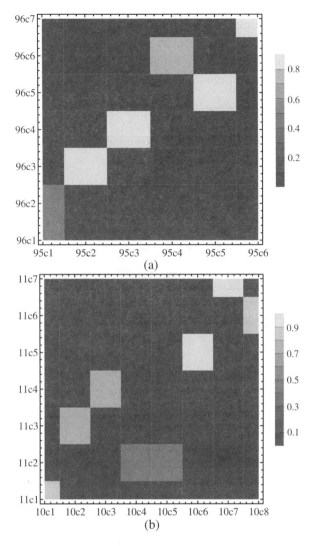

Fig. 8.29 The Jaccard indices between 1995 and 1996, and between 2010 and 2011. We define link relations for the pair of communities with the largest Jaccard index between adjoining years (from Ikeda et al. (2016)).

We define the link relation for a pair of communities with the largest Jaccard index between adjoining years. We observe clearly linked relationships for most of the communities. For instance, nodes 95c1 (community c_1 in 1995), 95c2, 95c3, 95c4, 95c5, and 95c6 are linked to nodes 96c1 (community c_1 in 1996), 96c3, 96c4, 96c6, 96c5, and 96c7, respectively. Similarly, nodes 10c1 (community c_1 in 2010), 10c4, 10c2, 10c3, 10c6, and 10c8 are linked to nodes 11c1 (community c_1 in 2011), 11c2, 11c3, 11c4, 11c5, and 11c6, respectively. We identified five **linked communities** between 1995 and 2011 as shown in Table 8.6 The identified link structure shows that a six-backboned structure exists in the international trade network.

The composition of the linked communities was analyzed in terms of the marginal rank of the trade volume in countries and industry sectors. The respective compositions of the first to the sixth linked communities are listed in Table 8.7 to Table 8.12. Features of the compositions of these linked communities are briefly described here.

Table 8.6 Linked communities

Year	1995	1996	1997	1998	1999	2000	2001	2002	2003
Linked comm1	c_1	c_1	c_1	c_2	c_1	c_1	c_2	c_1	c_1
Linked comm2	c_2	c_3	c_2	c_3	c_4	c_2	c_4	c_3	c_3
Linked comm3	c_3	c_4	c_3	c_4	c_2	c_3	c_3	c_2	c_2
Linked comm4	c_4	c_6	c_4	c_5	c_5	c_4	c_5	c_4	c_5
Linked comm5	c_5	c_5	c_5	c_7	c_7	c_5	c_6	c_5	c_6
Linked comm6	c_6	c_7	c_6	c_6	c_6	c_6	c_7	c_6	c_7

Year	2004	2005	2006	2007	2008	2009	2010	2011
Linked comm1	c_1	c_1	c_1	c_1	c_1	c_1	c_1	c_1
Linked comm2	c_3	c_2	c_3	c_2	c_2	c_3	c_4	c_2
Linked comm3	c_2	c_3	c_2	c_3	c_3	c_2	c_2	c_3
Linked comm4	c_4	c_4	c_4	c_4	c_4	c_4	c_3	c_4
Linked comm5	c_5	c_5	c_5	c_5	c_5	c_6	c_6	c_5
Linked comm6	c_6	c_6	c_6	c_6	c_6	c_7	c_8	c_6

Table 8.7 Linked community 1 (Total = US$ 24.360 trillion)

Rank	Share	Country	Industry Sector
Largest	5.07%	United States	Financial Intermediation
2nd	4.92%	United States	Renting of M&Eq and Other Business Activities
3rd	4.91%	Rest of World	Renting of M&Eq and Other Business Activities
4th	3.41%	United Kingdom	Renting of M&Eq and Other Business Activities
5th	3.00%	United Kingdom	Financial Intermediation

Table 8.8 Linked community 4 (Total = US$ 20.447 trillion)

Rank	Share	Country	Industry Sector
Largest	37.67%	Rest of World	Mining and Quarrying
2nd	6.06%	Rest of World	Coke, Refined Petroleum and Nuclear Fuel
3rd	5.11%	Russia	Mining and Quarrying
4th	4.38%	Canada	Mining and Quarrying
5th	3.04%	Australia	Mining and Quarrying

Table 8.9 Linked community 3 (Total = US$ 17.227 trillion)

Rank	Share	Country	Industry Sector
Largest	9.03%	Rest of World	Basic Metals and Fabricated Metal
2nd	8.33%	Germany	Basic Metals and Fabricated Metal
3rd	4.74%	Germany	Machinery, Nec
4th	3.58%	Japan	Basic Metals and Fabricated Metal
5th	3.46%	United States	Basic Metals and Fabricated Metal

Table 8.10 Linked community 2 (Total = US$ 16.520 trillion)

Rank	Share	Country	Industry Sector
Largest	9.27%	Rest of World	Chemicals and Chemical Products
2nd	7.04%	Germany	Chemicals and Chemical Products
3rd	5.88%	United States	Chemicals and Chemical Products
4th	3.49%	France	Chemicals and Chemical Products
5th	3.37%	United States	Wholesale Trade and Commission Trade, Except for Motor Vehicles and Motorcycles

Table 8.11 Linked community 5 (Total = US$ 15.096 trillion)

Rank	Share	Country	Industry Sector
Largest	13.95%	China	Electrical and Optical Equipment
2nd	13.22%	Rest of World	Electrical and Optical Equipment
3rd	11.80%	United States	Electrical and Optical Equipment
4th	9.15%	Japan	Electrical and Optical Equipment
5th	8.08%	Germany	Electrical and Optical Equipment

Table 8.12 Linked community 6 (Total = US$ 7.900 trillion)

Rank	Share	Country	Industry Sector
Largest	16.76%	Germany	Transport Equipment
2nd	16.01%	United States	Transport Equipment
3rd	11.08%	Japan	Transport Equipment
4th	7.38%	France	Transport Equipment
5th	5.85%	United Kingdom	Transport Equipment

The largest linked community is linked community 1 and its total amount of trade is US$ 24.360 trillion. This community consists of the *Financial Intermediation* and the *Renting of Machines and Equipment* sectors in the US and the UK. This industrial sector includes leasing of aircraft used by low cost carriers. For details, see Table 8.7. The second-largest linked community is linked community 4 and its total amount of trade is US$ 20.447

trillion. Linked community 4 consists of the *Mining and Quarrying* and the *Coke, Refined Petroleum and the Nuclear Fuel* sectors. This community includes Russia, Canada, and Australia, but the most noteworthy is the share of the *Rest of World* at more than 43 %. For details, see Table 8.8. The third-largest linked community is linked community 3 and its total amount of trade is US$ 17.227 trillion. Linked community 3 consists of the *Basic Metals and Fabricated Metal* and the *Machinery* sectors. This community includes Germany, Japan, and the US, mainly. The share in the *Rest of World* is considerably high. For details, see Table 8.9. The fourth-largest linked community is linked community 2 and its total amount of trade is US$ 16.520 trillion. Linked community 2 consists of the *Chemicals and Chemical Products* and the *Wholesale Trade and Commission Trade* sectors. This community includes Germany, United States, and France. For details, see Table 8.10. The fifth-largest linked community is linked community 5 and its total amount of trade is US$ 15.096 trillion. Linked community 5 consists of the *Electrical and Optical Equipment* sector. This community includes China, the *Rest of World*, the US, Japan, and Germany. Note that the share of China is as high as 14 %. For details, see Table 8.11. The sixth-largest linked community is linked community 6 and its total amount of trade is US$ 7.900 trillion. It consists of the *Transport Equipment* sector, which is mainly automobile manufacturing. This community includes Germany, the US, Japan, France, and UK. We note that the share of Germany is as high as 17 %. For details, see Table 8.12.

These characteristics of the community structure in Table 8.7 to Table 8.12 shows that international trade is actively transacted among the same or similar industry sectors, and as a result the structure of the community includes mainly the same or similar industry sectors in various countries.

We summarize our community analysis of the international trade network as follows:

(i) We apply conventional community analysis to each time slice of the international trade network data, the World Input–Output Database. This database contains the industry sector specific international trade data for 41 countries and 35 industry sectors from 1995 to 2011.

(ii) Once we had the community structure for each year, we identified the links between communities in adjoining years by using the Jaccard index as a similarity measure between communities in adjoining years. Furthermore, we confirmed the robustness of the observed community structure by quantifying the variation of information for a perturbed network structure. The identified linked communities show that six-backboned structures exist in the international trade network.

(iii) The largest linked community is the *Financial Intermediation* and the *Renting of Machines and Equipments* sectors in the US and UK. The second is the *Mining and Quarrying* sector in the *Rest of World*, Russia, Canada, and Australia. The third is the *Basic Metals and Fabricated Metal* sector in the *Rest of World*, Germany, Japan, and the US. These community structures show that international trade is actively transacted among the same or similar industry sectors.

8.7.6 Synchronization of international business cycles

We evaluated the phase time series $\theta_j(t)(j = 1, \cdots, 1435)$ for the growth rate of value added for the years 1995 to 2011 using the methodology described in Subsection 8.7.3.

Figure 8.30 shows the polar plot of the phase in 1997 for (a) all sectors, (b) community c_1. In these polar plots, the complex order parameter $u(t)$ is indicated by a black asterisk. The respective amplitude of the order parameters are 0.483, 0.359, and 0.534 for community c_1, c_2, c_3, respectively, while the amplitude for all sectors is 0.253. Figure 8.30 shows the polar plot of the phase in 2009 for (c) community c_2, and (d) community c_3. In these polar plots, the complex order parameter $u(t)$ is indicated by a black asterisk.

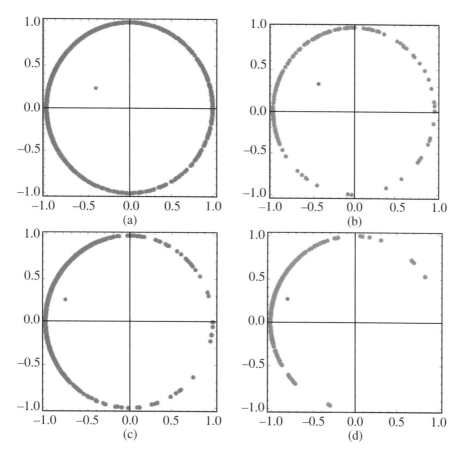

Fig. 8.30 Polar plot of the phase for (a) all sectors in 1997, (b) community c_1 in 1997, (c) all sectors in 2009, and (d) community c_1 in 2009. Amplitude $r(t)$ and average phase $\phi(t)$ for each community is equivalent to those quantities for all sectors. Note that the amplitudes and average phases in 2009 are larger than the quantities in 1997 (from Ikeda et al. (2016)).

The respective amplitude of the order parameters are 0.758, 0.512, 0.801 for community c_1, c_2, and c_3, respectively, while the amplitude for all sectors is 0.662. Amplitude $r(t)$ and average phase $\phi(t)$ for each community is equivalent to those quantities for all sectors. Note that the amplitudes and average phases in 2009 are larger than the quantities for 1997. This relation clearly indicates that coherence of the international business cycle has strengthened. However, we cannot say that the coherence in each community is stronger than that for all sectors, although we have denser links inside each community compared with the average density of links in the whole network.

Figure 8.31 shows the temporal change in amplitude for the order parameter $r(t)$ for the years 1996 to 2011. The phase coherence fell gradually in the late 1990s but increased sharply in 2001 and 2002. This temporal change might be related to the structural change in the international trade network discussed in Subsection 8.7.3. From 2002, the amplitudes for the order parameter $r(t)$ remain very high, although the years 2005 and 2009 are slightly lower.

Order parameters are estimated for the growth rate of the value added time series shuffled randomly while keeping the auto-correlation ((Iyetomi et al. (2011a,b)) and order parameter averaged over 1000 shuffled time series; they are plotted by the black curve in Figure 8.31. This means that the average order parameter is evidently larger than the systematic error of the analysis method. Therefore the high phase coherence observed for each linked communities is statistically significant.

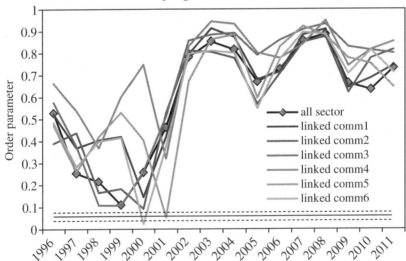

Fig. 8.31 The temporal change in amplitude for the order parameter $r(t)$ for the years 1996 to 2011. The phase coherence fell gradually in the late 1990s but increased sharply in 2001 and 2002. This temporal change might be related to the structural change in the international trade network discussed in Subsection 8.7.3. From 2002, the amplitudes for the order parameter $r(t)$ remain very high, although the years 2005 and 2009 are slightly lower (from Ikeda et al. (2016)) [see Color Plate at the end of book].

We recall that the estimated angular frequencies ω_i are almost identical for the business cycle in the six countries: Australia, Canada, France, the UK, Italy, and the US, as shown in Fig. 6.26. Based on this observation, we can assume that angular frequencies ω_i are almost identical for the business cycles in 35 industry sectors of 41 countries. The assumed equality of the angular frequencies and the observed high phase coherence suggest that the international business cycles are synchronized.

We summarize our analysis of the synchronization of international business cycles, as follows. We used the Hilbert transform to evaluate the phase time series of the growth rate of value added for 1435 nodes and then estimated the complex order parameters for communities. The temporal change in amplitude for the order parameter was studied for the years 1996 to 2011. We observed that phase coherence decreased gradually in the late 1990s but increased sharply in 2001 and 2002. From 2002, the amplitudes for the order parameter remained high except for the years 2005 and 2009. To confirm this synchronization, we estimated order parameters for the growth rate of the value added time series shuffled randomly while keeping the autocorrelation and order parameter averaged over 1000 shuffled time series. The comparison of order parameters for the original and randomly shuffled time series showed that the synchronization observed for each linked community was statistically significant.

8.7.7 Significant modes in economic crisis

In the analysis of the G7 Global Production Data, we obtained eigenvalue $\lambda^{(n)}$ and the corresponding eigenvector $\boldsymbol{V}^{(n)}$ for the complex correlation matrix C calculated using the theory of CHPCA described in Section 5.2 and selected statistically significant modes using the rotational random shuffling, as explained in Section 5.3.3. In rotational random shuffling, the growth rate of the production time series $x(t)$ was shuffled randomly while keeping the autocorrelation (Iyetomi et al. 2011a;) and then eigenvalues for the correlation matrix were calculated for the randomly shuffled time series. If an eigenvalue for the original time series is larger than the largest eigenvalue calculated for the randomly shuffled time series, the eigenvalue or eigen mode is regarded as statistically significant. The largest four eigenvalues of each period are shown in the upper panel of Fig. 8.32. The lower panel depicts temporal changes of the number of significant modes. Note that only a single eigen mode is significant for the periods that contain the sub-prime mortgage crisis of 2008. This means that production for all industry sectors in G7 countries behaves similarly during an **economic crisis**. This suggests that economic risk spread instantaneously to all industry sectors in G7 countries and all industries decrease their production simultaneously.

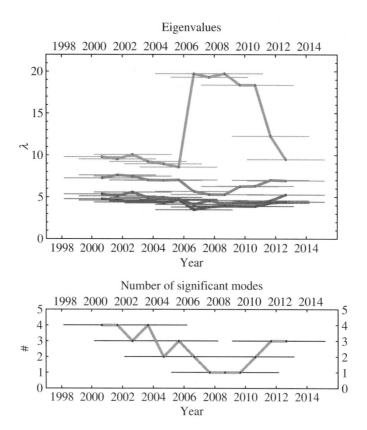

Fig. 8.32 The largest four eigenvalues of each period are shown in the upper panel. The lower panel depicts temporal change of the number of significant modes. Note that only a single eigen mode is significant for the periods that contain the sub-prime mortgage crisis of 2008. This means that production for all industry sectors in G7 countries behaves similarly during economic crisis (from Ikeda et al. (2016)).

8.7.8 Community structure of production network

We identified the community structure for each time slice of the global production network constructed using the methodology described in Section 8.7.3. Figure 8.33 shows examples of community structures obtained for (a) 2004, (b) 2007, (c) 2010, and (d) 2013. The average value of modularity Q_s is 0.302 during 2001 and 2013. The maximum and minimum modularity are 0.410 and 0.153, respectively. This means that the community structure is clearly identified for the global production network. The number of major communities varies between two and four. There were two major communities for the global economic crisis during 2007 and 2010. The observed community structure is consistent with the statistically significant modes obtained earlier. The number of major communities in the period of normal economy (2004 and 2013) is larger than for the period of an economic crisis (2007 and 2010). The small number of statistically significant modes in an economic crisis is observed as a small number of major communities in the global production network. This is again interpreted to mean that production for all industry sectors in G7 countries behaves similarly during economic crisis. The economic

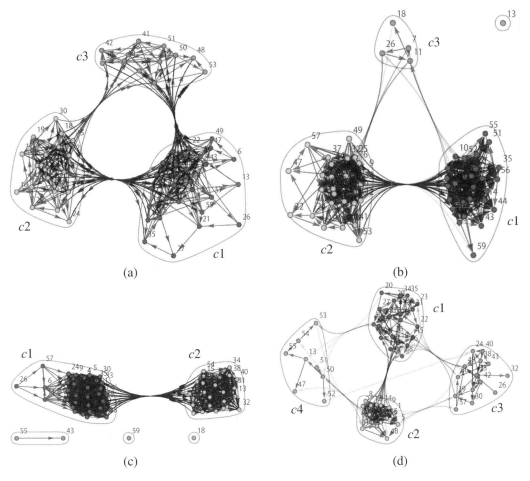

Fig. 8.33 Examples of community structures obtained for (a) 2004, (b) 2007, (c) 2010, and (d) 2013. Average value of modularity Q_s is 0.302 during 2001 and 2013. Maximum and minimum modularity are 0.410 and 0.153, respectively. This means that the community structure is clearly identified for the global production network. The number of major communities varies between two and four. There were two major communities for the global economic crisis during 2007 and 2010 (from Ikeda et al. (2016)).

risk spread instantaneously to all industry sectors in G7 countries and all industries look for new demand. Consequently, new links (trade relations) span beyond communities observed during normal economic period.

8.7.9 Linked communities of production networks

The temporal evolution of communities is characterized by the link relations for communities in adjoining years. The similarity of communities c_i and c_j in adjoining years was measured by the use of the Jaccard index $J(c_i, c_j)$. Figure 8.34 shows the Jaccard indices in adjoining years: (a) between 2003 and 2004, (b) between 2006 and 2007, (c) between 2009 and 2010, and (d) between 2012 and 2013. Communities were arranged in decreasing order of the number of industry sectors in each community. For instance community c_1 is the largest, and c_2 is the second largest.

We define the link relation for a pair of communities with the largest Jaccard index between adjoining year. Most of communities show clearly linked relationships with the communities of the adjoining year. For instance, Fig. 8.34 (a) shows that communities c_1, c_2, c_3, and c_4 in 2003 are linked to communities c_2, c_1, c_1, and c_3 in 2004, respectively. Similarly, Fig 8.34 (c) shows that communities c_1 and c_2 in 2009 are linked to communities c_1 and c_2 in 2010, respectively. These linked communities are summarized in Table 8.13, Table 8.14, and Table 8.15 for three periods: before the crisis, during the crisis, and after the crisis, respectively. The temporal changes in community structures, i.e., community dynamics is seen as an example of collective motion.

We obtained three linked communities before the crisis. Linked communities 1 to 3 are characterized by countries and corresponds to Europe, the US, and Canada, respectively. Japan distributed to three communities.

Then two linked communities (linked communities 4 and 5) were obtained for the period of the crisis and are characterized by sectors. For instance, linked community 4 is composed of the following sectors: *Steel products*, *Transportation equipments*, *Chemical products*, *Pulp and paper products*, *Computer and electronic product*, and others. Linked community 5 is composed of sectors: *Fabricated metal products*, *Precision machinery*, *Textile products*, and others.

We obtained four linked communities after the crisis. Linked communities 6 to 9 are characterized by countries as before. For instance, linked communities 6 to 9 corresponds to Canada, US, Japan, and Europe. Some European countries distributed to linked communities 6 and 7.

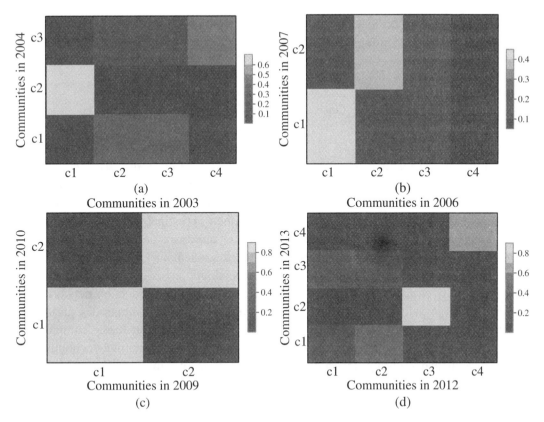

Fig. 8.34 Jaccard indices in adjoining years: (a) between 2003 and 2004, (b) between 2006 and 2007, (c) between 2009 and 2010, and (d) between 2012 and 2013. The temporal evolution of communities is characterized by the link relations for communities in adjoining years. The similarity of communities c_i and c_j in adjoining years was measured by the use of the Jaccard index $J(c_i, c_j)$ (from Ikeda et al. (2016)).

Table 8.13 Linked communities before the crisis

Year	2001	2002	2003	2004	2005	2006
Linked comm 1	c_1	c_2	$c_2 + c_3$	c_1	c_2	c_3
Linked comm 2	c_2	c_1	c_1	c_2	c_1	c_2
Linked comm 3	$c_3 + c_4$	c_3	c_4	c_3	c_3	c_1

Table 8.14 Linked communities during the crisis

Year	2008	2009	2010
Linked comm 4	c_2	c_1	c_1
Linked comm 5	c_1	c_2	c_2

Table 8.15 Linked communities after the crisis

Year	2012	2013
Linked comm 6	c_1	c_3
Linked comm 7	c_2	c_1
Linked comm 8	c_3	c_2
Linked comm 9	c_4	c_4

We summarize the features of collective motion during the economic crisis of 2008 by studying the G7 Global Production Network constructed using production index time series from January 1998 to January 2015 for G7 countries. The G7 Global Production Network was analyzed by using complex Hilbert principal component analysis, community analysis for a single-layer network and multiplex networks, and structural controllability. First, the complex Hilbert principal component analysis showed that only a single eigen mode was significant for the global economic crisis during 2007 and 2010. The community structure was clearly identified for the global production network with sufficiently large values of modularity. There were two major communities during the global economic crisis of 2007 and 2010, whereas there were four major communities during the normal period. We observe the structural change for the normal economic state and for the network during an economic crisis as collective motion of the G7 Global Production Network.

8.7.10 Communities in multiplex production networks

The dynamics of a community structure is observed in the temporal change of a community structure for a 13-layer multiplex network. We use the parameter $r_s = 1$ and inter slice coupling $C_{isr} = 0.8$. The color of each node corresponds to the identified community and each number indicates a layer from 2007 to 2013.

We compare community structures between a multiplex network and a single-layer network for 2004, 2007, 2010, and 2013. Figure 8.35 shows the results of the comparison. The left and right column of each panel corresponds to communities identified for a multiplex network and a single-layer network, respectively. Although the direction of links are ignored in the community analysis for a multiplex network, the two analysis agree reasonably well. Note that the number of communities obtained for a single-layer network is larger than that obtained for a multiplex network.

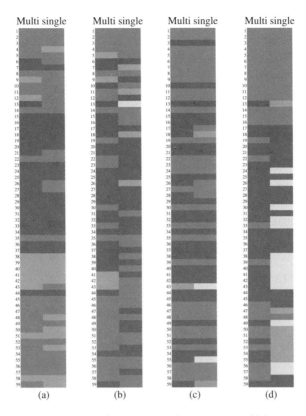

Fig. 8.35 We compare community structures between a multiplex network and a single-layer network for 2004, 2007, 2010, and 2013. The left and right column of each panel correspond to communities identified for a multiplex network and a single-layer network, respectively. Although the direction of links are ignored in the community analysis for a multiplex network, the two analysis agree reasonably well (from Ikeda et al. (2016)) [see Color Plate at the end of book].

8.8 Controllability of Production Network

8.8.1 Theory of structural controllability

The theory of **structural controllability** is applied to complex networks (Liu et al., 2011). The dynamics of the system is often approximately described by a linear equation,

$$\frac{d\boldsymbol{x}}{dt} = \tilde{A}\boldsymbol{x}(t) + \tilde{B}\boldsymbol{v}(t), \tag{8.85}$$

Here $\boldsymbol{x} = (x_1(t), \ldots, x_N(t))^T$ is the state vector of the system. The N × N matrix $\tilde{A}$ is identical to the transverse adjacent matrix and the N × M matrix $\tilde{B}$ identifies the driver node which is controlled from outside the system. If the N × NM matrix K,

$$K = (\tilde{B}, \tilde{A}\tilde{B}, \tilde{A}^2\tilde{B}, \ldots, \tilde{A}^{N-1}\tilde{B}) \tag{8.86}$$

has full rank, that is

$$\mathrm{rank}(K) = N, \tag{8.87}$$

the system is controllable.

The **driver node**s are identified by **maximum matching** in the bipartite representation of the network (Moore and Mertens, 2011). Maximum bipartite matching is written as an integer linear programming as follows,

$$\begin{aligned}&\underset{y}{\text{maximize}} \quad \mathbf{1}^T \mathbf{y} \\ &\text{subject to} \quad \mathbf{y} \geq 0, \ A^T \mathbf{y} \leq 1,\end{aligned} \tag{8.88}$$

where the M × N matrix A is called the incidence matrix whose component a_{ij} is

$$a_{ij} = \begin{cases} 1 \ (\text{if node } j \text{ is an endpoint of link } i) \\ 0 \ (\text{otherwise}) \end{cases} \tag{8.89}$$

and $\mathbf{y}$ is the variable

$$y_i = \begin{cases} 1 \ (\text{if i} \in \text{F}) \\ 0 \ (\text{if i} \notin \text{F}) \end{cases} \tag{8.90}$$

and $\mathbf{1}$ is a vector where all elements are equal to unity. The set F is matching if each node is incident to at most one link in F.

We define the **matchability** m to identify driver nodes as follows. In-degree $k_j^{(in)}$ and out-degree $k_j^{(out)}$ is calculated for each node j in the bipartite network after eliminating links with $y_i = 0$. If in-degree of node j is equal to 0, matchability of node j is equal to $m_j = 0$; on the other hand, if in-degree of node j is equal to 1, matchability of node j is equal to $m_j = 1/k_l^{(out)}$. Here node l is the origin node spanning a link toward node j in the bipartite network. If $m_j = 0$, node j is a pure driver node. On the other hand, if $m_j = 1$, node j is a pure controlled node. Thus a node with small matchability m is interpreted as a partial driver node. Note that the number of driver nodes is calculated by $N - \sum_j m_j$.

8.8.2 Illustrative example for identifying driver nodes

The first example for identifying driver nodes from Liu et al. (2011) is shown in Figure 8.36. The left panel is a network diagram and the right is a bipartite expression of the corresponding network. Both N and M are equal to three. The incidence matrix A is given by Eq. (8.91).

$$A = \begin{bmatrix} 0 & 1 & 0 \\ 0 & 0 & 1 \\ 0 & 0 & 1 \end{bmatrix} \tag{8.91}$$

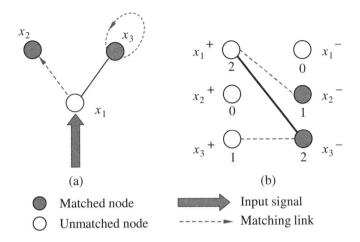

Fig. 8.36 An example for identifying driver nodes from Liu et al. (2011). The left panel is a network diagram and the right, a bipartite expression of the corresponding network.

The maximum bipartite matching using Eq. (8.88), Eq. (8.89), and Eq. (8.90) gives $y = (1, 0, 1)$. Matchability of each node is $m = (0, 1, 1)$ and therefore node 1 is the driver node.

Another example for identifying driver nodes from Liu et al. (2011) is shown in Fig. 8.37. The upper panels are network diagrams and the lower are bipartite expressions of the corresponding networks. Both N and M are equal to four. The incidence matrix A is given by Eq. (8.92).

$$A = \begin{bmatrix} 0 & 1 & 0 & 0 \\ 0 & 0 & 1 & 0 \\ 0 & 0 & 0 & 1 \\ 0 & 0 & 1 & 0 \end{bmatrix} \tag{8.92}$$

The maximum bipartite matching using Eq. (8.88), Eq. (8.89), and Eq. (8.90) gives $y = (1, 0, 1, 1)$. Matchability of each node is $m = (0, 1/2, 1, 1/2)$ and therefore node 1 is pure driver node and node 2 and 4 are partial driver nodes.

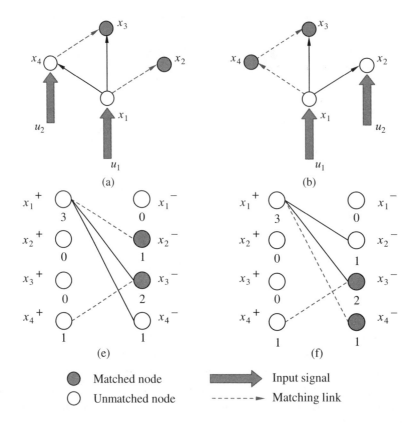

Fig. 8.37 Another example of identifying driver nodes from Liu et al. (2011). The upper panels are network diagrams and the lower are bipartite expressions of the corresponding networks.

8.8.3 Controllability of production networks

The number of driver nodes was identified by the use of Eq. (8.88) to Eq. (8.90). The matchability distribution $\bar{F}(m)$ is shown in Fig 8.38 for 2004, 2007, 2010, and 2013. It is clearly observed in panels (b) and (c) in Fig 8.38 that many nodes have a small value of matchability m during economic crises. Temporal changes of the number of driver nodes is shown in Fig. 8.39. Note that the number of driver nodes increased during the economic crisis from 2008 to 2010. During crises, the share of driver nodes n_D becomes about 80% of all the nodes, whereas n_D is about 60% during the normal period. This means that we cannot expect to control the global real economy by stimulating a relatively small number of nodes.

Partial driver nodes with matchability less than or equal to 0.2 ($m_j \leq 0.2$) are collected for 2004, 2007, 2010, and 2013 and shown in Table 8.16 to Table 8.19, respectively. During a normal period, we have 16 nodes (27.1%) and 15 nodes (25.4%) for 2004 and 2013, respectively. On the other hand, during an economic crisis, we have 36 nodes (61.0%) and 47 nodes (79.7%) for 2007 and 2010, respectively.

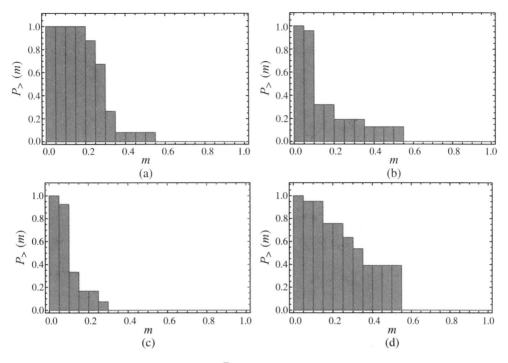

Fig. 8.38 Matchability distribution $\bar{F}(m)$ for 2004, 2007, 2010, and 2013. It is clearly observed in panels (b) and (c) that many nodes have a small value of matchability m during an economic crisis (from Ikeda et al. (2016)).

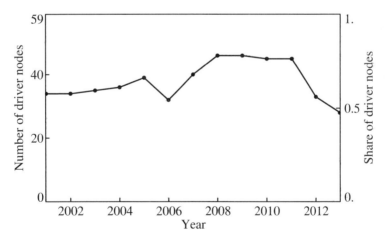

Fig. 8.39 Temporal changes of the number of driver nodes. Note that the number of driver nodes increased during the economic crisis from 2008 to 2010, their share of driver nodes n_D becomes about 80% of all the nodes, whereas n_D is about 60% during the normal period. This means that we cannot expect to control the global real economy by stimulating a relatively small number of nodes (from Ikeda et al. (2016)).

If we look at the country-wise distribution, we notice the following: In 2004, during the normal period, we have 0 node (0.0%), 1 node (25.0%), 2 nodes (50.0%), 3 nodes (75.0%), 2 nodes (50.0%), 5 nodes (18.8%), and 3 nodes (14.3%) for Canada, Germany, France, Great Britain, Italy, Japan, and the US, respectively. In 2013, during the normal period, we have 0 node (0.0%), 1 node (25.0%), 1 node (25.0%), 1 node (25.0%), 0 node (0.0%), 4 nodes (6.3%), and 8 nodes (38.1%) for Canada, Germany, France, Great Britain, Italy, Japan, and the US, respectively. In 2007, during an economic crisis, we have 3 nodes (50.0%), 3 nodes (75.0%), 2 nodes (50.0%), 2 nodes (50.0%), 1 node (25.0%), 12 nodes (12.5%), and 13 nodes (61.9%) for Canada, Germany, France, Great Britain, Italy, Japan, and the US, respectively. In 2010, during an economic crisis, we have 5 nodes (83.3%), 4 nodes (100.0%), 4 nodes (100.0%), 2 nodes (50.0%), 1 node (25.0%), 14 nodes (12.5%), and 17 nodes (81.0%) for Canada, Germany, France, Great Britain, Italy, Japan, and the US, respectively. Therefore, it is hard to find a country that dominates driver nodes for both normal and crises periods.

If we look at the sector-wise distribution, we notice the following: In 2004, during the normal period, we have 4 nodes (21.1%) and 12 driver nodes (30.0%)for durable/nondurable consumer goods sectors and capital/intermediate goods, respectively. In 2013, during the normal period, we have 4 driver nodes (21.1%) and 11 driver nodes (27.5%) for durable/nondurable consumer goods and capital/intermediate goods, respectively. In 2007, during an economic crisis, we have 6 driver nodes (31.6%) and 30 driver nodes (75.0%) for durable/nondurable consumer goods and capital/intermediate goods, respectively. In 2010, during an economic crisis, we have 12 driver nodes (63.2%) and 35 driver nodes (87.5%) for durable/nondurable consumer goods and capital/intermediate goods, respectively. Here we have 19 sectors classified under the category durable and non-durable consumer goods: JP10, JP11, JP14, US01, US02, US05, US18, US21, CA01, CA02, CA04, DE02, DE04, FR02, FR04, IT02, IT04, GB02, and GB04. The remaining 40 sectors are classified under the category of capital and intermediate goods. Therefore, we can say that capital/intermediate goods sectors are dominant over durable/nondurable consumer goods sectors for both normal and crises periods.

Table 8.16 Driver Nodes with $m \leq 0.2$ in 2004

Node	Description	Matchability	comm
FR02	Durable consumer goods	0.166	c_1
US06	Leather and allied products	0.166	c_1
JP03	Fabricated metal products	0.166	c_1
GB02	Durable consumer goods	0.2	c_1
GB01	Capital goods	0.2	c_1
IT01	Capital goods	0.2	c_1
US10	Petroleum and coal products	0.2	c_1
JP12	Miscellaneous	0.2	c_1

Contd...

Node	Description	Matchability	comm
JP10	Textile products	0.2	c_1
JP08	Plastic products	0.2	c_1
DE01	Capital goods	0.2	c_2
US16	Machinery	0.2	c_2
GB03	Intermediate goods	0.166	c_3
FR03	Intermediate goods	0.166	c_3
JP04	Transportation equipments	0.166	c_3
IT02	Durable consumer goods	0.2	c_3

Table 8.17 Driver Nodes with $m \leq 0.2$ in 2007

Node	Description	Matchability	comm
GB03	Intermediate goods	0.0625	c_1
DE03	Intermediate goods	0.0625	c_1
DE02	Durable consumer goods	0.0625	c_1
US15	Fabricated metal products	0.0625	c_1
JP16	Precision machinery	0.0625	c_1
JP15	General machinery	0.0625	c_1
GB01	Capital goods	0.0714	c_1
IT01	Capital goods	0.0714	c_1
FR01	Capital goods	0.0714	c_1
DE01	Capital goods	0.0714	c_1
CA03	Industrial production	0.0714	c_1
CA01	Goods-producing industries	0.0714	c_1
US19	Transportation equipment	0.0714	c_1
JP14	Electric appliances	0.0714	c_1
JP08	Plastic products	0.0714	c_1
JP05	Ceramic, stone and clay products	0.0714	c_1
JP04	Transportation equipments	0.0714	c_1
JP03	Fabricated metal products	0.0714	c_1
JP02	Nonferrous metal products	0.0714	c_1
JP01	Steel products	0.0714	c_1
FR03	Intermediate goods	0.0625	c_2
CA04	Non-durable manufacturing industries	0.0625	c_2
US22	Miscellaneous	0.0625	c_2
US17	Computer and electronic products	0.0625	c_2
US13	Nonmetallic mineral products	0.0625	c_2
US11	Chemical	0.0625	c_2
US04	Textile product mills	0.0625	c_2
US01	Food	0.0625	c_2
JP12	Miscellaneous	0.0625	c_2

Contd...

Node	Description	Matchability	comm
JP09	Pulp and paper products	0.0625	c_2
US18	Electrical equipment, appliance, and component	0.166	c_2
US16	Machinery	0.166	c_2
US12	Plastics and rubber products	0.166	c_2
US09	Printing and related support activities	0.166	c_2
US07	Wood products	0.166	c_2
JP06	Chemical products	0.166	c_2

Table 8.18 Driver Nodes with $m \leq 0.2$ in 2010

Node	Description	Matchability	comm
US10	Petroleum and coal products	0	c_1
GB01	Capital goods	0.0588	c_1
FR04	Non-durable consumer goods	0.0588	c_1
FR03	Intermediate goods	0.0588	c_1
FR01	Capital goods	0.0588	c_1
DE04	Non-durable consumer goods	0.0588	c_1
DE03	Intermediate goods	0.0588	c_1
CA02	Service-producing industries	0.0588	c_1
US19	Transportation equipment	0.0588	c_1
US17	Computer and electronic product	0.0588	c_1
US14	Primary metal	0.0588	c_1
US12	Plastics and rubber products	0.0588	c_1
US07	Wood products	0.0588	c_1
JP12	Miscellaneous	0.0588	c_1
JP05	Ceramic, stone and clay products	0.0588	c_1
JP01	Steel products	0.0588	c_1
GB03	Intermediate goods	0.111	c_1
US11	Chemical	0.111	c_1
US03	Textile mills	0.111	c_1
JP14	Electric appliances	0.111	c_1
JP09	Pulp and paper products	0.111	c_1
JP08	Plastic products	0.111	c_1
JP07	Petroleum and coal products	0.111	c_1
JP04	Transportation equipments	0.111	c_1
JP02	Nonferrous metal products	0.111	c_1
US08	Paper	0.2	c_1
US06	Leather and allied products	0.2	c_1
US01	Food	0.2	c_1
JP11	Food and tobacco	0.2	c_1
JP06	Chemical products	0.2	c_1

Contd...

Node	Description	Matchability	comm
DE02	*Durable consumer goods*	0.0588	c_2
JP16	*Precision machinery*	0.0588	c_2
IT01	*Capital goods*	0.0666	c_2
FR02	*Durable consumer goods*	0.0666	c_2
DE01	*Capital goods*	0.0666	c_2
CA05	*Durable manufacturing industries*	0.0666	c_2
CA04	*Non-durable manufacturing industries*	0.0666	c_2
CA03	*Industrial production*	0.0666	c_2
CA01	*Goods-producing industries*	0.0666	c_2
US22	*Miscellaneous*	0.0666	c_2
US21	*Furniture and related products*	0.0666	c_2
US18	*Electrical equipment,appliance,and component*	0.0666	c_2
US15	*Fabricated metal products*	0.0666	c_2
US09	*Printing and related support activities*	0.0666	c_2
US04	*Textile product mills*	0.0666	c_2
JP13	*Mining*	0.0666	c_2
JP10	*Textile products*	0.0666	c_2

Table 8.19 Driver Nodes with $m \leq 0.2$ in 2013

Node	Description	Matchability	comm
US19	*Transportation equipment*	0.125	c_1
US18	*Electrical equipment, appliance, and component*	0.125	c_1
US15	*Fabricated metal products*	0.125	c_1
US13	*Nonmetallic mineral products*	0.125	c_1
US12	*Plastics and rubber products*	0.125	c_1
US11	*Chemical*	0.125	c_1
US02	*Beverage and tobacco products*	0.125	c_1
GB04	*Non-durable consumer goods*	0.2	c_1
DE01	*Capital goods*	0.2	c_1
JP08	*Plastic products*	0.125	c_2
JP07	*Petroleum and coal products*	0.2	c_2
JP05	*Ceramic,stone and clay products*	0.2	c_2
JP03	*Fabricated metal products*	0.2	c_2
US10	*Petroleum and coal products*	0	c_3
FR04	*Non-durable consumer goods*	0	c_4

Degree distributions are shown in Fig. 8.40 for 2004, 2007, 2010, and 2013. The distributions have longer tail for the period of economic crisis as shown in Fig. 8.40 (b) and (c). The increase in the number of driver nodes during the economic crisis from 2008 to 2010 is discussed from the view point of heterogeneity in terms of degree distribution.

The average number of degree $\langle k \rangle$ is 13.7, 25.1, 25.0, and 9.42 for 2004, 2007, 2010, and 2013, respectively. The maximum degrees $k_{\max}$ are 26, 41, 37, and 19 for 2004, 2007, 2010, and 2013, respectively. If we assume the power-law degree distribution $\bar{F}(k) = k^{-\gamma'}$ for the entire region from 1 to $k_{\max}$, power exponent γ' are estimated to be 1.25, 1.09, 1.129, and 1.38, for 2004, 2007, 2010, and 2013, respectively. Here $\bar{F}$ is the cumulative probability. Thus, the power exponent γ for probability density $f(k) = k^{-\gamma}$ are 2.25, 2.09, 2.129, and 2.38, for 2004, 2007, 2010, and 2013, respectively. Therefore, we have $\langle k \rangle \approx 12$ and $\gamma \approx 2.3$ during the normal period. On the other hand, we have $\langle k \rangle \approx 20$ and $\gamma \approx 2.1$ during an economic crisis. With these values of $\langle k \rangle$ and γ, an analytical formulae (Liu et al., 2011) gives $n_D \approx 0.25$ and $n_D \approx 0.40$ for the normal period and crisis period, respectively. This means that an increase in the number of driver nodes during an economic crisis is explained qualitatively by heterogeneity in terms of degree distribution.

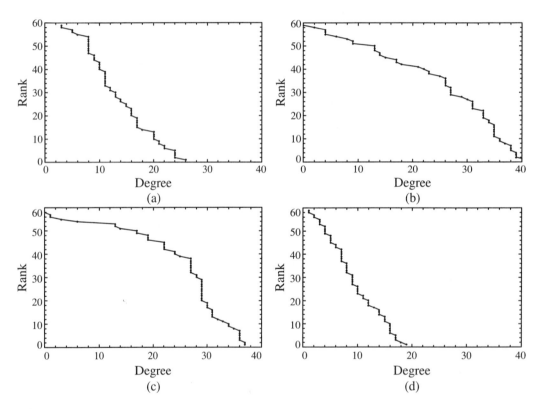

Fig. 8.40 Degree distributions for (a) 2004, (b) 2007, (c) 2010, and (d) 2013. The distributions have a longer tail for the period of an economic crisis. An increase in the number of driver nodes during the economic crisis from 2008 to 2010 is discussed from the view point of heterogeneity in terms of degree distributions (from Ikeda et al. (2016)).

Study of structural controllability shows that the number of driver nodes increased during the economic crisis from 2008 to 2010. During an economic crisis, the share of driver nodes n_D became about 80% of all the nodes, whereas n_D is about 60% during the normal period. The observed increase in the number of driver nodes during an economic crisis is explained qualitatively by the heterogeneity in terms of degree distribution. This means that we cannot expect to control the global real economy by stimulating a relatively small number of nodes; furthermore, it becomes more difficult to introduce some measure to control the state of the global economy during a crisis than during a normal period.

We observed various kinds of collective motions in the global economy under trade liberalization, as expected. Although many Japanese small and medium enterprises would achieve higher economic growth by free trade, we also need to pay attention to the fact that once a negative economic shock occurs in a regional economy, it will propagate to the rest of the world instantaneously and we as of now have no strong measure to control the economic crisis.

9
Systemic Risks

> Network theory provides a critique of standard wisdom on how to create a stable financial system.
>
> **Joseph E. Stiglitz**

Credit–debt relationships among economic agents comprise large-scale networks of the economic system on national and global scales. There are different layers in such networks even at the core of real economic and financial systems. One layer is the arena of the real economy, namely supplier–customer links among firms as nodes. Firm activities are financed by financial institutions as well as directly by financial markets. The layer of the supplier–customer network is thus linked to another layer of financial networks between firms and banks. Furthermore, banks are also creditors and debtors of themselves comprising another layer of inter-bank networks.

As a financial system, the inter-bank network resides at the core, which is connected with firms, via bank–firm networks, at the periphery of the system; the periphery is a large network of suppliers and customers comrising the engine of the real economy. These networks are actually further linked to financial markets, but one may depict the basic picture as in Fig. 9.1.

Systemic risk is a network effect caused by failures or financial deterioration of debtors and creditors through the credit–debt links to other nodes even in a remote part of the network. Systemic risk often has considerable consequences at a nation-wide scale, and sometimes at a world-wide extent, as one experiences today in repeated financial crises.

While an understanding of the inter-bank network at the core of the financial system is crucial, no less important is the propagation of risk from the core of banks to the periphery of firms, or vice versa, as well as the propagation of risk among firms. Unfortunately, empirical studies based on the real data of bank–firm networks or supplier–customer networks on a large scale is still lacking.

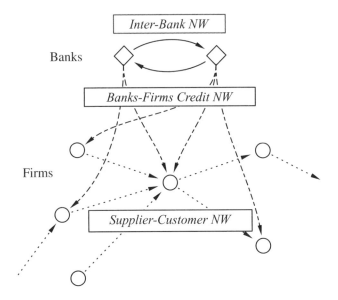

Fig. 9.1 Inter-bank, bank–firm credit, and supplier–customer networks. Financial institutions or banks (squares) and firms (circles) are creditors and debtors in the links of inter-bank credit (lines); lending–borrowing between banks–firms are shown as dashed lines, and supplier–customer links among firms are shown as dotted lines.

In this chapter, we shall study how one can quantify the systemic risk by using real data on a nation-wide scale for inter-firm production networks and firm–bank credit networks.

9.1 Nation-wide Production Network

Supplier–customer relationships among firms in the production network are the arenas where financial distress spreads from distressed debtors of customers to its creditors of suppliers. While the events of bankruptcies can be easily observed, the underlying contagion effect of financial distress can have considerable consequences such as a chain of bankruptcies.

DebtRank (Battiston et al., 2012) is a model aimed at quantifying the propagation of financial distress, and has been applied recently to the analysis and evaluation of systemic risk mainly for inter-bank contagion (di Iasio et al., 2013; Tabak et al., 2013; Poledna and Thurner, 2014; Fink et al., 2014; Puliga et al., 2014) as well as for the credit network between banks and firms by Aoyama et al. (2013a). and as well as for a credit network between banks and firms by Aoyama et al. (2013a). The typical size of the systems studied so far runs into the hundreds, while the production network in a nation comprises more than a million firms as nodes and millions of supplier–customer relationship as links, which is much larger than an inter-bank credit network. Owing to the lack of real data as well as computational challenges, little has been studied on the DebtRank analysis for such a large-scale production network.

This section studies the propagation of financial distress in production networks by employing the DebtRank methodology to model how financial distress spreads through supplier–customer relationships. Trade–credit has a crucial role in this model. Suppliers usually provide credit to their customers in trade anticipating payment in due time. If a customer delays or fails in making the payment, suppliers may lose expected sales and profits, depending on relative exposure, potentially causing a chain of failures and bankruptcies. We model this propagation of distress from downstream to upstream in the production network using the DebtRank. By assuming different initial configurations of idiosyncratic shocks on industrial sectors, and the model of distress propagation, we quantify financial distress, visualize the propagation, and evaluate the resulting ripple effect. The results have important implications in light of the fact that the current observation of the events of failures or bankruptcies underestimates the amount of financial stress in different parts of the network.

9.1.1 DebtRank method

The methodology of DebtRank was put forth by Battiston et al. (2012) to quantify systemic risk in credit networks among financial institutions. The network comprises banks as nodes and financial dependencies among them as links. The quantity is a measure of how financial distress at a single institution or in a set of institutions can potentially influence others along the links of financial dependencies, namely, exposure. The DebtRank and related methods have recently been studied to see how they can be applied to better evaluate systemic risk in such financial networks and also in credit networks between banks and firms (see, for example, di Iasio et al., 2013; Tabak et al., 2013; Aoyama et al., 2013a; Poledna and Thurner, 2014; Fink et al., 2014; Puliga et al., 2014).

Let us recapitulate the method of DebtRank in what follows. Consider a network with nodes $i = 1, 2, \cdots, N$ and with directed and weighted links. A link $j \Rightarrow i$ has a weight $0 < w_{ji} \leq 1$ that represents a relative exposure of i to j. At each time-step t in the computation, two variables of each node are updated:

- $h_i(t) \in [0, 1]$, the amount of financial distress of node i at time t.
- $s_i(t) \in \{U, D, I\}$, respectively, the state of "undistressed", "distressed", and "inactive" at time t.

As an initial configuration of distress at time $t = 0$, we assume that

$$h_i(t=0) = \begin{cases} h_{i0} & \text{if } i \in A, \\ 0 & \text{otherwise,} \end{cases} \quad (9.1)$$

starting from a set of distressed nodes, denoted by A (this can be a single node), and that

$$s_i(t=0) = \begin{cases} D & \text{if } i \in A, \\ U & \text{otherwise.} \end{cases} \quad (9.2)$$

We then update the distress according to

$$h_i(t) = \min\left[1, \ h_i(t-1) + \sum_{j:s_j(t-1)=D} w_{ji} h_j(t-1)\right], \quad (9.3)$$

where the summation is taken over all the is neighboring nodes j that are in the state of D at time $t-1$. The weight w_{ji} determines the strength of propagation. We denote the direction of propagation by $j \Rightarrow i$. Simultaneously, we update the state by

$$s_i(t) = \begin{cases} D & \text{if } h_i(t) > 0 \text{ and } s_i(t-1) \neq I, \\ I & \text{if } s_i(t-1) = D, \\ s_i(t-1) & \text{otherwise.} \end{cases} \quad (9.4)$$

One can see that a node with state D becomes I at the next time, and then does not affect any other node thereafter. This avoids an infinite number of repercussions in the propagation of distress. A node with state U becomes D, when distress reaches it, and then affects neighboring nodes the next time and becomes I. Note that a node with state I can continue to receive distress even as it does not affect any of the others.

After a finite number of time-steps, denoted by T, the propagation terminates resulting in a final configuration $h_i(t=T)$ for all the nodes. Finally, to define the total amount of distress in the entire network owing to the initial set A of distressed nodes, it is customary to take the average for all the $h_i(T)$ except for $i \in A$ as

$$D_A = \sum_{i \notin A} \hat{a}_i h_i(T) \quad (9.5)$$

where

$$\hat{a}_i = \frac{a_i}{\sum_{j \notin A} a_j} \quad (9.6)$$

and a_i is the size of the node i such as assets, sales and so forth. D_A is called the DebtRank for the system owing to the propagation of distress starting from the nodes in A. If A is a single node $\{i\}$, then it represents how much an individual node can affect the entire network. Note that the effect of the initially given distress is discarded by excluding it in the summation in Eq. (9.5), because one does not want to include the trivial effect owing to the given initial amount of distress.

Fig. 9.2 A photograph of the K computer system at the Advanced Institute of Computational Science, RIKEN, Kobe, used for DebtRank calculations (Courtesy of RIKEN)

Once the sets of nodes and links with weights w_{ji} are given as well as the attributes of nodes including the size a_i, it is easy to compute the DebtRank. In fact, the algorithm is quite similar to breadth-first search, an elementary graph algorithm (see Cormen et al., 2001). However, it takes an impractical amount of time, say more than a day, to compute each node's DebtRank values for our data comprising a million nodes. To overcome this difficulty, we employ one of the world's fastest supercomputers, called K computer (Fig. 9.2), and parallelize the computation on CPU cores in its facility of the RIKEN AICS so that the speed of computation increases greatly to within an hour or minutes (see Terai et al. (2016) for details). We believe that this advantage is important for simulations under a number of scenarios and initial conditions.

9.1.2 Bowtie structure and up/down streams of production

Our data for a production network is based on a survey done by Tokyo Shoko Research (TSR), one of the leading credit research agencies in Tokyo, with the Research Institute of Economy, Trade and Industry (RIETI). We use the three data sets of TSR Kigyo Jouhou, Kigyo Soukan Jouhou, and Kigyo Tousan Jouhou for basic information for more than a million firms, millions of supplier–customer and ownership links among firms, and a list of bankruptcies, respectively. The data were compiled in July 2011.

Let us denote a supplier–customer link as $i \to j$, where firm i is a supplier to another firm j, or equivalently, j is a customer of i. We extracted only the supplier–customer links for pairs of "active" firms to exclude inactive and failed firms by using a flag in the basic information. Eliminating self-loops and parallel edges (duplicate links recorded in the data), we have a network of firms as nodes and supplier–customer links as edges. When viewed as an undirected network, it has the largest connected component (99% in terms of the number of firms), which we shall now study. Denoting the number of nodes and links by N and M respectively, we have

$$N = 986,185, \tag{9.7}$$

$$M = 4,402,270. \tag{9.8}$$

as a result of our pre-processing. The largest connected component is often called a giant weakly connected component (GWCC). Numbers of firms in a standard classification of industrial sectors are given in Table 9.1.

Table 9.1 Classification of industrial sectors (Japan Standard Industrial Classification, Rev. 12, November 2007). Numbers of firms (third column) and fractional numbers (fourth column) are based on the divisions according to the primary industry of each firm.

ID	Divisions	#firms	#firms(%)
A	Agriculture, forestry	6,821	0.69
B	Fishing	1,001	0.10
C	Mining and quarrying	1,332	0.14
D	Construction	337,206	34.19
E	Manufacturing	156,847	15.90
F	Electricity, gas, heat, water supply	648	0.07
G	Information and communications	23,441	2.38
H	Transportation and storage	33,246	3.37
I	Wholesale and retail trade	249,610	25.31
J	Finance and insurance activities	6,054	0.61
K	Real estate activities	34,325	3.48
L	Professional, scientific, technical	33,757	3.42
M	Accommodation and food service	14,617	1.48
N	Arts, entertainment and recreation	16,015	1.62
O	Education/learning support	3,651	0.37
P	Human health and social work	21,004	2.13
Q	Compound services	5,586	0.57
R	Other service activities	41,021	4.16
	Total	986,185	100.00

Let us examine how firms are located in the upstream and downstream of the entire network. To do so, we regard the network as a directed graph and find the so-called "bowtie" structure. A GWCC can be decomposed into its parts defined as follows (see Fig. 9.3):

NW The whole network.

GWCC (Giant weakly connected component): This is the largest connected component when viewed as an undirected graph. An undirected path exists for an arbitrary pair of firms in the component.

DC (Disconnected components): These are connected components other than the GWCC.

GSCC (Giant strongly connected component): This is the largest connected component when viewed as a directed graph. A directed path exists for an arbitrary pair of firms in the component.

Table 9.2 Bowtie structure: Sizes of different components for a million firms in Japan

Component	#firms	Note
GWCC	986,185	100%
GSCC	488,347	50% × GWCC
IN	179,127	18% × GWCC
OUT	282,331	29% × GWCC
TE	36,380	4% × GWCC
Total	$N = 986,185$	equal to GWCC

IN This refers to the firms from which the GSCC is reached via a directed path.

OUT This refers to the firms that are reachable from the GSCC via a directed path.

TE ("Tendrils"); This is the rest of the GWCC (note that TEs may not look like tendrils).

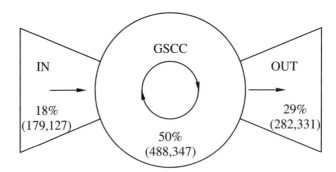

Fig. 9.3 Bowtie structure for the production network including GSCC (giant strongly connected component) in which any pair of firms is mutually connected by a directed path; IN and OUT components comprised firms in the GSCC's upstream and downstream sides respectively. See the main text for a full explanation (from Fujiwara et al. (2016).)

See Table 9.2 for the sizes of each components. It follows from the definitions that

$$\text{NW} = \text{GWCC} + \text{DC} \tag{9.9}$$

$$\text{GWCC} = \text{GSCC} + \text{IN} + \text{OUT} + \text{TE} \tag{9.10}$$

The shortest-path lengths (distances) from the GSCC and firms in the IN and OUT are given by:

Distance from GSCC to IN

Distance	#firms
1	172,526
2	6,368
3	221
4	12
Total	179,127

Distance from GSCC to OUT

Distance	#firms
1	269,555
2	12,414
3	350
4	12
Total	282,331

By comparing the over- and under-presence of industrial divisions in each of these components, we can see that in the portion of IN, the numbers of firms in the sectors of real estate (K), agriculture and forestry (A), and information and communications (G) are larger when compared with the corresponding sectors in the SCC. In the portion of OUT, human health and social work (P), accommodation and food service (M), and education and learning support (O) are more abundant. This is reasonable, because these industries are mainly located either in the upstream or in the downstream. Nevertheless, all industries are basically embedded in the SCC with entanglement. We also analyze the community structure of the network and obtain results that are similar to those found by previous works (Fujiwara and Aoyama, 2010).

The diameter of a graph is the maximum length for all ordered pairs (i, j) of the shortest path from i to j. The average distance is the average length for all those pairs (i, j). We found that the average distance is 4.59 while the diameter is 22. This implies that the computation for the DebtRank will terminate at most within the time-steps corresponding to the diameter.

9.1.3 Calculation of DebtRank using the K computer

Assumption

Supplier–customer links are regarded as creditor–debtor relationships. A supplier depends on its customers for sales and profits. If one of the customers has financial distress, it may delay or even be unable to make payment, which results in the financial distress of the supplier. We assume that this is the most important channel for the propagation of distress. In fact, there is empirical and theoretical evidence for this assumption (see Battiston et al., 2007; Fujiwara and Aoyamam 2010 and the references therein). Thus, if there is a supplier–customer relationship, $i \to j$, from firm i as a supplier and to firm j as a customer, it is assumed that the direction of the distress propagation in Eq. (9.3), $j \Rightarrow i$, is *opposite* of the direction of supplier–customer link $i \to j$.

It would be ideal to have information about the strength w_{ji} in Eq. (9.3) from the amount of trade, for example, which is not available in our data. We assume that the strength w_{ji} in Eq. (9.3), or relative exposure of the customer i to its suppliers js is given by

$$w_{ji} = \frac{1}{\text{number of customers } is \text{ of supplier } j}. \tag{9.11}$$

Because the data were collected from the nomination of suppliers and customers that are most important to a particular firm under investigation, one could take into account the order of importance, but we simply assume that this is a reasonable approximation.

As for the attributes of firms, namely firm-size a_i in Eq. (9.5) and Eq. (9.6), we employed the amount of sales and the number of employees, which gave us qualitatively similar results, as far as statistical properties are concerned.

DebtRank for individual firms

We computed the DebtRank values $x = D_{\{i\}}$ starting from each node i under distress. We assume that the initial value of the distress, $h_{i0} = 1$ in Eq. (9.1), is the maximum value of distress in the model. The cumulative distribution $\bar{F}(x)$ is shown in Fig. 9.4. One can observe that the distribution obeys a power law

$$\bar{F}(x) \propto x^{-\alpha}, \tag{9.12}$$

in the region for large x, where the value of α is given by

$$\alpha = 1.28 \pm 0.01, \tag{9.13}$$

estimated by the maximum likelihood or Hill method (error at 99% significance level).

We note that the distributions for the in-degree (number of suppliers) and out-degree (number of customers) of a firm also have power-laws with quantitatively similar exponents (see Fujiwara and Aoyama, 2010). In fact, there is a significant correlation between the value of DebtRank and degrees for each firm. This is important, because a big firm usually has a number of suppliers and customers, and the amount of financial distress coming from these firms tend to be large, even if the weight tends to be small. On the other hand, a small firm typically depends on a limited number of customers, say one or two firms, and is easily influenced by others. Note that the resulting value of DebtRank tends to be small, because it is a weighted average by firm-size.

To examine the relation between the DebtRank and firm-size, we show the scatter plots in Fig. 9.5 for the amount of sales for firm-size, and also in Fig. 9.6 for the number of employees as an alternative measure of size.

Systemic Risks

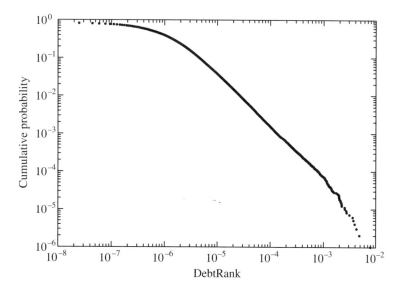

Fig. 9.4 Cumulative distribution $\bar{F}(x)$ for the DebtRank value x of individual firms (weighted by firm-size of sales). The distribution obeys a power-law, $\bar{F}(x) \propto x^{-\alpha}$, in the region for large x, with exponent $\alpha = 1.28 \pm 0.01$ (from Fujiwara et al. (2016).)

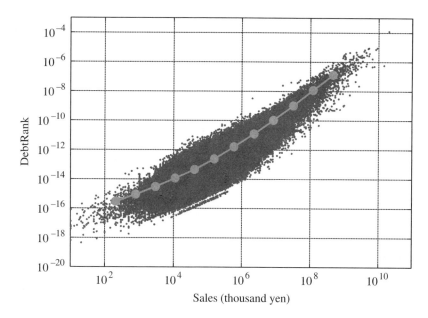

Fig. 9.5 Scatter plot for the pair of values of DebtRank and the amount of sales for individual firms. Also shown is the conditional average of DebtRank for firms present in each slice, corresponding to a particular size of firm (represented by each point in the red line) (from Fujiwara et al. (2016).)

We also show the conditional averages of DebtRank for firms with given ranges of firm-size, because the scatter plot is densely populated by points and can mislead the interpretation of the density. As obvious from the line for the conditional averages, it has an interesting non-linearity in the sense that the value of DebtRank becomes much larger than what is expected by a linear relation between the DebtRank and firm-size. This implies that big firms can have a larger impact on the entire system than what one can naively expect from the numbers of suppliers and customers. Namely, the "higher-order" structure of a network rather than the degrees, certainly plays an important role.

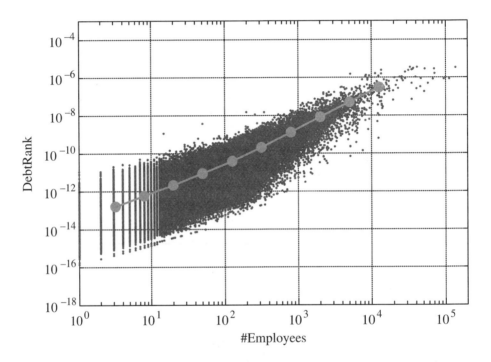

Fig. 9.6 Scatter plot for the pair of values of DebtRank and the number of employees for individual firms. Also shown is the conditional average of DebtRank for firms present in each slice, corresponding to a particular size of firm (represented by each point in the red line) (from Fujiwara et al. (2016).)

DebtRank for Sectors

Let us turn our attention to the industrial sectors by computing the DebtRank D_A, starting from firms in the sector of A. We assume that the initial distress is given by $h_{i0} = 0.1$ in Eq. (9.1) for $i \in A$, and 0 otherwise. By this configuration, we suppose that a relatively weak but simultaneous shock occurs in a particular sector. Let us assume that the level of sectoral classification is given as shown in Table 9.1, i.e., divisions from A to R which contain relatively large number of firms.

We calculate D_A according to the definition in Eq. (9.5) and Eq. (9.6). We also compute the total sum of firm-sizes for firms in sector A so that one can estimate the size of the initial configuration of financial distress:

$$S_A = \sum_{i \in A} a_i. \tag{9.14}$$

We show the scatter plot for the 18 sectors (A to R) in Fig. 9.7. One can see the power-law relation between S_A and D_A:

$$D_A \propto S_A^{-\beta}, \tag{9.15}$$

where the exponent is estimated as $\beta = 0.92$ using the minimum square estimate for the logarithmic values of the variables. Remember that we exclude the trivial effect of the initial stress in the quantification of DebtRank.

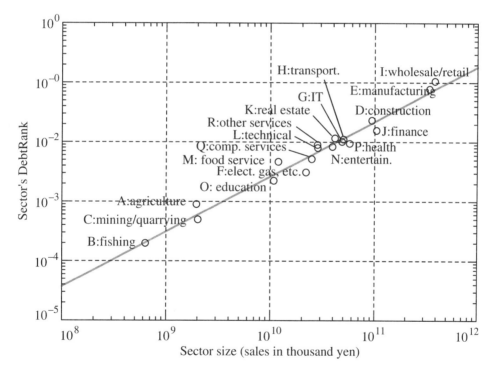

Fig. 9.7 The size of each sector S_A (horizontal) and each sector's value of DebtRank D_A (vertical). The straight line is a power-law fit, $D_A \propto S_A^{-\beta}$, where $\beta = 0.92$ (MSE).

The linear relationship between the logarithms of sector's size and DebtRank guides us to pay attention to deviations from it based on individual sectors.

The sector of construction (D) has a smaller size than that for the sector of finance and insurance activities (J), but the former's impact is larger than the latter's. This implies that one or more hops from the sector of construction in the upstream occupies a greater part of the network producing a bigger impact. Similarly, while the sector of accommodation and food service (M) has a size comparable to that of education/learning support (O), it has nearly double the impact on the whole system. By taking a close look at each sector, one can see that the deviation from Eq. (9.14) corresponds to the location of the sector in the upstream and downstream portions of the network, larger for upstream and smaller for downstream than what is expected by Eq. (9.14).

One is able to further divide the sectors into smaller ones, and examine the relation between S_A and D_A in order to understand the relationship between the size of the stress and the location of the firms in the sector.

Vulnerability of sectors

The financial distress brought to each sector can be used as a measure of vulnerability of firms in the sector under the condition that the initial distress starts from a single sector A.

To do so, one simply decomposes the DebtRank into different components, that is

$$D_A = \sum_g D_{A_g}, \qquad (9.16)$$

where A_g is the sector from $g = 1, \ldots, G$. The quantity D_{A_g} is simply the decomposition of the sum given by Eq. (9.5) into the various sectors, except the initially distressed sector, and represents how much distress is propagated into the sector S_g. The larger the quantity is, the more vulnerable the corresponding sector is. It is a measure of the vulnerability of each sector owing to the initial configuration of distress.

Figure 9.8 shows the matrix of such vulnerabilities. Each row represents the initially distressed sector. Each column is a measure of the vulnerability.

The result shows that depending on which sector is initially distressed, there is a heterogeneous propagation of distress into different sectors resulting in different levels of vulnerability. We see that this method of examining vulnerability can be employed to identify the likelihood of failures of the firms in the more vulnerable sectors.

Discussion

As one can see from the methodology of DebtRank, the model is based on abstract quantification of financial stress. It is important to know how the model is related to the financial state of individual firms in terms of stock and flow variables in balance sheets and profit-and-loss statements. With regard to the application of DebtRank to financial institutions, there are two recent works, those by Battiston et al. (2015) and Bardoscia et al. (2015). They attempted to clarify the link between financial distress defined and modeled by balance-sheet dynamics and the DebtRank and found that the dynamics of

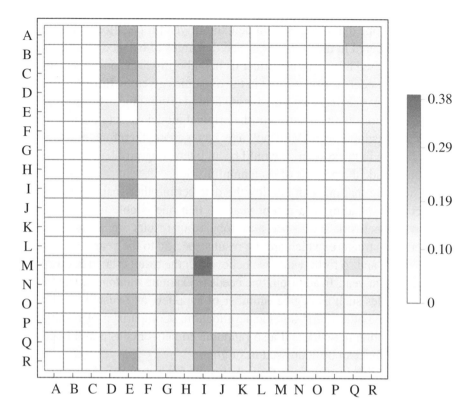

Fig. 9.8 Vulnerability of each sector owing to another sector's financial distress. Each row represents the initially distressed sector, from which financial distress propagates to other sectors in columns (from Fujiwara et al. (2016)).

DebtRank is naturally interpreted by the dynamics of balance-sheets defining variables of distress, such as debt and capital ratios.

In the application of production networks, while there are works such as those by Hazama and Uesugi (2012); Mizuno et al. (2014); Goto et al. (2015), which found various aspects of systemic risks in production networks, there is little work that simulates the entire system by using the dynamics based on the actual balance-sheets of individual firms and the model of DebtRank and its extension to the actual network. It would be valuable to relate these works to the simulation.

Secondly, in the current model, we focus on the propagation of distress from downstream (customers) to upstream (suppliers), but not on the opposite direction from upstream to downstream. The latter is relevant in a typical case of the influence of price changes. When the price of raw materials goes up, the prices of commodities in the downstream eventually increase, potentially affecting the financial state of those firms. Another case is external shock arising from the supply side and propagating in a similar direction, as in the event of a natural disaster, for example. Although these problems are not in the current scope that is focused on demand-driven propagation of financial distress, they could be considered in the aforementioned dynamics of balance-sheets.

Thirdly, a distressed firm does not affect its neighboring firms once it enters an inactive state in the current model, as exemplified in the dynamics given by Eq. (9.3) and Eq. (9.4). Since it is usually the case that such a distressed firm may continue in business, affecting the neighbors, we may consider a variation of the original model so as to include amplification by such firms. Our estimation of DebtRank may be regarded, therefore, as a lower bound of the financial distress in the system. It would be an interesting problem to extend the model so that one can include the process of such amplification.

Fourthly, we note that one can employ the data set in other snapshots available in the RIETI project corresponding to a recent year, so that one is able to compare the results for more than one network and to see how robust our results are, what are the possible changes in network structure, which are the results specific to the year 2011, right after the East Japan Disaster, and so forth, while we believe that the results stated in this paper do not depend on a particular year. In addition, one needs to examine *random networks*, preserving macroscopic variables such as degree (number of suppliers and customers) but otherwise random, as a null hypothesis for statistical validation. Comparison of results from other snapshots of production network and also randomized networks as a null hypothesis is an important problem that can be addressed in future.

Concerning possible implications for policies related to small and medium enterprises (SMEs), we state the following: Current SME policies include safety-net guarantee programs. This program supports SMEs whose business stability is threatened by external factors, such as a major customer's restricted operations or application for bankruptcy, by making additional credit guarantees available. One of the strategies of this program is aimed at mitigating the possibility of chained bankruptcies of SMEs, each of which provides a credit of accounts receivable to the bankrupted customer of a large-sized firm. The policy of making additional credit guarantees available should be based on a certain evaluation for the propagation of financial stress in the system, because of the budget constraint for credit guarantees. The current model can serve as a benchmark for such evaluation.

9.1.4 Summary

We apply the methodology of DebtRank (Battiston et al., 2012), to the propagation of financial distress along the supplier–customer links from the downstream of customers and to the upstream of suppliers. If a customer does not fulfill the payment to its suppliers owing to a financial distress, then its suppliers are also possibly under financial distress potentially causing propagation of distress.

Assuming that the propagation takes place in the opposite direction to the supplier–customer relation $i \to j$, namely from customers to suppliers, $j \Rightarrow i$, and that the strength of propagation W_{ji} is given by the inverse of the number of customers js for firm i, we perform the simulation and computation of the DebtRank on a million firms and millions of links by supercomputers, including the world's fastest K computer. Such computation under many different scenarios has been difficult in practical computation time.

We show that the distribution of DebtRank for individual firms obeys a power-law in a significant correlation between the DebtRank and size for each firm. This fact is not trivial in the sense that big firms are affected by many connected but less-depending firms, while smaller firms are strongly influenced by distress. There is an interesting non-linearity, namely that the DebtRank becomes much larger than what is expected by a linear relationship between the DebtRank and firm-size. This implies that the role of big firms are usually underestimated.

By calculating the DebtRank of individual sectors, we show that there is a linear relationship between the logarithms of a sector's size and DebtRank, but also that there is a deviation owing to the location of the sector in the upstream and downstream portions of the network. This implies that the linear relationship between the logarithms of a sector's size and DebtRank guides us to pay attention to deviations from it depending on individual sectors.

Finally, we show that one can measure vulnerability in the methodology of DebtRank which can be potentially useful to identify the likelihood of failures of firms in those more vulnerable sectors. One will be able to use simulations on supercomputers under many scenarios, models of financial distress propagation, and various initial configurations.

9.2 Bank–Firm Credit Network

Let us now analyze the lending/borrowing relationship between Japanese banks and firms, which form bipartite credit networks. We introduce distress to some initial node(s) (banks or firms) and let it propagate and contaminate other nodes in this network according to their relative exposure. First, by choosing the initial node to be a bank and taking the weighted average of the resulting distress distribution, with the weights proportional to the size (total assets) of each node, we identify the bank's importance to the whole network at the time of a crisis. It leads to a nonlinear relationship between the importance and the size of the bank, which implies that mergers with same-sized partners would result in most increase in importance. Secondly, by introducing the initial distress to firms in certain industrial sector(s), we evaluate the vulnerability of banks and firms in other sectors owing to the distress in the initial sectors (Aoyama et al., 2013b).

9.2.1 Japanese bank–firm data

The database we analyze is an annual list of bank-loans (both long-term and short-term) to firms in Japan, which form a weighted bipartite network (Fujiwara et al., 2009; De Masi et al., 2011). See also recent works by Marotta et al. (2015, 2016) for more studies.

It covers all banks (including 'savings and loans' type of monetary institutions, called "regional banks") and large firms, most of which are listed. Their total numbers for the years 1980 to 2011 are plotted in Fig. 9.9. A network visualization is given in Fig. 9.10, where we see that city banks (redsquares) are in the center of the upper layer, meaning that they have many links (loans) to firms in the lower layer, and therefore play major roles, while regional banks are around the peripheral, indication that they play peripheral roles in this system, although they are numerous in numbers.

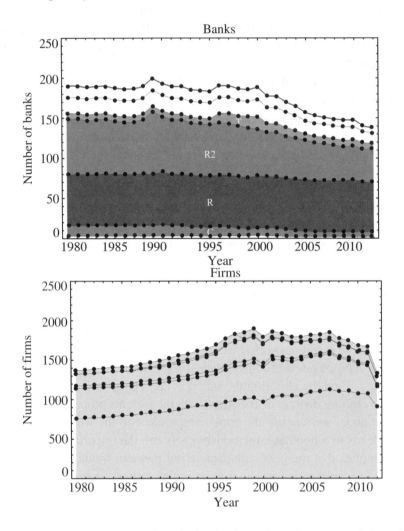

Fig. 9.9 Number of banks and firms in the database (from Aoyama et al. (2013b)).

We denote[1] banks and firms by Greek letters β ($\mu = 1, 2, \ldots, B$) and Latin letters f ($f = 1, \ldots, F$) respectively, and B is the number of banks, and F is that of firms. An edge between a bank β and a firm f is defined to be present if there is a credit relationship between them.

---

[1]The notation differs from Fujiwara et al. (2009), so that the name (bank or firm), the index (β and f) and the total numbers (B and F) match.

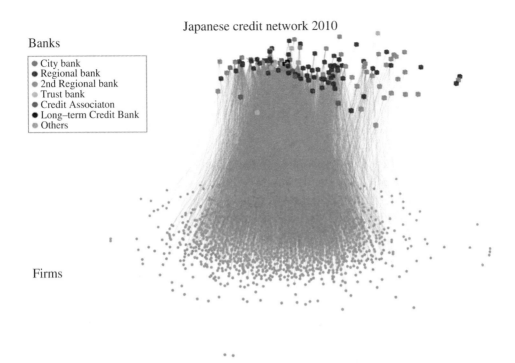

Fig. 9.10 Network formed by banks (upper layer) and firms (lower layer) in 2010, where edges are bank-loans unweighted by the amount (from Aoyama et al. (2013b)) [see Color Plate at the end of book].

We note that there are no inter-bank edges or inter-firm edges in our system. Technically, this is simply owing to the fact that they are not available to us currently. However, one might argue that they are not relevant for our current purpose, which is to examine banks' importance and vulnerability at the time of a crisis: Any distress or failure of a bank might affect the whole economic system, as other banks absorb its damage through government intervention and not because of any pre-crisis inter-bank relationship. Firms may go through chain-bankruptcy owing to bankruptcy of other firms with whom they trade, for which we need information of the inter-firm trading network. However, its effect on the whole system may go through their strongest ties to the banks, as they fail to repay the loans. For these reasons, we believe that the current analysis with bank–firm bipartite network data would give us a good clue as to the importance and vulnerability of bank(s).

Lending patterns of financial institutions have similarity arising from geographical regions. It is obvious that regional banks often lend to a limited but overlapping set of firms in the same local regions, and also that city and mega banks provide credit to many and overlapping sets of firms. As we have shown, some results of the DebtRank analysis can be interpreted in terms of the similarity of lending, which we elaborate here.

Let us define a distance between a pair of banks based on their lending patterns. A lending pattern of a bank β is a vector $\boldsymbol{x}_\beta$ of dimension F, the number of firms. Each component is given by

$$(\boldsymbol{x}_\beta)_f \propto \begin{cases} 1 & \text{if } \beta \text{ lends to } f \\ 0 & \text{otherwise} \end{cases}, \tag{9.17}$$

for $f = 1, \cdots, F$, namely 1 or 0 according to the presence or absence of the credit relation between bank β and firm f. Alternatively, one could use the information of lending weight, but it is sufficient to use the information of 0/1 for our purpose. Then, we normalize the vector so that $|\boldsymbol{x}_\beta| = 1$, and define the distance between β and α as

$$d(\beta, \alpha) = |\boldsymbol{x}_\beta - \boldsymbol{x}_\alpha| = \sqrt{2(1 - \boldsymbol{x}_\beta \cdot \boldsymbol{x}_\alpha)}. \tag{9.18}$$

We employ the well-known method of minimum spanning tree (MST) as a compact representation of the entire map of similarities among banks. An MST is an undirected and tree graph of nodes $\beta = 1, \cdots, B$ and $(B - 1)$ edges. Assume that the edge for a pair of nodes (β, α) has a weight equal to $d(\beta, \alpha)$. One can construct such a tree arbitrarily, and the MST is the one way the total sum of weights of resulting edges in the tree is a minimum. By this construction, a pair of nodes whose distance is small is more likely to be located with a smaller shortest-path in the tree. Therefore the banks are likely to be closely located in the MST.

Figure 9.11 reflects the data in the year 2010. We can identify geographical regions in the MST as sets of banks having similar patterns specific to those regions, as shown by circles with colors corresponding to eight regions in Japan. Moreover, in the bottom and left portion, a set of city mega banks is present implying that they provide credit to many and overlapping firms.

Let us use the DebtRank concept elaborated earlier in Section 9.1.1, to evaluate network effects in this system.

The amount $C_{\beta f}$ associated with the edge is the amount of the credit (total lending by the bank β to the firm f), as illustrated in Fig. 9.12. The propagation matrix element w_{ji} is defined to the relative exposure;

$$w_{f\beta} := \frac{C_{\beta f}}{C_\beta}, \tag{9.19}$$

$$w_{\beta f} := \frac{C_{\beta f}}{C_f}. \tag{9.20}$$

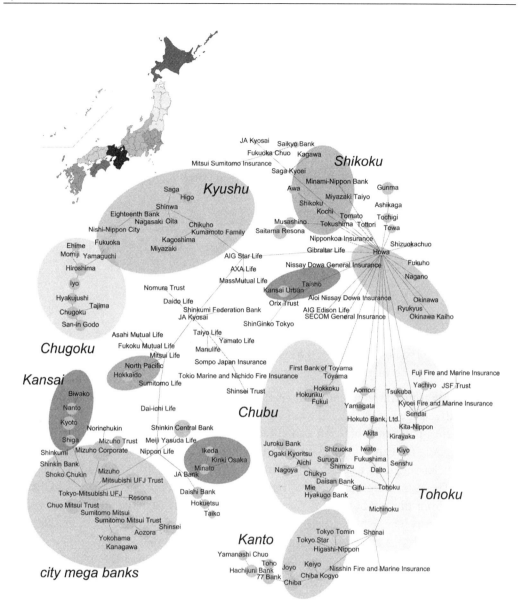

Fig. 9.11 Minimum spanning tree for the nodes of banks and the edges with weights of distances between banks. Distance of a pair of nodes is smaller if the banks have similar lending to firms. The node size is proportional to the logarithm of the bank's asset. Circles with colors are identified as geographical regions and city mega banks. (Upper-left) Japan and eight geographical regions; Hokkaido (red), Tohoku (yellow), Kanto including Tokyo (green), Chubu (cyan), Kansai including Osaka and Kyoto (blue), Chugoku (orange), Shikoku (purple), Kyushu including Okinawa (gray) (from Aoyama et al. (2013b)) [see Color Plate at the end of book].

where C_β is the total amount of lending by bank β;

$$C_\beta := \sum_f C_{\beta f}, \qquad (9.21)$$

and likewise

$$C_f := \sum_\beta C_{\beta f}. \qquad (9.22)$$

The CDFs of C_β and C_f are plotted in Fig. 9.13 from 1980 to 2012.

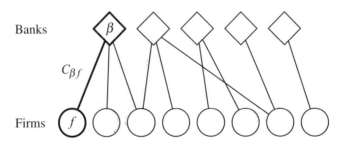

Fig. 9.12 The quantity $C_{\beta f}$ is the sum of the long-term loans and the short-term loans in the year of discussion from bank β to firm f (from Aoyama et al. (2013b)).

The propagation matrix element, Eq. (9.19), reflects the fact that once the firm f is in distress, it affects the banks β, from whom the firm f has borrowed the amount $C_{\beta f}$, through delayed interest payments, or, even a total failure to repay the borrowed money. We model this by using relative exposure of the bank to the firm and *not* the absolute amount, because if the bank is lending a lot more to other firms, the effect of the firm f would be small, and vice versa. The latter weight, Eq. (9.20) rises from the fact that once a bank β is in distress, it affects the firms that are borrowing from it through an increase in the interest rate, request for further security, etc., which are again relative exposure of the firm to the bank. Note that these are identical to A and B in Fujiwara et al. (2009), respectively.

As stated in the previous section, after all the distress propagation is completed at time-step $t = T$, we define two weighted averages of the distress: The DebtRank on the bank layer;

$$D_A^{(\text{banks})} = \sum_{\beta=1, \notin A}^{B} \hat{a}_\beta h_\beta(T), \quad \hat{a}_\beta := a_\beta \bigg/ \sum_{\beta'=1, \notin A}^{B} a_{\beta'}, \qquad (9.23)$$

where a_β is the total asset of the bank β, so that the larger the node is, the larger its distress is counted in. Note that we *exclude* the node in the initial set A in the sums, so that the resulting DebtRank is not a simple reflection of the initial node(s): without this, a large initial node contributes (initial) $h = 1$ with large weight, thus resulting in a large DebtRank. In other words, by excluding the initial node(s), our DebtRank is a measure of how the initial node(s) affect *other* nodes in the network.

Systemic Risks

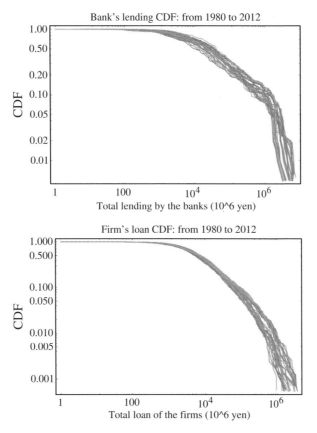

Fig. 9.13 CDF (cumulative distribution function) of C_β (left) and C_f (right) from the year 1980 to 2012 (from Aoyama et al. (2013b)).

Similarly, the DebtRank of the firm layer;

$$D_A^{(\text{firms})} = \sum_{f=1}^{F} \hat{a}_f h_f(T), \quad \hat{a}_f := a_f \Big/ \sum_{f'=1}^{B} a_{f'}, \quad (9.24)$$

where a_f is the total asset of the firm f. This DebtRank on the firm layer $D_A^{(\text{firms})}$ is a measure of the distress caused to firms.

By using these two DebtRanks, $D^{(\text{banks})}$ and $D^{(\text{firms})}$, we shall quantify importance and vulnerability of the node(s) A.

9.2.2 'Too big to fail?' and other questions

We first choose the set $A = \{\beta_0\}$ to evaluate the bank β_0's importance in the system. The result for the year 2010 is given in Fig. 9.14 in linear-scale (left) and in log-scale (right). The dashed line in the latter plot is the best-fit power-law,

$$D^{(\text{firms})} = 1.76 \, D^{(\text{banks})\,1.13}, \quad (9.25)$$

which hold as a good relationship as an average and any deviation from this is a measure of their characteristics in lending practice: Large banks such as Tokyo Mitsubishi UFJ and Mizuho Corporate show that their effect on firms are larger, while Mizuho and Risona have more of an effect on other firms. Trust banks in general have a higher effect on firms than the average, which is a natural consequence of the fact that their role is to support firms in financing and management services (Trust Companies Association of Japan, 2013).

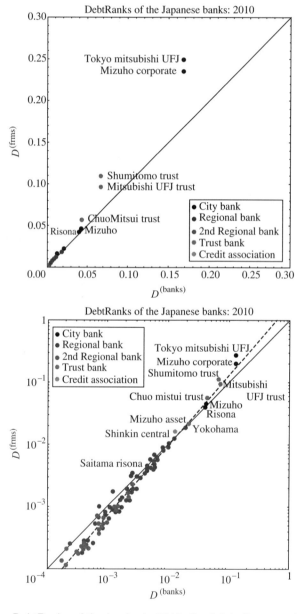

Fig. 9.14 The DebtRanks of the banks in 2010, the left in linear scale, the right in log scale (from Aoyama et al. (2013b)) [see Color Plate at the end of book].

This plot and the average behavior explained earlier have several implications, which we shall elaborate on now.

Figure. 9.15 is the distribution of the banks on the plane of (Asset, The total DebtRank $(D^{(\text{banks})} + D^{(\text{firms})})$). We find here a general trend that the larger the bank is, the larger is the total DebtRank, with the best-fit

$$D^{(\text{banks})} + D^{(\text{firms})} = 6.55 \times 10^{-19} S^{1.50}, \qquad (9.26)$$

where S is the size (total asset) of the bank.

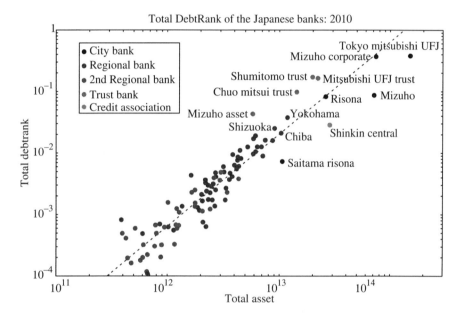

Fig. 9.15 Total asset vs total DebtRank in 2010 (from Aoyama et al. (2013b)) [see Color Plate at the end of book].

Too big to fail? Nonlinearity, the power behavior in Eq. (9.26), evidently shows the importance of the size of the bank. This, no matter how trivial it sounds, is relevant in the context of DebtRank, as our DebtRank *excludes* the bank in question (see Eq. (9.23)) and measures its importance to *other* banks and firms. Therefore, we obtain an independent and objective conclusion that big banks are important *in average*. On the other hand, if one looks at the deviations from this average behavior, we find that this statement is not always true. The biggest bank, Tokyo Mitsubishi UFJ has total assets that are twice as large as those of Mizuho Corporate but has about the same level of DebtRank. Similarly, Mizuho's total assets is about 2.7 times as large as that of Risona and yet they have the same level of DebtRank, and so forth. Most of these deviations from the average behavior comes from difference in their lending practice; in general, trust banks, as noted before, stress their role in lending to firms and this explains their high DebtRanks. Shikin Central

Bank is another case in point: it plays the role of a central bank for all the trust banks, and its importance is underrated from this analysis of lending to firms.

Nonlinearity and merger The fact that the exponent is larger than one ($\simeq 1.50$ in Eq. (9.26)) is a significant result: it means that if the bank becomes, say, twice as big, its total DebtRank becomes 2.82 times as before. Therefore, it means that more than simply stating that "big banks are important", one should say "big banks are *far more* important than small banks." This provides a strong motivation for mergers. Let us think of bank mergers in a more practical situation. Imagine two banks of size S_1 and S_2. Since

$$S_1^\alpha + S_2^\alpha < (S_1 + S_2)^\alpha, \tag{9.27}$$

for $\alpha > 1, S_1 \neq 0, S_1 \neq 0$, their merger will result in a total DebtRank that is larger than the sum of their DebtRank. In fact, their ratio is a function of S_1/S_2;

$$\frac{(S_1 + S_2)^\alpha}{S_1^\alpha + S_2^\alpha} = R\left(\frac{S_1}{S_2}\right), \tag{9.28}$$

which has a maximum at $S_1/S_2 = 1$ *regardless of* the value of the exponent α as long as it is greater than 1 (Fig. 9.16), with the peak value $R(1) = 2^{\alpha-1}$. Therefore, we conclude that by merging with an equal-sized partner, they achieve the maximum increase in their importance.[2]

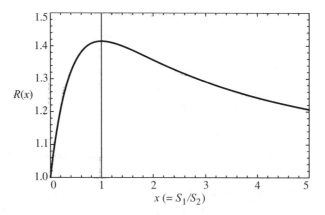

Fig. 9.16 Plot of the importance ratio function $R(x) = (x+1)^\alpha/(x^\alpha+1)$ in Eq. (9.28) for $\alpha = 1.50$ (from Aoyama et al. (2013b)).

We note that this is a general discussion on mergers and individual cases differ from this because of several reasons, some of which are; (1) the deviation of the agents from the average curve, (2) merger is associated with separation of divisions from

---

[2]It is curious to note that if $\alpha < 1$, the total DebtRank is always less than the sum of the total DebtRanks before the merger, whose ratio is minimum at $S_1 = S_2$.

either agents and asset management/reduction, as they try to cope with a difficult transition, and (3) the aforementioned discussion ignores changes in the lending structure by the merger. More specifically, the propagation matrix elements, Eqs. (9.19) and (9.20), are not homogeneous in the amount of the lending $C_{\beta f}$.

For example, Sanwa Bank and Tokai Bank merged to form the UFJ bank in 2002. Before the merger, their DebtRanks were 0.23 and 0.14, respectively and after the merger, it was 0.33, slightly less than the simple sum of the DebtRanks before the merger, contrary to general theory. Their total assets, however, were reduced by 20%, which disqualifies the direct application of the general theory.

9.2.3 Vulnerability

The distress $h_{f,\beta}$ can be used as a measure of vulnerability of nodes in this network.

Let us put firms in certain industrial sectors in distress, by choosing them to be in the initial distress set A, and all the other firms and banks $\notin A$. Then, the resulting distress h_β (after all the propagation) is the measure of vulnerability of bank β to the failure or distress in those sectors.

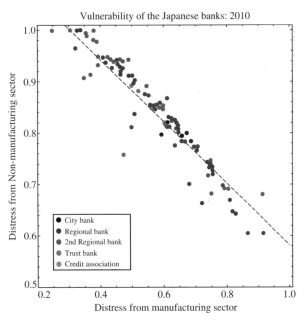

Fig. 9.17 Vulnerability of banks to distress in the manufacturing sectors and the distress in the non-manufacturing sectors (from Aoyama et al. (2013b)) [see Color Plate at the end of book].

First, we choose A to consist of all the firms in the manufacturing sector and measure the vulnerability of banks. Then, we do the same for the non-manufacturing (service) sector. Figure. 9.17 is the plot of the resulting vulnerabilities, where the dashed line is the best-fit linear function, $y = 1.18 - 0.60x$ with x and y being the horizontal and the vertical coordinates, respectively. We observe here that the city banks are well balanced, while regional and second regional banks are widely distributed, which is in agreement with the fact that city banks are large and lend to a wide spectrum of firms, while the (second) regional banks tend to be small and their lending may be limited to a small set of firms. On the other hand, we see that this method of studying vulnerability is a powerful one to identify small firms with a limited lending practice that makes them quite vulnerable to various systemic crisis.

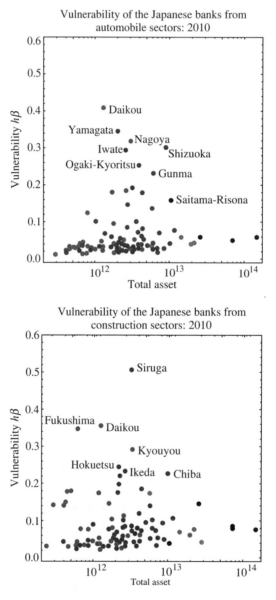

Fig. 9.18 Vulnerability of banks to distress in the automobile sector (left) and the construction sector (right) (from Aoyama et al. (2013b)).

Another useful aspect to study is the vulnerability of banks to distress in the automobile sector, and in the construction sector, since they are most affected by external shocks, as was true at the time of the Lehman shock.[3] The results are shown in Fig. 9.18 together with their total assets.

Doing this kind of analysis for all 33 sectors (see Table 9.3), we find the resulting distress for all 159 banks shown in Fig. 9.19. In the top matrix view, we observe the bright (white to red) column, which is Sector no. 23, "Credit & Leasing", followed by Sector no. 22, "Retail Trade" and then by Sector no. 21 "Wholesale Trade", which we find are important sectors for the stability of monetary institutions.

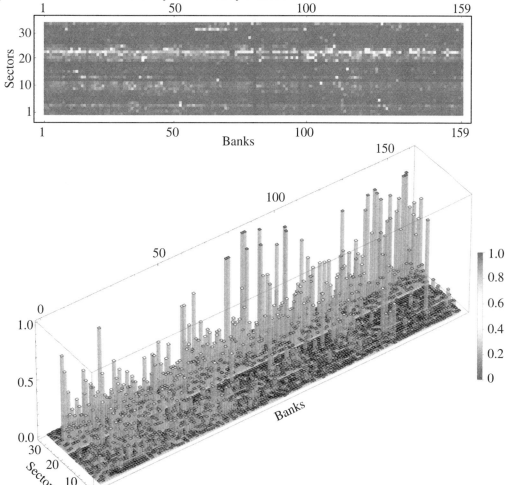

Fig. 9.19 Vulnerability (distress) of 159 banks caused by each of the 33 sectors in 2010, in matrix view (top) and 3D view (bottom) (from Aoyama et al. (2013b)).

---

[3] As is explained before, the sub-prime mortgage crisis is called "Lehman shock" (not even "Lehman Brothers shock") in Japan. We therefore use this term here as we are discussing Japanese economics.

Table 9.3 All 33 sectors specified in the Nikkei database. Code in 100s (No.1–17) refer to the manufacturing sector and those in 200s (No.18–33) to the non-manufacturing sector (from Aoyama et al. (2013b)).

No.	Nikkei code	Sector
1	101	Foods
2	103	Textile Products
3	105	Pulp & Paper
4	107	Chemicals
5	109	Drugs
6	111	Petroleum
7	113	Rubber Products
8	115	Stone, Clay & Glass Products
9	117	Iron & Steel
10	119	Non ferrous Metal & Metal Products
11	121	Machinery
12	123	Electric & Electronic Equipment
13	125	Shipbuilding & Repairing
14	127	Motor Vehicles & Auto Parts
15	129	Transportation Equipment
16	131	Precision Equipment
17	133	Other Manufacturing
18	235	Fish & Marine Products
19	237	Mining
20	241	Construction
21	243	Wholesale Trade
22	245	Retail Trade
23	252	Credit & Leasing
24	253	Real Estate
25	255	Railroad Transportation
26	257	Trucking
27	259	Sea Transportation
28	261	Air Transportation
29	263	Warehousing & Harbor Transportation
30	265	Communication Services
31	267	Utilities - Electric
32	269	Utilities - Gas
33	271	Other Services

Let us look at the vulnerability of one sector to another, which we measure by means of h_f for the firms in the former sector, caused by propagation of distress of the firms in the latter sector. This is shown in Fig. 9.20. Here again we observe the bright column apparent in the matrix view (left), which is Sector no. 23, "Credit & Leasing", in keeping with the aforementioned observation.

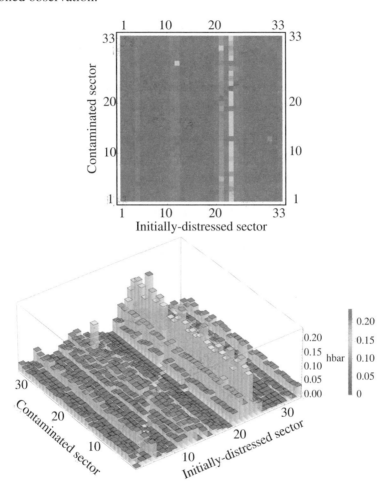

Fig. 9.20 Vulnerability (distress) of sectors caused by other sectors, in matrix view (left) and 3D view (right) (from Aoyama et al. (2013b)).

9.2.4 Summary

We have shown that DebtRank approach is a powerful approach that provides one with a measure of importance of the nodes as well as vulnerability of the nodes at times of crisis.

We defined two DebtRanks, one for the bank-layer and the other for the firm-layer, and found that they are almost equal with slight non-linearity, as in Eq. (9.25). Their sum, the total Debt-Rank, is a good measure of their importance in the network. On average, it has a good correlation with the bank's size (total asset) as in Eq. (9.26), whose exponent 1.50 implies that a merger with a same-sized partner is the optimal solution for an increase of the total DebtRank.

Vulnerability is measured by how much the node(s) in question receives distress from a particular set of node(s). This has a wide range of applications for investigating the weak spots to particular types of crisis. In this paper, we have identified banks that are quite vulnerable to crisis in the automobile sector. We have found that the distress in the "Credit & Leasing sector" affects all sectors most, which implies that it is fairly important to keep them in good financial standing.

In the current research, we have reported mainly the results of the analysis of 2010. The data we have, however, covers 1980 to 2012 and therefore comparing the results of each year would be important.

There are several missing links in our consideration, which are interbank (bank–bank) and inter-firm (firm–firm) interactions. The latter can be extracted from the trading network discussed earlier in this chapter. Such mergers of database is required to advance this line of research.

Another important improvement of the DebtRank approach would be to incorporate in some way the constraint on capital-to-asset ratio for banks by the Basel Capital Accords. This may act as a threshold to the distress variable, so that once the ratio is below a pre-set mark, the distress variable may be enhanced. We may look into this modification of the current method in the near future.

Appendix A
Computer Programs for Beginners

> People think that computer science is the art of geniuses but the actual reality is the opposite, just many people doing things that build on each other, like a wall of mini stones.
>
> Donald Knuth

A.1 *Mathematica* Codes for Finance

If you want to start the study of macro-econophysics, economics, finance, physics, statistics, and mathematics, through analysis of real data, *Mathematica* is a useful tool. *Mathematica* provides data and tools. We list a few useful commands for finance here.

```
 In[1]:= SetDirectory[NotebookDirectory[]]
 In[2]:= FinancialData["Classes"]
 In[3]:= FinancialData["Exchanges"]
 In[4]:= FinancialData["NYSE*"]
 In[5]:= Take[FinancialData["NYSE*"], 20]
 In[6]:= Take[FinancialData["NASDAQ*"], 20]
 In[7]:= Take[FinancialData["^*"], 20]
 In[8]:= FinancialData["NASDAQ:AAPL", "Properties"]
 In[9]:= FinancialData["NASDAQ:AAPL", "Name"]
In[10]:= FinancialData["AAPL", "Exchange"]
```

```
In[11] := FinancialData["^DJI", "Name"]
In[12] := FinancialData["SP500", "Name"]
In[13] := FinancialData["^GSPC", "Name"]
In[14] := FinancialData["NASDAQ:AAPL", "OHLCV"]
In[15] := Take[FinancialData["NASDAQ:AAPL", "OHLCV",All],20]
In[16] := DateListPlot[FinancialData["NASDAQ:AAPL", All], PlotRange -> All]
In[17] := Export["Apple.csv", FinancialData["NASDAQ:AAPL", "OHLCV", All]]
In[18] := DateListPlot[FinancialData["NASDAQ:AAPL", "Return", All], PlotRange
             -> All]
In[19] := data = FinancialData["NASDAQ:AAPL", "OHLCV", All];
          logret = Table[{data[[i, 1]], Log[data[[i, 2, 4]]/data[[i, 2,
          1]]]}, {i, Length[data]}];
          Take[logret, 20]
          DateListPlot[logret, PlotRange -> All]
In[20] := data = FinancialData["NYSE:*"];
          Take[data, 10]
          n = Length[data]
In[21] := For[i = 1, i < 4, i++,
          Export[StringReplace[data[[i]], "NYSE:" -> ""] <> ".csv",
          FinancialData[data[[i]], "OHLCV", All]]]
In[22] := For[i = 1, i < n + 1, i++,
          Export[StringReplace[data[[i]], "NYSE:" -> ""] <> ".csv",
          FinancialData[data[[i]], "OHLCV", All]]]
```

A.2 Tools for Network Analysis

Many tools for network analysis and visualization are available today. Readers are able to find them using a web search. Here we list a limited number of them based on a somewhat biased selection.

Pajek (http://mrvar.fdv.uni-lj.si/pajek)

Netminer (http://www.netminer.com)

UCINET (https://sites.google.com/site/ucinetsoftware)
  These three are originally developed in sociology for social network analysis, while recent development enables faster computation for larger networks.

Appendix 339

NetworkX (https://networkx.github.io)
> Python library; network analysis and visualization, flexible as a script language; also applicable to small-scale visualization. `anaconda`, all scientific libraries in a single package, includes this and the required libraries, and is better for easy installation.

igraph (http://igraph.org)
> R library; network analysis and visualization, flexible as a script language with abundant tools of R, statistical computing and graphics; easy to install. Python and C versions are also available.

Gephi (https://gephi.org)
> Network visualization with basic tools of network analyses included; applicable to relatively large-scale networks.

JUNG (http://jung.sourceforge.net)
> Network visualization applicable to large-scale networks.

Net Workbench (http://nwb.cns.iu.edu)
> Toolbox of network analyses and visualization originally developed in computer science and complex networks.

Boost graph library (http://www.boost.org)
> C++ library for data structure of graphs; generic programming with efficient computation.

LEDA/AGD (http://www.algorithmic-solutions.com/leda)
> C++ library for data structure and algorithms for graphs; commercially available.

yEd (http://www.yworks.com/products/yed)
> Graph editor, frequently used for figures in a paper.

Graphviz (http://www.graphviz.org)
> Visualization of graphs with basic tools of graph algorithms; long history with fast computation and also handy visualization up to middle-scale networks.

Mathematica (http://www.wolfram.com)
> This includes a nice toolbox for graph theoretical analyses, network analyses and visualization. See also Section A.4.

A.3 Python Codes for Basic Graph Algorithms

In Section 8.1.1, we explained elementary graph-search algorithms in order to use them in simple applications, potentially useful to economic networks, which are not fully explained elsewhere.

It is not easy to understand why these algorithms work for generic graphs, since they involve genius tricks, somewhat similar to those in *chess* or *go*, for non-experts. Readers are highly recommended to appreciate how they work by doing experiments for as many examples as possible, rather than attempting to prove mathematically why they work. For a full exposition, see Cormen et al. (2001); Mehlhorn and Naher (1999); Siek et al. (2002) and also Newman (2010, Part III) for a concise guide.

In the following, we encourage the readers by providing computer programs in Python, which one can write and execute easily, not depending on any other library nor requiring Compile[4]. We assume the reader's familiarity with the language of Python, but it should not be difficult to understand the source codes even if you are not familiar with it at all.

Program A.1: Read a list of nodes and a list of links

```
1  import getopt
2  import sys
3
4  try:
5      opt, argv = getopt.getopt(sys.argv[1:], 'd')
6  except getopt.GetoptError, ge:
7      sys.stdout.write(ge.msg + "\n")
8      exit()
9
10 if len(argv) != 2:
11     sys.stdout.write("args: [-d] nodes.dat edges.dat\n")
12     exit()
13
14 fn_n = argv[0]
15 fn_e = argv[1]
16
17 nodes = []
18 adj = {}
19 fin = open(fn_n)
20 for line in fin:
21     n = line.strip()
22     nodes.append(n)
23     adj[n] = []
24 fin.close()
25
26 fin = open(fn_e)
27 for line in fin:
28     line = line.strip()
29     f = line.split(' ', 2)
30     m = f[0]
31     n = f[1]
32     adj[m].append(n)
33     adj[n].append(m) if ('-d', '') not in opt else None
34 fin.close()
35
36 print(nodes)
37 print(adj)
```

---

[4] Thanks to Yudai Fujiwara for converting the programs originally written in Ruby to Python.

Note that the nodes are represented by a variable nodes, which is a list, i.e., an array of nodes. The links are expressed as an adjacency list by a variable adj. This is a dictionary, i.e., an associative array mapping a "key" to a "value", where a key is a node and a value is a list of nodes that are the opposite ends of the links emanating from the node of the key.

Given a list of nodes (left column in the following) and a list of links (middle column) for the undirected graph of Fig. 8.6, this program is executed and generates the output as shown[5] (right column).

```
# ex1.n     # ex1.e     $ python readgraph.py ex1.n ex1.e
r           r v
s           s w         ['r', 's', 't', 'u', 'v', 'w', 'x', 'y']
t           s r         {'r':['v', 's'],
u           w t          's':['w', 'r'],
v           w x          't':['w', 'u', 'x'],
w           t u          'u':['t', 'y', 'x'],
x           t x          'v':['r'],
y           u y          'w':['s', 't', 'x'],
            x u          'x':['w', 't', 'u', 'y'],
            x y          'y':['u', 'x']}
```

For a directed graph, we put the option -d in execution. An example for the directed graph in Fig. 8.2 is shown here.

```
# ex2.n     # ex2.e     $ python readgraph.py -d ex2.n ex2.e
a           a b
b           a d         ['a', 'b', 'c', 'd', 'e', 'f']
c           b d         {'a':['b', 'd'],
d           b e          'b':['d', 'e'],
e           c e          'c':['e', 'f'],
f           c f          'd':[],
            e d          'e':['d'],
            f f          'f':['f']}
```

Note that for the undirected graph, the link (r,v) appears in adj twice as 'r':['v','s'] and 'v':['r']. For the directed graph, the link (a,b) appears once in 'a':['b', 'd'].

Now let us assume that we have read the data of nodes and adj by using the Program A.1 to implement the breadth-first search algorithm given by Algorithm 8.5 explained in Section 8.1.1. The implementation is given in Program A.2.

Program A.2: Breadth-first search

```
1  # bfs.py
2  color = {}
3  distance = {}
```

---
[5]The order of lists in the output of print(adj) was manually sorted for clarity.

```
 4  predecessor = {}
 5
 6  # Initialize
 7  for v in nodes:
 8      color[v] = "WHITE"
 9      distance[v] = -1      # for infinity
10      predecessor[v] = v    # for nil
11  color[s] = "GRAY"
12  distance[s] = 0
13
14  Q = []
15  Q.append(s)
16  print("Q= {0}".format(Q))
17  while Q:
18      v = Q.pop(0)      # pop
19      print("{0}: distance= {1}".format(v, distance[v]))
20      for w in adj[v]:
21          if color[w] == "WHITE":
22              print("({0},{1}:white): a tree edge".format(v, w))
23              color[w] = "GRAY"
24              distance[w] = distance[v] + 1
25              predecessor[w] = v
26              Q.append(w)
27          elif color[w] == "GRAY":
28              if distance[w] == distance[v] + 1:
29                  msg = "a cross edge /dist.diff.= 1"
30              elif distance[w] == distance[v]:
31                  msg = "a cross edge /dist.diff.= 0"
32              else:
33                  msg = "this case should be impossible!"
34              print("({0},{1}:gray): {2}".format(v, w, msg))
35          else:
36              print("({0},{1}:black): already done".format(v, w))
37      print("Q= {0}".format(Q))
38      color[v] = "BLACK"
```

For the example of undirected graph given earlier, which was also shown in Fig. 8.7 in Section 8.1.1, we have the output as follows. Note that actually one needs to insert Program A.1 in the beginning of Program A.3.

```
$ python bfs.py ex1.n ex1.e

Q= ['s']
s: distance= 0
(s,w:white): a tree edge
(s,r:white): a tree edge
Q= ['w', 'r']
w: distance= 1
(w,s:black): already done
(w,t:white): a tree edge
(w,x:white): a tree edge
Q= ['r', 't', 'x']
r: distance= 1
(r,v:white): a tree edge
(r,s:black): already done
Q= ['t', 'x', 'v']
t: distance= 2
(t,w:black): already done
(t,u:white): a tree edge
(t,x:gray): a cross edge /dist.diff.= 0
```

```
Q= ['x', 'v', 'u']                    u: distance= 3
x: distance= 2                        (u,t:black): already done
(x,w:black): already done             (u,y:gray): a cross edge /di. t.diff.= 0
(x,t:black): already done             (u,x:black): already done
(x,u:gray): a cross edge /dist.diff.= 1   Q= ['y']
(x,y:white): a tree edge              y: distance= 3
Q= ['v', 'u', 'y']                    (y,u:black): already done
v: distance= 2                        (y,x:black): already done
(v,r:black): already done             Q= []
Q= ['u', 'y']
```

Similarly, the depth-first search algorithm given by Algorithm 8.1 and Algorithm 8.2 explained in Section 8.1.1 can be readily implemented as shown in Program A.3.

Program A.3:   Depth-first search

```
1   # dfs.py
2
3   # DFSvisit
4   def dfs_visit(v):
5       global color, discover, finish, predecessor, adj, time, tab
6       color[v] = "GRAY"
7       time += 1
8       discover[v] = time
9       tab += "  |"
10      print("{0}{1} discovered at time= {2}".format(tab, v, discover[v]))
11      for w in adj[v]:
12          if color[w] == "WHITE":
13              print("{0}({1},{2}:white): a tree edge".format(tab, v, w))
14              predecessor[w] = v
15              dfs_visit(w)
16          elif color[w] == "GRAY":
17              if predecessor[v] != w:
18                  print("{0}({1},{2}:gray): a back edge".format(tab, v, w))
19              else:
20                  if discover[v] > discover[w]:
21                      msg = "a cross edge"
22                  else:
23                      msg = "a forward edge"
24                  print("{0}({1},{2}:black): {3}".format(tab, v, w, msg))
25      color[v] = "BLACK"
26      time += 1
27      finish[v] = time
28      print("{0}{1} finished at time= {2}".format(tab, v, finish[v]))
29      tab = tab[3:]
30
31  # DFS
32  color = {}
33  discover = {}
34  finish = {}
35  predecessor = {}
```

```
36  tab = ""      # for log
37
38  # Initialize
39  for v in nodes:
40      color[v] = "WHITE"
41      discover[v] = -1
42      finish[v] = -1
43      predecessor[v] = v
44
45  time = 0
46  for v in nodes:
47      if color[v] == "WHITE":
48          dfs_visit(v)
```

For the example of directed graph given earlier, which was also shown in Fig. 8.3 in Section 8.1.1, we have the output as follows. Note that actually one needs to insert Program A.1 in the beginning of Program A.3. The option -d is put in the command line to run the program of dfs.py for a directed graph.

```
$ python dfs.py -d ex2.n ex2.e

|a discovered at time= 1
|(a,b:white): a tree edge
|   |b discovered at time= 2
|   |(b,d:white): a tree edge
|   |   |d discovered at time= 3
|   |   |d finished at time= 4
|   |(b,e:white): a tree edge
|   |   |e discovered at time= 5
|   |   |(e,d:black): a cross edge
|   |   |e finished at time= 6
|   |b finished at time= 7
|(a,d:black): a forward edge
|a finished at time= 8
|c discovered at time= 9
|(c,e:black): a cross edge
|(c,f:white): a tree edge
|   |f discovered at time= 10
|   |(f,f:gray): a back edge
|   |f finished at time= 11
|c finished at time= 12
```

Programs for further algorithms and examples including

- Enumerate the neighbours of neighbours with no duplication. More generally, count the number of nth neighbours at n links away from a node or a set of nodes.
- How to find the so-called bow-tie structure to identify "upstream" and "downstream" for a directed network.

- DebtRank algorithm as an extension of breadth-first search.

as well as the ones in this appendix are downloadable from the URL:

http://www.econophysics.jp/book_macroeconophysics

A.4 *Mathematica* Codes for Network Analysis

We summarize *Mathematica* codes useful for network analysis.

```
In[1]:= SetDirectory[NotebookDirectory[]]
In[2]:= ExampleData["NetworkGraph"]
In[3]:= g = ExampleData[{"NetworkGraph", "ZacharyKarateClub"}]
In[4]:= FindGraphPartition[g, {0.38, 0.62}]
         HighlightGraph[g, Table[Subgraph[g, i], {i, %}]]
         N[GraphAssortativity[g, %%]]
In[5]:= FindGraphPartition[g]
         HighlightGraph[g, Table[Subgraph[g, i], {i, %}]]
         N[GraphAssortativity[g, %%]]
In[6]:= FindGraphPartition[g,3]
         HighlightGraph[g, Table[Subgraph[g, i], {i, %}]]
         N[GraphAssortativity[g, %%]]
In[7]:= FindGraphPartition[g,4]
         HighlightGraph[g, Table[Subgraph[g, i], {i, %}]]
         N[GraphAssortativity[g, %%]]
In[8]:= FindGraphCommunities[g, Method -> "Hierarchical"]
         HighlightGraph[g, Subgraph[g, #] & /@%, GraphHighlightStyle ->
      "DehighlightHide"]
         N[GraphAssortativity[g, %%]]
In[9]:= FindGraphCommunities[g, Method -> "Modularity"]
         HighlightGraph[g, Subgraph[g, #] & /@%, GraphHighlightStyle ->
      "DehighlightHide"]
         N[GraphAssortativity[g, %%]]
In[10]:= FindGraphCommunities[g, Method -> "Centrality"]
         HighlightGraph[g, Subgraph[g, #] & /@%, GraphHighlightStyle ->
      "DehighlightHide"]
```

```
        N[GraphAssortativity[g, %%]]
In[11]:= FindGraphCommunities[g, Method -> "CliquePercolation"]
        HighlightGraph[g, Subgraph[g, #] & /@%, GraphHighlightStyle ->
    "DehighlightHide"]
        N[GraphAssortativity[g, %%]]
In[12]:= FindGraphCommunities[g, Method -> "Spectral"]
        HighlightGraph[g, Subgraph[g, #] & /@%, GraphHighlightStyle ->
    "DehighlightHide"]
        N[GraphAssortativity[g, %%]]
In[13]:= FindClique[g]
        HighlightGraph[g, Subgraph[g, %]]
In[14]:= DegreeCentrality[g]
        HighlightGraph[g, VertexList[g], VertexSize -> Thread[VertexList[g]
    -> Rescale[%]]]
In[15]:= ClosenessCentrality[g]
        HighlightGraph[g, VertexList[g], VertexSize -> Thread[VertexList[g]
    -> Rescale[%]]]
In[16]:= BetweennessCentrality[g]
        HighlightGraph[g, VertexList[g], VertexSize -> Thread[VertexList[g]
    -> Rescale[%]]]
In[17]:= RadialityCentrality[g]
        HighlightGraph[g, VertexList[g], VertexSize -> Thread[VertexList[g]
    -> Rescale[%]]]
In[18]:= EccentricityCentrality[g]
        HighlightGraph[g, VertexList[g], VertexSize -> Thread[VertexList[g]
    -> Rescale[%]]]
In[19]:= PageRankCentrality[g, 0.1]
        HighlightGraph[g, VertexList[g], VertexSize -> Thread[VertexList[g]
    -> Rescale[%]]]
In[20]:= EdgeBetweennessCentrality[g]
        coloring = {EdgeList[g], Map[ColorData["TemperatureMap"],
    Rescale[%]]};
        HighlightGraph[g, Style[Style @@@ Transpose[coloring], Thick]]
In[21]:= EigenvectorCentrality[g]
        HighlightGraph[g, VertexList[g], VertexSize -> Thread[VertexList[g]
    -> Rescale[%]]]
```

Solution to Exercises

2.1 In the case of the original Yule model with **Y1** and **Y2**, one can see that the parameter of α corresponding to Simon's model is given by

$$\alpha = \frac{\gamma N_g}{\sigma N_s + \gamma N_g} \tag{A.1}$$

which changes in time as follows. We have already shown that the number of genera grows exponentially as

$$N_g = e^{\gamma t}. \tag{A.2}$$

The total number of species, N_s, can be calculated by

$$N_s(t) = e^{\sigma t} + \int_0^t du\, \gamma \cdot N_g(u) \cdot e^{\sigma(t-u)} = \frac{\gamma}{\gamma - \sigma} e^{\gamma t} + \frac{\sigma}{\sigma - \gamma} e^{\sigma t}. \tag{A.3}$$

Here, in the right-hand side of the first equality, the first term corresponds to the number of species contained in the original genus, while the second term represents the species contained in the new genera generated by bifurcation with the rate of birth, $\gamma\, du$, in the interval $(u, u+du)$ (integrated over time).

For the case $\gamma > \sigma$, after a sufficiently large time, one has

$$\alpha \simeq 1 - \sigma/\gamma, \tag{A.4}$$

which is inserted into Eq. (2.23) to obtain exactly Eq. (2.5).

For the case $\gamma < \sigma$, one finds that

$$\alpha = A\, N_s^{-(1-\gamma/\sigma)}, \tag{A.5}$$

where A is a constant that depends on σ and γ. By using the master equations with the temporally changing parameter $\alpha = \alpha(s)$, Eq. (2.16) and Eq. (2.17), and also

$$dW(s) = \alpha(s)\, dN(s), \tag{A.6}$$

which can be derived from them, one has the stationary distribution, Eq. (2.20), after a sufficiently large time as easily shown. Then one arrives at Eq. (2.5) as well.

2.2 For the extension of Simon's model with the probability given by Eq. (2.28), the master equations read

$$n_k(s+1) - n_k(s) = (1-\alpha)\left[\frac{(k-1+c)\,n_{k-1}(s)}{N(s)+cW(s)} - \frac{k\,n_k(s)}{N(s)+cW(s)}\right]$$

$$(\text{for } k > k_0), \tag{A.7}$$

$$n_{k_0}(s+1) - n_{k_0}(s) = \alpha - (1-\alpha)\frac{(k_0+c)\,n_{k_0}(s)}{N(s)+cW(s)}. \tag{A.8}$$

At time step s, the number of vocabularies (different words) and the total number of words are respectively

$$W(s) := \sum_{k \geq k_0} n_k(s), \tag{A.9}$$

$$N(s) := \sum_{k \geq k_0} k\, n_k(s). \tag{A.10}$$

Sum Eq. (A.7) over $k > k_0$ and add Eq. (A.8) to have

$$W(s+1) - W(s) = \alpha. \tag{A.11}$$

Muliply Eq. (A.7) by k and sum over $k > k_0$, and add it to a multiple of Eq. (A.8) by k_0. One has

$$N(s+1) - N(s) = 1 + \alpha(k_0 - 1). \tag{A.12}$$

From these equations, for a sufficiently large time step s

$$\{1 + \alpha(k_0 - 1)\}\, W(t) = \alpha\, N(t), \tag{A.13}$$

which is Eq. (2.29).

Then assume the stationary distribution:

$$p_k(s) := \frac{n_k(s)}{W(s)}, \tag{A.14}$$

and insert the ansatz into the master equations. By using Eq. (2.29), the master equations lead to the following:

$$p_k = \frac{k-1+c}{k-1+c+(k_0+c)\alpha(1-\alpha)^{-1}} p_{k-1} \quad (k > k_0), \tag{A.15}$$

$$p_{k_0} = \frac{1-\alpha+\alpha(k_0+c)}{1-\alpha+k_0+c}. \tag{A.16}$$

Note that the factor given by Eq. (2.31) appears in the above equations. Recursive calculation shows that

$$p_k = \frac{k-1+c}{k-1+c+\mu} \cdot \frac{k-2+c}{k-2+c+\mu} \cdots \frac{k_0+c}{k_0+c+\mu} p_{k_0} \tag{A.17}$$

$$= \frac{\Gamma(k+c)}{\Gamma(k_0+c)} \cdot \frac{\Gamma(k_0+c+\mu)}{\Gamma(k+c+\mu)} p_{k_0} \tag{A.18}$$

$$= \frac{B(k+c, \mu)}{B(k_0+c, \mu)} p_{k_0}. \tag{A.19}$$

This is the stationary distribution given by Eq. (2.30) that was to be obtained.

2.3 Entropy of the system may be defined as

$$S = -k_B \int_{-\infty}^{\infty} f(v) \ln f(v) dv. \tag{A.20}$$

Maximization of the entropy under the following constraints leads to the Maxwell distribution:

$$\int_{-\infty}^{\infty} f(v) dv = 1, \tag{A.21}$$

$$\int_{-\infty}^{\infty} v f(v) dv = u, \tag{A.22}$$

$$\int_{-\infty}^{\infty} \frac{1}{2} m v^2 f(v) dv = \langle \epsilon \rangle. \tag{A.23}$$

Note that the averaged energy $\langle \epsilon \rangle$ of the gas molecule to be conserved consists of two components:

$$\langle \epsilon \rangle = \frac{1}{2} m u^2 + \frac{1}{2} k_B T, \tag{A.24}$$

where the first term is the translational kinetic energy and the second term is the thermal energy related to T through the equipartition law of thermal energy.

2.4 From $s = e^x = g(x)$, we have $dg/dx = s$, $d^2g/dx^2 = s$, $dg/dt = 0$. The last equation means that the function $g(x)$ does not depend on time explicitly. Using Ito's lemma, we write a stochastic differential equation to describe function $g(x)$ as

$$ds = dg = \frac{dg}{dx}dx + \frac{dg}{dt}dt + \frac{1}{2}\sigma^2 \frac{d^2g}{dx^2}dt. \quad (A.25)$$

By substituting the aforementioned three relations into the derivative of $g(x)$ and the standard Brownian process, Eq. (2.140), into dx, we obtain the stochastic differential equation to describe the stock price $s = e^x$

$$ds = \left(\mu + \sigma^2/2\right) s dt + \sigma s dz. \quad (A.26)$$

Therefore, stock price s obeys the geometrical Brownian process.

2.5 In Fig. A.1, we have $r = 1/4$, $a = 2$. Therefore, fractal dimension is $D_s = 0.5$.

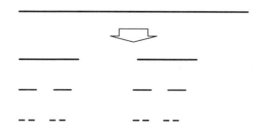

Fig. A.1 Diagram obtained by dividing a line segment into four and removing every other line segment

2.6 From Eq. (2.164), we have

$$x_t = (1-B)^{-d} \frac{\pi(B)}{\phi(B)} \xi_t. \quad (A.27)$$

By binomial expansion of operator $(1-B)^{-d}$, we have

$$(1-B)^{-d} = \sum_{k=0}^{d} \binom{-d}{k}(-B)^k = \sum_{k=0}^{d} \frac{\Gamma(d+k)}{\Gamma(d)\Gamma(k+1)}(B)^k. \quad (A.28)$$

Here we used the property of the negative binomial coefficient

$$\binom{-d}{k} = (-1)^k \binom{d+k-1}{d-1} = (-1)^k \frac{\Gamma(d+k)}{\Gamma(d)\Gamma(k+1)}. \quad (A.29)$$

As a result, we have

$$x_t = \frac{1}{\Gamma(d)} \frac{\pi(B)}{\phi(B)} \sum_{k=0}^{d} \frac{\Gamma(d+k)}{\Gamma(k+1)} \xi_{t-k}. \quad (A.30)$$

2.7 We use the first order approximation of Stirling's approximation

$$\Gamma(n) = e^{-n} n^{n-\frac{1}{2}} \sqrt{2\pi} \left(1 + \frac{1}{12n} + \frac{1}{288n^2} + \cdots \right). \tag{A.31}$$

From $\tau \gg 1$ and $\tau \gg d$, we obtain

$$\frac{\Gamma(\tau + d)}{\Gamma(\tau + 1 - d)} = \tau^{2d-1}. \tag{A.32}$$

By substituting this relation into t autocorrelation function, Eq. (2.167), we obtain relation Eq. (2.168).

3.1 When $a = \langle x \rangle$, $c_0 = -\sigma^2$, $c_1 = 0$, and $c_2 = 0$, Eq. (3.12) becomes

$$\frac{df(x)}{dx} = -\frac{(x - \langle x \rangle) f(x)}{\sigma^2}. \tag{A.33}$$

The solution of Eq. (A.33) is

$$f(x) = K \exp\left[-\frac{(x - \langle x \rangle)^2}{2\sigma^2}\right], \tag{A.34}$$

where K is a normalization constant determined by the normalization condition:

$$\int_{-\infty}^{+\infty} f(x) = 1. \tag{A.35}$$

Then, we obtain

$$f(x) = \frac{1}{\sqrt{2\sigma^2 \pi}} \exp\left[-\frac{(x - \langle x \rangle)^2}{2\sigma^2}\right]. \tag{A.36}$$

This is the probability density function of the normal (or Gaussian) distribution with mean $\langle x \rangle$ and variance σ^2.

3.2 When $a = 0$, $c_0 = -\lambda$, $c_1 = 0$, and $c_2 = 0$, Eq. (3.12) becomes

$$\frac{dp(x)}{dx} = -\frac{p(x)}{\lambda}. \tag{A.37}$$

By solving Eq. (A.37), we obtain

$$p(x) = \frac{1}{\lambda} \exp\left[-\frac{x}{\lambda}\right]. \tag{A.38}$$

This is the probability density function of the exponential distribution and corresponds to the type X Pearson distribution shown in Table 3.1.

3.4 1. The CDF is given by the following:
$$\bar{F}(x) = \theta(a - x), \qquad (A.39)$$
whose shape is shown in Fig. A.2.

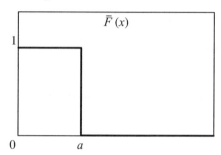

Fig. A.2 CDF for $f(x) = \delta(x - a)$.

2. Using the CDF in (A.39), we obtain,

$$\langle X_N^{(\max)} \rangle = -\int_0^\infty x \frac{d}{dx} \left(1 - \theta(a - x)\right)^N dx \qquad (A.40)$$

$$= -\int_0^\infty x \frac{d}{dx} \left(1 - \theta(a - x)\right) dx \qquad (A.41)$$

$$= \int_0^\infty x \, \delta(x - a) dx = a. \qquad (A.42)$$

Here we used Eq. (3.89) on the first line. Since the stochastic variable X takes the value a only, the maximum value is a, which agrees with the result obtained in Eq. (A.42).

3.5 This equation is correct for $N = 2$ evidently. Next, let us assume that this equation is correct at N, and calculate it at $N + 1$:

$$G_{N+1}(x) = \int_0^x dx_1 p(x_1) \frac{1}{(N-1)!} \left(\int_0^{x_1} p(x') dx'\right)^{N-1}$$

$$= \int_0^x dx_1 \frac{d}{dx_1} \frac{1}{N!} \left(\int_0^{x_1} p(x') dx'\right)^N$$

$$= \frac{1}{N!} \left(\int_0^{x_1} p(x') dx'\right)^N \bigg|_0^x$$

$$= \frac{1}{N!} \left(\int_0^x p(x') dx'\right)^N. \qquad (A.43)$$

Therefore, Eq. (3.92) is correct at arbitrary N.

3.6 Straightforward calculation leads to the following:

$$G_1(t) = e^t \, t \, \Gamma(0, t) = -t \log t - \gamma_E \, t - t^2 \log t + O(t^2). \tag{A.44}$$

From this, we find that $C_1 = -\gamma_E$, where $\gamma_E = 0.577216\ldots$ is the Euler constant. $\Gamma(p, z)$ is the incomplete gamma function.

4.1 An elasticity of substitution for a general production function Φ is defined by

$$\frac{1}{\sigma} := \left. \frac{d \ln |dK/dL|}{d \ln(K/L)} \right|_{\Phi_{\text{CD}}}. \tag{A.45}$$

For the case of CD production function, we obtain

$$\left. \frac{dK}{dL} \right|_{\Phi_{\text{CD}}} = -\frac{\beta}{\alpha} \frac{K}{L}, \tag{A.46}$$

therefore

$$\sigma = 1. \tag{A.47}$$

4.2 The elasticity of substitution is defined by Eq. (A.45). For the case of CES production function

$$\left. \frac{dK}{dL} \right|_{\Phi_{\text{CES}}} = -\frac{\gamma L^{cp-1}}{(1-\gamma)K^{cp-1}}, \tag{A.48}$$

therefore,

$$\sigma = \frac{1}{cp - 1}. \tag{A.49}$$

4.3 In the limit of $p \to 0$, the logarithm of CES production function becomes

$$\ln \Phi_{\text{CES}}(L, K; \gamma, c, p) = \ln A + \frac{1}{p} \ln \{\gamma L^{cp} + (1-\gamma) K^{cp}\} \tag{A.50}$$

$$= \ln A + \frac{1}{p} \ln \left[1 + cp \{\delta \log L + (1-\delta) \log K\} + O(p^2) \right] \tag{A.51}$$

$$= \ln A + c \{\gamma \ln L + (1-\gamma) \ln K\} + O(p) \tag{A.52}$$

$$\simeq \ln A K^{(1-\gamma)c} L^{\gamma c} \tag{A.53}$$

$$= \ln \Phi_{\text{CD}}(L, K; \alpha = (1-\gamma)c, \beta = \gamma c). \tag{A.54}$$

4.4 From Eq. (4.64), one can readily confirm $G(0) = 0$ and

$$G(2) = G(1+1) = G(1) + G(1) = 2G(1). \tag{A.55}$$

Repeated use of Eq. (4.64) thus leads to

$$G(n) = nG(1), \tag{A.56}$$

with any positive integer n. This equation is generalized for an arbitrary number x as

$$G(nx) = nG(x). \tag{A.57}$$

Setting $c = 1/n$ in Eq. (A.57) gives

$$G\left(\frac{1}{n}\right) = \frac{1}{n}G(1). \tag{A.58}$$

Furthermore, setting $c = 1/m$ with any integer m in Eq. (A.57) yields

$$G\left(\frac{n}{m}\right) = nG\left(\frac{1}{m}\right) = \frac{n}{m}G(1). \tag{A.59}$$

This result shows that $G(c)$ is at least linear for *rational* numbers c. If the continuity of $G(c)$ with respect to c is assumed, $G(c)$ must be a linear function of *real* number c:

$$G(c) \propto c. \tag{A.60}$$

Note that we should not have assumed $G(c)$ was differentiable in c to obtain this solution. One can derive Eq. (4.65) from Eq. (A.60) in a straightforward way.

4.5 Begin with taking logarithm of Eq. (4.85):

$$\ln f(n;g) = \ln g! - \ln(g-n)! - n\ln g. \tag{A.61}$$

Using the Stirling formula, Eq. (2.74), one can approximate Eq. (A.61) as

$$\ln f(n;g) \simeq g\ln g - (g-n)\ln(g-n) - n - n\ln g, \tag{A.62}$$

where $g \gg n \gg 1$ is assumed. If the second term on the right-hand side of Eq. (A.62) is expanded in $n/g (\ll 1)$ and the terms up to its second order are retained, the Gaussian approximation form, Eq. (4.85), is obtained.

5.1 This is a simple exercise in integration on a complex plane. Using the expressions;

$$\cos\omega t = \frac{e^{i\omega t} + e^{-i\omega t}}{2}, \quad \sin\omega t = \frac{e^{i\omega t} - e^{-i\omega t}}{2i}, \tag{A.63}$$

and by closing the contour on the upper plane or the lower plane depending on the convergence at infinity, the desired results are obtained.

5.2 Use the following integration formula:

$$\int_a^b \frac{\sqrt{(b-x)(x-a)}}{x}dx = \frac{\pi}{2}\left(\sqrt{b}-\sqrt{a}\right)^2, \quad (A.64)$$

with $0 < a < b$.

5.3 The autocorrelation function for $\omega(t)$ is identical to that for $w(t)$ given by

$$\psi(\tau) = \frac{\langle w(\tau+t)w(t)\rangle}{\langle w(t)^2\rangle}. \quad (A.65)$$

The numerator in the right-hand side of this equation is expressed in terms of $x(t)$ as

$$\langle w(\tau+t)w(t)\rangle = \langle (x(t+\tau+1) - x(t+\tau))(x(t+1) - x(t))\rangle$$
$$= 2\zeta(\tau) - \zeta(\tau+1) - \zeta(\tau-1), \quad (A.66)$$

with

$$\zeta(\tau) = \langle x(\tau+t)x(t)\rangle. \quad (A.67)$$

Using Eq. (5.105), one can calculate the correlation function $\zeta(\tau)$ as

$$\zeta(\tau) = \frac{\langle \epsilon^2\rangle}{1-\phi^2}\phi^\tau \quad (\tau \geq 0). \quad (A.68)$$

Substitution of Eq. (A.68) in Eq. (A.66) gives

$$\langle w(\tau+t)w(t)\rangle = -\frac{1-\phi}{1+\phi}\langle \epsilon^2\rangle\phi^{\tau-1}, \quad (A.69)$$

where $\tau \geq 1$ is assumed. The denominator in the right-hand side of Eq. (A.65) is likewise computable as

$$\langle w(t)^2\rangle = 2(\zeta(0) - \zeta(1)) = \frac{2\langle \epsilon^2\rangle}{1+\phi}. \quad (A.70)$$

Combining Eqs. (A.69) and (A.70) derives the desired formula.

Epilogue

SALVIATI: Yesterday took us into such good digressions on Prof. Feynman's words and his van that I do not know whether I shall be able to go ahead without your assistance in putting me back on the track of macro-econophysics.

SAGREDO: You do not need to worry, Salviati. Together with Simplicio, I read this book to my pleasure. I found it to contain many beautiful considerations which are novel and impressive. ... You shake your head, Simplicio and smile as if I uttered some absurdity.

SIMP.: I merely smile, but believe me. I have an impression that all matters of economy are not covered and the authors do not have answers on business cycles, not to mention how to predict an economic crisis and how to deal with it.

SALV.: I may remind you that we do not yet have a basic understanding of the underlying dynamics of anything at the micro-level: Anyone can write out models of interactions between economic agents like firms and financial institutions, and simulate the dynamics on simulated networks. But there are so many parameters and so many artificial networks.

SAGR.: Ahuh... They have real credit and trade networks, don't they, which can serve as a basis for all the new directions of work?

SALV.: Doubtless there are. We need all the analytical and mathematical tools and modeling ideas exposed in this book to strive for a real understanding of the nature of economic dynamics.

SIMP.: You are simply talking about agent modeling and everyone knows that they are so limited.

SALV.: You may know mathematics well, but you don't know science by just looking at real world. Non-linear equations for macro-variables, exact solutions, stability and chaos may be fun to play with. But they are made-up plays only.

SAGR.: I agree with Salviati. Think of molecules. They interact with each other, optimizing action, which may be comparable to the notion of utility function in economics. With complete understanding of such interactions at micro-level and introducing concepts of statistical physics on how to handle the large number of such interacting heterogeneous objects comes understanding of the dynamics of macro-matter.

SALV.: Such was elaborated upon in Chapter 1.

SIMP.: But it takes so much effort and time to deal with it. For it, don't you think it is better to fully articulate artificial economic systems that can serve as laboratories in which one may examine economic matters of concern such as business cycles?

SAGR.: That is a dangerous thought, Simplicio. You may be as articulate as anyone can possibly be, but your "system" would have nothing to do with reality.

SALV.: This is where Prof. Feynman's words comes in. Do not think of costs and time. Do what we can do what we can and improve upon it for a better future.

SAGR.: And think of our old friend Johannes Kepler. His three laws of planetary motion are surprising findings out of a huge number of handwritten numbers from observations. Think what they initiated. Newtonian classical mechanics, quantum physics, statistical physics with a beautiful understanding of phase transitions, particle physics which tells us that we are in a 10-dimensional space-time. ...

SALV.: Mr. Galilei will be so glad to hear about all of them. And I understand your whole point. We are standing here and now as Mr. Kepler or even some of his predecessors. Let us not believe in our "articulateness" but work on real data with all we have. We may go on to apply AI with all the deep-learning on these matters too.

SAGR.: Yes, the day and age of AI is starting. Let us not argue fruitlessly; instead let us proceed with good hopes for a better economic society for all humans,

SIMP.: I concur with your opinion. Farewell then.

SALV.: Farewell with hopes for more books from these authors.

Bibliography

Aitchison, J. and J. A. C. Brown. 1957. *The Lognormal Distribution.* Cambridge: Cambridge University Press.

Akaike, H. 1973. 'Information Theory and an Extension of the Maximum Likelihood Principle', in *2nd International Symposium on Information Theory,* edited by B. N. Petrov and F. Csaki, 267–81. Budapest: Akademiai Kiado.

———. 1974. 'A New Look at the Statistical Model Identication'. *IEEE Transactions on Automatic Control* 19 (6): 716–23.

Allen, M. P. and D. J. Tildesley. 1987. *Computer Simulation of Liquids.* Oxford, UK: Oxford University Press.

Amaral, L. A. N., S. V. Buldyrev, S. Havlin, H. Leschhorn, P. Maass, M. A. Salinger, H. E. Stanley, and M. H. R. Stanley. 1997. 'Scaling Behavior in Economics: I. Empirical Results for Company Growth'. *Journal de Physique I France* 7 (4): 621–33.

———, M. A. Salinger, and H. E. Stanley. 1998. 'Power Law Scaling for a System of Interacting Units with Complex Internal Structure'. *Physical Review Letters* 80 (7): 1385–8.

Amoroso, L. 1925. 'Ricerche intorno alla curva dei redditi'. *Annali di matematica pura ed applicata* 2 (1): 123–59.

Aoki, M. 2002. *Modeling Aggregate Behavior and Fluctuations in Economics.* Cambridge: Cambridge University Press.

Aoki and H. Yoshikawa. 2002. 'Demand Saturation-Creation and Economic Growth'. *Journal of Economic Behavior & Organization* 48(2): 127–54.

———. 2006. *Reconstructing Macroeconomics: A Perspective from Statistical Physics and Combinatorial Stochastic Processes.* Cambridge: Cambridge University Press.

———. 2012. 'Non-Self-Averaging in Macroeconomic Models: A Criticism of Modern Micro-Founded Macroeconomics'. *Journal of Economic Interaction and Coordination* 7 (1): 1–22.

Aoyama, H. and J. Constable. 1999. 'Word Length Frequency and Distribution in English: Part I. Prose.' *Literary and Linguistic Computing* 14, 339–58.

———, H. Iyetomi, and H. Yoshikawa. 2015. 'Equilibrium Distribution of Labor Productivity: A Theoretical Model'. *Journal of Economic Interaction and Coordination* 10 (1): 57–66.

———, H. Iyetomi, W. Souma, and H. Yoshikawa. 2015. 'Application of the Concept of Entropy to Equilibrium in Macroeconomics.' *RIETI Discussion Papers 15-E-070.*

———, S. Battiston, and Y. Fujiwara. 2013. 'DebtRank Analysis of the Japanese Credit Network'. Discussion Paper 13-E-087. 1–19. Research Institute of Economy, Trade & Industry.

———, W. Souma, and Y. Fujiwara. 2003. 'Growth and Fluctuations of Personal and Company's Income'. *Physica A: Statistical Mechanics and its Applications* 324 (1): 352–8.

———, Y. Nagahara, M. P. Okazaki, H. Takayasu, and M. Takayasu. 2000. 'Paretos law for income of individuals and debt of bankrupt companies'. *Fractals* 8 (03): 293–300.

———, Y. Fujiwara, Y. Ikeda, H. Iyetomi, and W. Souma. 2010. *Econophysics and Companies: Statistical Life and Death in Complex Business Networks*. Cambridge: Cambridge University Press.

Arai, Y. and H. Iyetomi. 2012. 'Numerical Study of Generalized Random Correlation Matrices: Autocorrelation Effects'. *Progress of Theoretical Physics Supplement* 194: 84–93.

———, T. Yoshikawa, and H. Iyetomi. 2015. 'Complex Principal Component Analysis of Dynamic Correlations in Financial Markets'. *Intelligent Decision Technologies, Frontiers in Articial Intelligence and Applications* 255: 111–19.

———, T. Yoshikawa, and H. Iyetomi. 2015. 'Dynamic Stock Correlation Network.' *Procedia Computer Science* 60: 1826–35.

Arnold, B.C. 2015. *Pareto Distribution.* Wiley Online Library.

Arrow, K. J. 1967. 'Samuelson Collected'. *Journal of Political Economy* 75 (5): 730–7.

———, H. B. Chenery, B. S. Minhas, and R. M. Solow. 1961. 'Capital-Labor Substitution and Economic Efciency'. *Review of Economics and Statistics* 43: 225–50.

Atoda, N., T. Suruga, and T. Tachibanaki. 1988. 'Statistical Inference of Functional Forms for Income Distribution'. *The Economic Studies Quarterly* 39(1): 14–40.

Bandourian, R., J. McDonald, and R. S. Turley. 2002. 'A Comparison of Parametric Models of Income Distribution across Countries and over Time'. *Luxembourg Income Study Working Paper No. 305.*

Bank of Japan. 2014. 'Outline of Statistics and Statistical Release Schedule'. https://www.boj.or.jp/en/statistics/outline/exp/pi/excgpi02.htm/.

Barabási, A.-L. and R. Albert. 1999. 'Emergence of Scaling in Random Networks'. *Science* 286: 509–12.

Bardoscia, M., S. Battiston, F. Caccioli, and G. Caldarelli. 2015. 'DebtRank: A Microscopic Foundation for Shock Propagation'. Mimeo.

Barigozzi, M., G. Fagiolo, and G. Mangioni. 2011. 'Identifying the Community Structure of the International-Trade Multi-Network'. *Physica A: Statistical Mechanics and its Applications* 390 (11): 2051–66.

Barnes, J. and P. Hut. 1986. 'A Hierarchical O(N log N) Force-Calculation Algorithm'. *Nature* 324 (6096): 446–49.

Barnett, T. P. 1983. 'Interaction of the Monsoon and Pacic Trade Wind System at Interannual Time Scales Part I: The Equatorial Zone'. *Mon. Wea. Rev.* 111: 756–73.

Bartels, C.P.A. 1977. *Economic Aspects of Regional Welfare*. Springer.

Battista, G. D., P. Eades, R. Tamassia, and I. G. Tollis. 1998. *Graph Drawing: Algorithms for the Visualization of Graphs*, 1st ed. Upper Saddle River, NJ, USA: Prentice Hall PTR.

Battiston, S., D. Delli Gatti, M. Gallegati, B. Greenwald, and J. E. Stiglitz. 2007. Credit Chains and Bankruptcy Propagation in Production Networks'. *Journal of Economic Dynamics & Control* 31: 261–2084.

———, G. Caldarelli, M. D'Errico, and S. Gurciullo. 2015. 'Leveraging the Network: A Stress-Test Framework Based on DebtRank'. Mimeo.

———, M. Puliga, R. Kaushik, P. Tasca, and G. Caldarelli. 2012. 'DebtRank: too Central to Fail? Financial Networks, the FED and Systemic Risk'. *Scientic Reports* 2: 541.

Bendat, J. S. and A. G. Piersol. 2011. *Random Data: Analysis and Measurement Procedures*. Wiley Series in Probability and Statistics. Wiley.

Benini, Rodolfo, 1906. *Principii di Statistica Metodologica*. Torino: UTET.

Blanchard, O. J. and D. Quah. 1989. 'The Dynamic Effects of Aggregate Demand and Supply Disturbances'. *American Economic Review* 79(4): 655–73.

Blondel, V. D., J.-L. Guillaume, R. Lambiotte, and E. Lefebvre. 2008. 'Fast Unfolding of Communities in Large Networks'. *Journal of Statistical Mechanics: Theory and Experiment* 2008 (10): P10008.

Bollerslev, T. 1986. 'Generalized Autoregressive Conditional Heteroskedasticity'. *Journal of Econometrics* 31: 307–27.

Bordley, R. F. and J. B. McDonald. 1993. 'Estimating Aggregate Automotive Income Elasticities from the Population Income-Share Elasticity'. *Journal of Business & Economic Statistics* 11 (2): 209–214.

———, and A. Mantrala. 1997. 'Something New, Something Old: Parametric Models for the Size of Distribution of Income'. *Journal of Income Distribution* 6(1): 91–103.

Borg, I. and P. J. F. Groenen. 2005. *Modern Multidimensional Scaling: Theory and Applications.* Springer.

Borghesi, C., M. Marsili, and S. Miccich. 2007. 'Emergence of Time-Horizon Invariant Correlation Structure in Financial Returns by Subtraction of the Market Mode'. *Physical Review E* 76: 026104.

Bottazzi, G. and A. Secchi. 2003. 'Why are Distributions of rm Growth Rates Tent-Shaped?.' *Economics Letters* 80 (3): 415–20.

Box, G. E. P. and G. M. Jenkins. 1970. *Time Series Analysis, Forecasting and Control.* San Francisco: Holden-Day.

Brandes, U. 2001. 'A Faster Algorithm for Betweenness Centrality'. *Journal of Mathematical Sociology* 25(2): 163–77. http://ella.slis.indiana.edu/ katy/L579/brandes.pdf.

——— and C. Pich. 2007. 'Eigensolver Methods for Progressive Multidimensional Scaling of Large Data.' *Proceedings of 14th Symposium on Graph Drawing (GD).* pp. 42–53. Lecture Notes in Computer Science 4372.

Buja, A. and N. Eyuboglu. 1992. 'Remarks on Parallel Analysis'. *Multivariate Behavioral Research* 27: 509–40.

Bureau van Dijk Electronic Publishing. Brussels, Belgium. http://www.bvdinfo.com/en-gb/home.

Burns, A. F. and W. C. Mitchell. 1946. *Measuring Business Cycles.* New York: NBER.

Burr, I. W. 1942. 'Cumulative Frequency Functions'. *The Annals of Mathematical Statistics* 13 (2): 215–32.

Butler, R. J. and J. B. McDonald. 1989. 'Using Incomplete Moments to Measure Inequality'. *Journal of Econometrics* 42 (1): 109–19.

Champernowne, D. G. 1952. 'The Graduation of Income Distributions'. *Econometrica: Journal of the Econometric Society* 591–615.

———. 1953. 'A Model of Income Distribution.' *Economic Journal* 63(250): 318–51.

———. 1973. *The Distribution of Income between Persons.* Cambridge: Cambridge University Press.

Chandrasekhar, S. 1943. 'Stochastic Problems in Physics and Astronomy'. *Rev. Mod. Phys.* 15: 1–89.

Clauset, A., C. R. Shalizi, and M. E. J. Newman. 2009. 'Power-Law Distributions in EMPIRICAL Data'. *SIAM Review* 51(4): 661–703.

———, M. E. J. Newman, and C. Moore. 2004. 'Finding Community Structure in Very Large Networks'. *Physical Review E* 70(6): 66111.

Cobb, C. W. and P. H. Douglas. 1928. 'A Theory of Production'. *American Economic Review, Supplement* 18: 139–65.

Cormen, T. H., C. E. Leiserson, R. L. Rivest, and C. Stein. 2001. *Introduction to Algorithms*. Cambridge, MA: MIT Press.

Costantini, D. and U. Garibaldi. 1989. 'Classical and Quantum Statistics as nite Random Processes'. *Foundations of Physics* 19(6): 743–54.

CRD Association. Tokyo, Japan. http://www.crd-ofce.net/CRD/english/index.htm.

Creedy, J. 1977. 'Notes and Memoranda Pareto and the Distribution of Income'. *Review of Income and Wealth* 23 (4): 405–11.

Dagum, C. 1977. 'New Model of Personal Income-Distribution-Specication and Estimation'. *Economie Appliqué* 30(3): 413–37.

———. 1980. 'Sistemas Generadores de Distribucion del Ingreso y la Ley de Pareto,'. *El Trimestre Económico* 47(188): 877–917.

———. 1990. 'Generation and Properties of Income Distribution Functions', in *Income and Wealth Distribution, Inequality and Poverty*, edited by C. Dagum and M. Zenga, 1–17. Springer.

———. 2008. 'A New Model of Personal Income Distribution: Specication And Estimation' in *Modeling Income Distributions and Lorenz Curves*, edited by D. Chotikapanich, 3–25. Springer.

De Masi, G., Y. Fujiwara, M. Gallegati, B. Greenwald, and J.E. Stiglitz. 2011. 'An Analysis of the Japanese Credit Network'. *Evolutionary and Institutional Economic Review* 7 (2): 209–32.

de S. Price, D. J. 1965. 'Networks of Scientic Papers'. *Science* 149: 510–15.

di Iasio, G., S. Battiston, L. Infante, and F. Pierobon. 2013. 'Capital and Contagion In nancial Networks'. MPRA Paper 52141, Munich Personal RePEc Archive.

Diebold, F. X. and G. D. Rudebusch. 1990. 'A Nonparametric Investigation of Duration Dependence in the American Business Cycle'. *Journal of Political Economy* 98(3): 596–616.

Drăgulescu, A. and V. M. Yakovenko. 2001. 'Evidence for the Exponential Distribution of Income in the USA'. *The European Physical Journal B: Condensed Matter and Complex Systems* 20 (4): 585–9.

———. 2001. 'Exponential and Power-Law Probability Distributions of Wealth and Income in the United Kingdom and the United States'. *Physica A: Statistical Mechanics and its Applications* 299(1–2): 213–21.

Engle, R. F. 1982. 'Autoregressive Conditional Heteroskedasticity with Estimates of the Variance of United Kingdom Ination'. *Econometrica* 50: 987–1007.

Eurostat. 'Eurostat'. http://ec.europa.eu/eurostat/data/database/.

Fair, R. C. 1989. 'Book Review of R. E. Lucas, *Models of Business Cycles*'. *Journal of Economic Literature* 27(1): 104–5.

Feldman, M. 2011. 'Hilbert Transform in Vibration Analysis'. *Mechanical Systems and Signal Processing* 25 (3): 735–802.

Feller, W. 1968. *An Introduction to Probability Theory and Its Applications,* vol. I. 3rd ed. John Wiley & Sons.

Feynman, R. P., R. B. Leighton, and M. Sands. 1964. *The Feynman Lectures on Physics*, vol. II. Addison-Wesley.

Fink, K., U. Kruger, B. Meller, and L. H. Wong. 2014. 'BSLoss A Comprehensive Measure for Interconnectedness'. Discussion Paper. Deutsche Bundesbank.

Fisk, P. R. 1961. 'Estimation of Location and Scale Parameters in a Truncated Grouped sech Square Distribution'. *Journal of the American Statistical Association* 56(295): 692–702.

———. 1961. 'The Graduation of Income Distributions'. *Econometrica: Journal of the Econometric Society* 171–185.

Foerster, A. T., P. D. G. Sarte, and M. W. Watson. 2008. 'Sectoral vs. Aggregate shocks: A Structural Factor Analysis of Industrial Production'. Technical Report, National Bureau of Economic Research.

Foley, D. 1994. 'A Statistical Equilibrium Theory of Markets'. *Journal of Economic Theory* 62: 321–45.

———. 1996. 'Statistical Equilibrium in a Simple Labor Market'. *Metroeconomica* 47: 125–47.

Fortunato, S. 2010. 'Community Detection in Graphs'. *Physics Reports* 486 (3): 75–174.

——— and M. Barthelemy. 2007. 'Resolution Limit in Community Detection'. *Proceedings of the National Academy of Sciences* 104(1): 36–41.

Francis, N. and V. A. Ramey. 2005. 'Is the Technology-Driven Real Business Cycle Hypothesis Dead? Shocks and Aggregate uctuations Revisited'. *Journal of Monetary Economics* 52(8): 1379–99.

Franklin, S. B., D. J. Gibson, P. A. Robertson, J. T. Pohlmann, and J. S. Fralish. 1995. 'Parallel Analysis: A Method for Determining Signicant Principal Components'. *Journal of Vegetation Science* 6: 99–106.

Frenkel, D. and B. Smit. 2002. *Understanding Molecular Simulation: From Algorithms to Applications*, 2nd ed. San Diego, CA, USA: Academic Press.

FRS. 'Federal Reserve System'. http://www.federalreserve.gov/releases/g17/table1_2.htm/.

Fu, D., F. Pammolli, S. V. Buldyrev, M. Riccaboni, K. Matia, K. Yamasaki, and H. E. Stanley. 2005. 'The Growth of Business Firms: Theoretical Framework and Empirical Evidence'. *Proceedings of the National Academy of Sciences* 102 (52): 18801–6.

Fujita, Y., Y. Fujiwara, and W. Souma. 2016. 'Large Directed Graph Layout and its Application to a Million-rms Economic Network'. *Evolutionary and Institutional Economic Review*.

Fujiwara, Y. 2009. 'Visualizing a Large-Scale Structure of Production Network by N-body Simulation'. *Progress of Theoretical Physics Supplement* 179: 167–177.

——— and H. Aoyama. 2010. 'Large-Scale Structure of a Nation-Wide Production Network'. *European Physical Journal B* 77(4): 565–80.

———, C. Di Guilmi, H. Aoyama, M. Gallegati, and W. Souma. 2004. 'Do ParetoZipf and Gibrat Laws Hold True? An Analysis with European Firms'. *Physica A: Statistical Mechanics and its Applications* 335(1): 197–216.

———, H. Aoyama, C. Di Guilmi, W. Souma, and M. Gallegati. 2004. 'Gibrat and Pareto–Zipf Revisited with European Firms'. *Physica A: Statistical Mechanics and its Applications* 344 (1): 112–6.

———, Y. Ikeda, H. Iyetomi, and W. Souma. 2009. 'Structure and Temporal Change of the Credit Network between Banks and Large Firms in Japan'. *Economics: The Open-Access, Open-Assessment E-Journal* 3: 7.

———, M. Terai, Y. Fujita, and W. Souma. 2016. 'DebtRank Analysis of Financial Distress Propagation on a Production Network in Japan'. *RIETI Discussion Papers* 16-E-046: 1–16.

———, W. Souma, H. Aoyama, T. Kaizoji, and M. Aoki. 2003. 'Growth and Fluctuations of Personal Income'. *Physica A: Statistical Mechanics and its Applications* 321(3): 598–604.

Gabaix, X. 1999. 'Zipf's Law for Cities: An Explanation'. *Quarterly Journal of Economics* 739–67.

———. 2008. 'The Granular Origins of Aggregate Fluctuations'. SSRN eLibrary. Working Paper Series, http://ssrn.com/paper=1111765.

Gabor, D. 1946. 'Theory of Communication'. *J. Inst. Electr. Eng.–Part III, Radio Commun. Eng.* 93: 429–57.

———. 1946. 'Theory of Communication. Part 1: The Analysis of Information'. *Electrical Engineers-Part III: Radio and Communication Engineering, Journal of the Institution of* 93 (26): 429–41.

Galton, F. 1879. 'The Geometric Mean, in Vital and Social Statistics'. *Proceedings of the Royal Society of London* 29 (196–199): 365–7.

Gardiner, C. 2009. *Stochastic Methods A Handbook for the Natural and Social Sciences*, 4th ed. Springer-Verlag, 2009.

Garibaldi, U. and E. Scalas. 2010. *Finitary Probabilistic Methods in Econphysics*. Cambridge: Cambridge University Press.

Gastwirth, J. L. 1972. 'The Estimation of the Lorenz Curve and Gini Index'. *The Review of Economics and Statistics* 306–16.

——— and J. T. Smith. 1972. 'A New Goodness-of-Fit Test'. *ASA Proceedings of the Business and Economic Statistics Section* 320, 322.

Gibrat, R. 1931. *Les Inégalités Économiques* Paris: Librairie du Recueil Sirey.

Girvan, M. and M. E. J. Newman. 2002. Community Structure in Social and Biological Networks'. *Proceedings of the National Academy of Sciences* 99(12): 7821–6.

Goodwin, R. M. 1951. 'The Nonlinear Accelerator and the Persistence of Business Cycles'. *Econometrica: Journal of the Econometric Society* 1–17.

Goto, H., H. Takayasu, and M. Takayasu. 2015. 'Empirical Analysis of Firm-Dynamics on Japanese Interrm Trade Network' in *Proceedings of the International Conference on Social Modeling and Simulation, plus Econophysics Colloquium 2014*, edited by H. Takayasu, N. Ito, I. Noda, and M. Takayasu, 195–204. Springer Proceedings in Complexity. Springer.

Granger, C. W. J. 1966. 'The Typical Spectral Shape of an Economic Variable'. *Econometrica* 34 (1): 150–61.

——— and M. Hatanaka. 1964. *Spectral Analysis of Economic Time Series.*, Princeton Univ. Press.

——— and R. Joyeaux. 1980. 'An Introduction to Long-Memory Time Series Models and Fractional Differencing'. *Journal of Time Series Analysis* 1: 15–20.

Haberler, G. 1946. *Prosperity and Depression*, vol. 24. Transaction Publishers.

———. 1964. *Prosperity and Depression*. New edition. First published by the League of Nations, 1937. Cambridge, Mass: Harvard University Press.

Hall, R. 1988. 'Comment on Shapiro and Watson'. *NBER Macroeconomics Annual* 3: 148–51.

Hannachi, A., I. T. Jolliffe, and D. B. Stephenson. 2007. 'Empirical Orthogonal Functions and Related Techniques in Atmospheric Science: A Review'. *International Journal of Climatology* 27 (9): 1119–52.

Hansen, J. P. and I. R. McDonald. 2006. *Theory of Simple Liquids*, 3rd ed. London: Academic Press.

Harrison, A. 1979. 'The Upper Tail of the Earnings Distribution: Pareto or Lognormal?'. *Economics Letters* 2 (2): 191–5.

———. 1981. 'Earnings by Size: A Tale of Two Distributions'. *The Review of Economic Studies* 48(4): 621–31.

Hatanaka, M. 1964. *Spectral Analysis of Economic Time Series*. Princeton, New Jersey: Princeton University Press.

Hazama, M. and I. Uesugi. 2012. 'Measuring the Systemic Risk in Interrm Transaction Networks'. Working Papers Series 20. Center for Interrm Network, Institute of Economic Research, Hitotsubashi University.

He, J. and M. W. Deem. 2010. 'Structure and Response in the World Trade Network'. *Physical Review Letters* 105(19): 198701.

Heider, F. 1944. 'Social Perception and Phenomenal Causality'. *Psychological Review* 51(6): 358.

———. 1946. 'Attitudes and Cognitive Organization'. *The Journal of Psychology* 21(1): 107–12.

———. 2013. *The Psychology of Interpersonal Relations*. Oxford: Taylor & Francis.

———. 1950. *A Contribution to the Theory of Trade Cycle*. Oxford University Press.

Hilbert, D. 1912. *Grundzüge einer Allgemeinen Theorie der Linearen Integralgleichungen*. Druck und Verlag con B. G. Teubner.

Horel, J. D. 1984. 'Complex Principal Component Analysis: Theory and Examples'. *J. Appl. Meteor.* 23: 1660–73.

Horn, J. L. 1965. 'A Rationale and Test for the Number of Factors in Factor Analysis'. *Psychometrica* 30: 179–85.

Houthakker, H. S. 1955. 'The Pareto Distribution and the Cobb-Douglas Production Function in Activity Analysis'. *The Review of Economic Studies* 23(1): 27–31.

Hu, Y. 2006. 'Efcient, High-Quality Force-Directed Graph Drawing'. *The Mathematica Journal* 10: 37–71.

Hurst, H. E. 1951. 'Long-Term Storage Capacity of Reservoirs'. *Transactions of the American Society of Civil Engineers* 116: 770–808.

Huygens, C. 1966. *Horologium Oscillatorium*. London: Dawsons of Pall Mall.

Iino, T. and H. Iyetomi. 2015. 'Community Structure of a Large-Scale Production Network in Japan' in *The Economics of Interrm Networks* edited by T. Watanabe, I. Uesugi and A. Ono. Japan: Springer.

———, K. Kamehama, H. Iyetomi, Y. Ikeda, T. Ohnishi, H. Takayasu, and M. Takayasu. 2010. 'Community Structure in a Large-Scale Transaction Network and Visualization'. *Journal of Physics: Conference Series* 221: 012013.

Ijiri, Y. and H. A. Simon. 1977. *Skew Distributions and the Sizes of Business Firms*. Amsterdam: North-Holland.

Ikeda, Y., H. Aoyama, and H. Yoshikawa. 2013. 'Synchronization and the Coupled Oscillator Model in International Business Cycles'. *RIETI Discussion Papers*.

———, H. Iyetomi, and H. Yoshikawa. 2013. 'Direct Evidence for Synchronization in Japanese Business Cycles'. *Evolutionary and Institutional Economics Review* 10 (2): 315–27.

———, Y. Fujiwara, H. Iyetomi, K. Ogimoto, W. Souma, and H. Yoshikawa. 2012. 'Coupled Oscillator Model of the Business Cycle with Fluctuating Goods Markets'. *Progress of Theoretical Physics Supplement* 194: 111–21.

———, H. Iyetomi, T. Mizuno, T. Ohnishi, and T. Watanabe. 2014. 'Community Structure and Dynamics of the Industry Sector-Specic International-Trade-Network' in *Proceedings of the Tenth International Conference on Signal-Image Technology and Internet-Based Systems (SITIS)*, 2014 Tenth International Conference on IEEE, pp. 456–61.

———, T. Ohnishi, T. Watanabe, H. Aoyama, T. Mizuno, and Y. Sakamoto. 2016. 'Econophysics Point of View of Trade Liberalization: Community Dynamics, Synchronization'. *RIETI Discussion Papers*.

———, W. Souma, H. Aoyama, Y. Fujiwara, and H. Iyetomi. 2010. 'Analysis of Labor Productivity Using Large-Scale Data of Firms Financial Statements'. *The European Physical Journal B* 76(4): 491–9.

Iyetomi, H. 2012. 'Labor Productivity Distribution with Negative Temperature'. *Progress of Theoretical Physics, Supplement* 194: 135–43.

———, H. Aoyama, Y. Fujiwara, Y. Ikeda, and W. Souma. 2012. 'A Paradigm Shift from Production Function to Production Copula: Statistical Description of Production Activity of Firms'. *Quantitative Finance* 12(9): 1453–66.

———, Y. Nakayama, H. Aoyama, Y. Fujiwara, Y. Ikeda, and Wataru Souma. 2011. Fluctuation Dissipation Theory of Input-Output Interindustrial Relations'. *Physical Review E* 83(1): 016103.

———, H. Yoshikawa, H. Aoyama, Y. Fujiwara, Y. Ikeda, and W. Souma. 2011. 'What Causes Business Cycles? Analysis of the Japanese Industrial Production Data'. *Journal of the Japanese and International Economies* 25(3): 246–72.

Jaccard, P. 1912. 'The Distribution of the Flora in the Alpine Zone'. *New Phytologist* 11(2): 37–50.

Johnson, N. O. 1937. 'The Pareto Law'. *The Review of Economics and Statistics* 19(1): 20–6.

Johnson, N. L., S. Kotz, and N. Balakrishnan. 1995. *Continuous Univariate Distributions*. New York: John Wiley & Sons.

Kaldor, N. 1940. 'A Model of the Trade Cycle'. *The Economic Journal* 78–92.

Kalecki, M. 1945. 'On the Gibrat Distribution'. *Econometrica: Journal of the Econometric Society* 161–70.

Kapteyn, J. C., M. J. van Uven. 1916. *Skew Frequency Curves in Biology and Statistics*. Groningen: Hoitsema Brothers.

Karrer, B., E. Levina, and M. E. J. Newman. 2008. 'Robustness of Community Structure in Networks'. *Physical Review E* 77(4): 046119.

Kawai, A. and T. Fukushige. 2006. '$158/GFLOPS Astrophysical N-body Simulation with Recongurable Add-in Card and Hierarchical Tree Algorithm' in *Proceedings of the 2006 ACM/IEEE Conference on Supercomputing*, edited by B. Horner-Miller, 48. New York: ACM.

Kesten, H. 1973. 'Random Difference Equations and Renewal Theory for Products of Random Matrices'. *Acta Mathematica* 131(1): 207–48.

Keynes, J. M. 1936. *General Theory of Employment, Interest, and Money.* London: Macmillan.

Kichikawa, Y., H. Iyetomi, H. Aoyama, and H. Yoshikawa. 2016. Dynamical Linkage among Economic Conditions, Exchange Rates, and Prices. *RIETI Discussion Papers* 16J-046. In Japanese.

Kim, D. H. and H. Jeong. 2005. 'Systematic Analysis of Group Identication in Stock Markets'. *Physical Review E* 72: 046133.

Kirman, A. 1992. 'Whom or What Does the Representative Individual Represent?'. *Journal of Economic Perspectives* 6(2): 117–36.

Kittel, C. and H. Kroemer. 1980. *Thermal Physics*, 2nd ed. San Francisco: W. H. Freeman.

Klenow, P. J. and B. A. Malin. 2011. 'Microeconomic Evidence on Price-Setting', in *Handbook of Monetary Economics*, vol. 3A, edited by Benjamin H. Friedman and Michael Woodford, Chapter 6. Amsterdam: North-Holland.

Kloek, T. and H. K. van Dijk. 1978. 'Efcient Estimation of Income Distribution Parameters'. *Journal of Econometrics* 8(1): 61–74.

Kmietowicz, Z. W. 1984. 'The Bivariate Lognormal Model for the Distribution of Household Size and Income'. *The Manchester School* 52 (2): 196–210.

——— and H. Ding. 1993. 'Statistical Analysis of Income Distribution in the Jiangsu Province of China'. *The Statistician* 107–21.

Krapivsky, P. L. Antal T. and S. Redner. 2006. 'Social Balance on Networks: The Dynamics of Friendship and Enmity'. *Physica D: Nonlinear Phenomena* 224(1): 130–6.

Krugman, P. 1996. *The Self-Organizing Economy.* New York: John Wiley & Sons.

———, K. M. Dominquez, and K. Rogoff. 1990. 'Its Baaack: Japan's Slump and the Return of the Liquidity Trap'. *Brookings Papers on Economic Activity* 1998(2): 137–205. Brookings Institution Press.

Kuramoto, Y. 2012. *Chemical Oscillations, Waves, and Turbulence*, vol. 19. New York: Springer Science & Business Media.

Kydland, F. E. and E. C. Prescott. 1982. 'Time to Build and Aggregate Fluctuations'. *Econometrica: Journal of the Econometric Society* 1345–70.

Laloux, L., P. Cizeau, J. P. Bouchaud, and M. Potters. 1999. 'Noise Dressing of Financial Correlation Matrices'. *Physical Review Letters* 83(7): 1467–70.

Landau, L. D. and E. M. Lifshitz. 1980. *Statistical Physics, Part 1*, 3rd ed. Oxford: Elsevier Butterworth-Heinemann.

Levy, M. 2003. 'Are Rich People Smarter?'. *Journal of Economic Theory* 110(1): 42–64.

Liu, Y. Y., J. J. Slotine, and A. L. Barabási. 2011. 'Controllability of Complex Networks'. *Nature* 473(7346): 167–73.

Lomax, K. S. 1954. 'Business Failures: Another Example of the Analysis of Failure Data'. *Journal of the American Statistical Association* 49 (268): 847–52.

Lorenz, H. W. 1993. *Nonlinear Dynamical Economics and Chaotic Motion*, vol. 334. New York: Springer.

Lucas Jr., R. E. 1972. 'Expectations and the Neutrality of Money'. *Journal of Economic Theory* 4 (2): 103–24.

———. 1987. *Models of Business Cycles*. Oxford: Basil Blackwell.

———. 2003. 'Macroeconomic Priorities'. *American Economic Review* 93(1): 1–14.

Lydall, H. F. 1959. 'The Distribution of Employment Incomes'. *Econometrica: Journal of the Econometric Society* 110–15.

———. 1968. *The Structure of Earnings*. Clarendon Press.

Macgregor, D. H. 1936. 'Pareto's Law'. *The Economic Journal* 46(181): 80–7.

Majumder, A. and S. R. Chakravarty. 1990. 'Distribution of Personal Income: Development of a New Model and its Application to US Income Data'. *Journal of Applied Econometrics* 5(2): 189–96.

Makino, J. and M. Taiji. 1998. *Scientic Simulations with Special-Purpose Computers*. Chichester, UK: John Wiley and Sons.

Mandelbrot, B. 1960. 'The Pareto-Levy Law and the Distribution of Income'. *International Economic Review* 1(2): 79–106.

———. 1961. 'Stable Paretian Random Functions and the Multiplicative Variation of Income'. *Econometrica: Journal of the Econometric Society* 517–43.

———. 1983. *The Fractal Geometry of Nature*. New York: W. H. Freeman.

———, R. L. Hudson, and E. Grunwald. 2005. 'The (mis) behaviour of markets'. *The Mathematical Intelligencer* 27(3): 77–9.

Mankiw, N. G. 1989. 'Real Business Cycles: A New Keynesian Perspective'. *Journal of Economic Perspectives* 79–90.

Mantegna, R. N. 1999. 'Hierarchical Structure in Financial Markets'. *European Physical Journal B* 11(1): 193–7.

March, L. 1898. 'Quelques Exemples de Distribution des Salaires'. *Journal de la Societe de Statistique de Paris* 39: 193–206.

Marotta, L., S. Miccichè, Y. Fujiwara, H. Iyetomi, H. Aoyama, M. Gallegati, and R. N. Mantegna. 2015. 'Bank-Firm Credit Network in Japan. An Analysis of a Bipartite Network'. *PLOS ONE* 10: e0123079.

———. 2016. 'Backbone of Credit Relationships in the Japanese Credit Market'. *EPJ Data Science* 5(10).

Marsili, M. and Y. C. Zhang. 1998. 'Interacting Individuals Leading to Zipf's Law'. *Physical Review Letters* 80(12): 2741–4.

Marčenko, V. A. and L. A. Pastur. 1967. 'Distribution of Eigenvalues for Some Sets of Random Matrices'. *Mathematics of the USSR-Sbornik* 1(4): 457–83.

Mavroeidis, S., M. Plagborg-Møller, and J. H. Stock. 2014. Empirical Evidence on Ination Expectations in the New Keynesian Phillips Curve.' *Journal of Economic Literature* 52(1): 124–88.

McAlister, D. 1879. 'The Law of the Geometric Mean'. *Proceedings of the Royal Society of London* 29: 367–76.

McDonald, J. B. 1984. 'Some Generalized Functions for the Size Distribution of Income'. *Econometrica: Journal of the Econometric Society* 647–63.

——— and A. Mantrala. 1995. 'The Distribution of Personal Income: Revisited'. *Journal of Applied Econometrics* 10(2): 201–4.

——— and M. R. Ransom. 1979. 'Functional Forms, Estimation Techniques and the Distribution of Income'. *Econometrica: Journal of the Econometric Society* 1513–25.

——— and Y. J. Xu. 1995. 'A Generalization of the Beta Distribution with Applications'. *Journal of Econometrics* 66(1): 133–52.

McNeil, A. J., R. Frey, and P. Embrechts. 2005. *Quantitative Risk Management: Concepts, Techniques and Tools*. Princeton: Princeton University Press.

Mehlhorn, K. and S. Näher. 1999. *LEDA: A Platform for Combinatorial and Geometric Computing*. Cambridge: Cambridge University Press. http://www.mpi-inf.mpg.de/mehlhorn/LEDAbook.html.

Mehta, M. L. 2004. *Random Matrices*. Amsterdam: Elsevier.

Merton, R. K. 1968. 'The Matthew Effect in Science'. *Science* 159: 56–63.

METI, Japan. 2015. 'Indices of Industrial Production, Producer's Shipments, Producer's Inventory of Finished Goods and Producers Inventory Ratio of Finished Goods'. http://www.meti.go.jp/english/statistics/tyo/iip/index.html.

Metzler, L. A. 1941. 'The Nature and Stability of Inventory Cycles'. *The Review of Economics and Statistics* 23(3): 113–29.

Mitchell, W. C. 1951. *What Happens During Business Cycles: A Progress Report*. National Bureau of Economic Research.

Mitzenmacher, M. 2004. 'A Brief History of Generative Models for Power Laws and Lognormal Distributions'. *Internet Mathematics* 1: 226–51.

Mizuno, T., W. Souma, and T. Watanabe. 2014. 'The Structure and Evolution of Buyer-Supplier Networks'. *PLOS ONE* 9(7): e100712.

Molloy, M. and B. Reed. 1998. 'The Size of the Giant Component of a Random Graph with a Given Degree Sequence'. *Combinatorics, Probability and Computing* 7(03): 295–305.

Montroll, E. 1981. 'On the Entropy Function in Sociotechnical Systems'. *Proceedings of National Academy of Sciences, U.S.A.* 78(12): 7839–93.

Montroll, E. W. 1978. 'Social Dynamics and the Quantifying of Social Forces'. *Proceedings of National Academy of Sciences, U.S.A.* 75(10): 4633–7.

———. 1987. 'On the Dynamics and Evolution of Some Socio-Technical Systems'. *Bulletin (New Series) of American Mathematical Society* 16(1): 1–46.

Moore, C. and S. Mertens. 2011. *The Nature of Computation*. Oxford: OUP.

Mortensen, D. T. 2011. 'Markets with Search Friction and the DMP Model'. *American Economic Review* 101: 1073–91.

Mucha, P. J., T. Richardson, K. Macon, M. A. Porter, and J. P. Onnela. 2010. 'Community Structure in Time-Dependent, Multiscale, and Multiplex Networks'. *Science* 328(5980): 876–8.

Musmeci, N., T. Aste, and T. Di Matteo. 2015. 'Relation between Financial Market Structure and the Real Economy: Comparison between Clustering Methods'. *PLOS ONE* 10(3): e0116201.

Nelsen, R. B. 2006. *An Introduction to Copulas*. New York: Springer.

———. 2004. 'Analysis of Weighted Networks'. *Physical Review E* 70(5): 056131.

———. 2004. 'Fast Algorithm for Detecting Community Structure in Networks'. *Physical Review E* 69: 066133.

———. 2005. 'Power Laws, Pareto Distributions and Zipf's Law'. *Contemporary Physics* 46: 323–51.

———. 2006. 'Finding Community Structure in Networks Using the Eigenvectors of Matrices'. *Physical Review E* 74(3): 036104.

———. 2010. *Networks: An Introduction*. Oxford: Oxford University Press.

———. 2012. 'Communities, Modules and Large-Scale Structure in Networks'. *Nature Physics* 8 (1): 25–31.

——— and M. Girvan. 2004. 'Finding and Evaluating Community Structure in Networks. *Physical Review E* 69(2): 026113.

———, S. H. Strogatz, and D. J. Watts. 2001. 'Random Graphs with Arbitrary Degree Distributions and their Applications'. *Physical Review E* 64 (2): 026118.

Nikkei Digital Media, Inc. Tokyo, Japan. http://www.nikkei.co.jp/digitalmedia/.

Nikkei Media Marketing, Inc. 2008. www.nikkeimm.co.jp/english/index.html.

Nirei, M. and W. Souma. 2006. 'Income Distribution and Stochastic Multiplicative Process with Reset Events' in *The Complex Dynamics of Economic Interaction: Essays in Economics and Econophysics*, edited by M. Gallegati, A. P. Kirman, and M. Marsili, 161–8. New York: Springer-Verlag.

——— and W. Souma. 2007. 'A Two Factor Model of Income Distribution Dynamics'. *Review of Income and Wealth* 53(3): 440–59.

Okuyama, K., M. Takayasu, and H. Takayasu. 1999. 'Zipf's Law in Income Distribution of Companies'. *Physica A* 269(1): 125–31.

Pareto, V. 1895. 'La legge Della Domanda'. *Giornale Degli Economisti* pp. 59–68.

———. 1897. *Cours d'économie Politique: Professe a Luniversité de Lausanné* – tone second. Lausanne F. Rouge.

Parker, S. C. 1997. 'The Distribution of Self-employment Income in the United Kingdom, 1976–1991'. *The Economic Journal* 107(441): 455–66.

Paul, D. and A. Aue. 2014. 'Random Matrix Theory in Statistics: A Review'. *Journal of Statistical Planning and Inference* 150: 1–29.

Pearson, K. 1895. 'Contributions to the Mathematical Theory of Evolution. II. Skew Variation in Homogeneous Material'. *Philosophical Transactions of the Royal Society of London A: Mathematical, Physical and Engineering Sciences* 186: 343–414.

Pfalzner, S.1996. *Many-Body Tree Methods in Physics*. New York: Cambridge University Press.

Phillips, A. W. 1958. 'The Relation between Unemployment and the Rate of Change of Money Wage Rates in the United Kingdom, 1861–1957'. *Economica* 25(100): 283–99.

Piccardi, C. and L. Tajoli. 2012. 'Existence and Signicance of Communities in the World Trade Web'. *Physical Review E* 85(6): 066119.

Piketty, T. 2014. *Capital in the Twenty-First Century*. Cambridge: Harvard University Press.

Pikovsky, A., M. Rosenblum, and J. Kurths. 2003. *Synchronization: A Universal Concept in Nonlinear Sciences*, vol. 12. Cambridge: Cambridge University Press.

Pisarenko, V. F. and D. Sornette. 2012. 'Robust Statistical Tests of Dragon-Kings Beyond Power Law Distributions'. *The European Physical Journal Special Topics* 205(1): 95–115.

Pissarides, C. A. 2011. 'Equilibrium in the Labor Market with Search Frictions'. *American Economic Review* 101: 1092–1105.

Plerou, V., P. Gopikrishnan, B. Rosenow, L. A. N. Amaral, and H. E. Stanley. 1999. 'Universal and Nonuniversal Properties of Cross Correlations In nancial Time Series'. *Physical Review Letters* 83(7): 1471–4.

———, T. Guhr, and H. E. Stanley. 2002. 'Random Matrix Approach to Cross Correlations in Financial Data'. *Physical Review E* 65(6): 066126.

Poledna, S. and S. Thurner. 2014. 'Elimination of Systemic Risk in Financial Networks by Means of a Systemic Risk Transaction Tax'. http://arxiv.org/abs/1401.8026.

Press, W. H., S. A. Teukolsky, W. T. Vetterling, and B. P. Flannery. 2007. *Numerical Recipes: The Art of Scientic Computing*, 3rd ed.. New York: Cambridge University Press.

Puliga, M., G. Caldarelli, and S. Battiston. 2014. 'Credit Default Swaps Networks and Systemic Risk'. *Scientic Reports* 4: 6822.

Ransom, M. R. and J. S. Cramer. 1983. 'Income Distribution Functions with Disturbances. *European Economic Review* 22(3): 363–72.

Rasmusson, E. M., P. A. Arkin, W. Y. Chen, and J. B. Jalickee. 1981. 'Biennial Variations in Surface Temperature over the United States as Revealed By Singular Decomposition'. *Mon. Wea. Rev.* 109: 587–98.

Redner, S. 1998. 'How Popular is Your Paper?'. *European Physical Journal B* 4:131–4.

Rosenblum, M. G., A. S. Pikovsky, and J. Kurths. 1996. 'Phase Synchronization of Chaotic Oscillators'. *Physical Review Letters* 76(11): 1804.

Rutherford, R. S. G. 1955. 'Income Distributions: A New Model'. *Econometrica: Journal of the Econometric Society* 277–94.

Salem, A. B. Z. and T. D. Mount. 1974. 'A Convenient Descriptive Model of Income Distribution: The Gamma Density'. *Econometrica: Journal of the Econometric Society* 1115–27.

Samuelson, P. A. 1939. 'Interaction between the Multiplier Analysis and the Principle of Acceleration'. *The Review of Economics and Statistics* 21: 75–8.

Sargent, T. J. 2015. 'Robert E. Lucas Jr's Collected Papers on Monetary Theory'. *Journal of Economic Literature* 53 (1): 43–64.

——— and C. A. Sims. 1977. 'Business Cycle Modeling without Pretending to have Too Much A Priori Economic Theory'. *New Methods in Business Cycle Research* 1: 145–68.

Sato, K. 1975. *Production Functions and Aggregation*. Amsterdam: North Holland Pub.

Scalas, E. and U. Garibaldi. 2009. 'A Dynamic Probabilistic Version of the Aoki-Yoshikawa Sectoral Productivity Model'. *Economics: The Open-Access, Open-Assessment E-Journal* 3(15): 1–10.

Shapiro, M. D. and M. W. Watson. 1988. 'Sources of Business Cycle Fluctuations'. *NBER Macroeconomics Annual* 3: 111–48.

Shirras, G. F. 1935. 'The Pareto Law and the Distribution of Income'. *The Economic Journal* 45(180): 663–81.

Siek, J., L. Q. Lee, and A. Lumsdaine. 2002. *The Boost Graph Library: User Guide and Reference Manual*. New York: Addison-Wesley.

Simkin, M. V. and V. P. Roychowdhury. 2006. 'Re-inventing Willis'. arXiv:physics/0601192.

Simon, H. A. 1955. 'On a Class of Skew Distribution Functions'. *Biometrika* 42: 425–40.

Singh, S. K. and G. S. Maddala. 1976. 'A Function for Size Distribution of Incomes'. *Econometrica: Journal of the Econometric Society* 963–70.

Sklar, A. 1959. 'Fonctions de Repartition a n Dimensions et Leurs Marges'. *Publ. Inst. Statist. Univ. Paris* 8: 229–31.

Solomon, S. and M. Levy. 1996. 'Spontaneous Scaling Emergence in Generic Stochastic Systems'. *International Journal of Modern Physics C* 7(05): 745–51.

Sornette, D. 2003. *Why Stock Markets Crash: Critical Events in Complex Financial Systems*. Princeton: Princeton University Press.

———. 2006. *Critical Phenomena in Natural Sciences*, 2nd ed. New York: Springer-Verlag.

Souma, W. 2001. 'Universal Structure of the Personal Income Distribution'. *Fractals* 9(04): 463–70.

———. 2002. 'Physics of Personal Income' in *Empirical Science of Financial Fluctuations*, edited by H. Takayasu, 343–52. Tokyo: Springer.

———. 2007. 'Networks of Firms and the Ridge in the Production Space', in *Econophysics of Markets and Business Networks: Proceedings of the Econophys-Kolkata III* edited by B. K. Chakrabarti and A. Chatterjee, 149–58. Milano: Springer.

——— and M. Nirei. 2005. 'Empirical Study and Model of Personal Income' in *Econophysics of Wealth Distributions*, edited by A. Chatterjee, S. Yarlagadda, B. K. Chakrabarti, 34–42. Milano: Springer, Milano.

———, Y. Ikeda, H. Iyetomi, and Y. Fujiwara. 2009. 'Distribution of Labor Productivity in Japan over the Period 1996–2006'. *Economics: The Open-Access, Open-Assessment E-Journal* 3: 2009–14.

Stanley, M. H. R., L. A. N. Amaral, S. V. Buldyrev, S. Havlin, H. Leschhorn, P. Maass, M. A. Salinger, and H. E. Stanley. 1996. 'Scaling Behaviour in the Growth of Companies'. *Nature* 379: 804–6.

Statcan. Statistics Canada. http://www5.statcan.gc.ca/cansim/.

Statistics Bureau. 2014. 'Consumer Price Index'. http://www.stat.go.jp/english/data/cpi/index.htm.

Stein, K., A. Timmermann, and N. Schneider. 2011. 'Phase Synchronization of the El Niño-Southern Oscillation with the Annual Cycle'. *Phys. Rev. Lett.* 107: 128501.

Steindl, J. 1965. *Random Processes and the Growth of Firms: A Study of the Pareto Law*. London: Grifn.

Sterman, J. D. and E. Mosekilde. 1994. 'Business Cycles and Long Waves: A Behavioral Disequilibrium Perspective, in *Business Cycles: Theory and Empirical Methods*, edited by W. Semmler, 13–51. New York: Springer.

Steyn, H. S. 1959. 'A Model for the Distribution of Incomes'. *South African Journal of Economics* 27(2): 149–56.

———. 1966. 'On the Departure from the Logarithmic Normal Distribution for Income'. *South African Journal of Economics* 34 (3): 225–32.

Stiglitz, J. E. 2010. 'Needed: A New Economic Paradigm'. *Financial Times*

Stock, J. H. and M. W. Watson. 1998. 'Diffusion Indices'. NBER Working Paper 6702.

———. 2002. 'Forecasting Using Principal Components from a Large Number of Predictors'. *Journal of the American Statistical Association* 97(460): 1167–79.

———. 2005. 'Understanding Changes in International Business Cycle Dynamics'. *Journal of the European Economic Association* 3(5): 968–1006.

Strogatz, S. H. 2000. 'From Kuramoto to Crawford: Exploring the Onset of Synchronization in Populations of Coupled Oscillators'. *Physica D: Nonlinear Phenomena* 143(1): 1–20.

Summers, L. H. 1986. 'Some Skeptical Observations on RBC Theory'. *Federal Reserve Bank of Minneapolis Quarterly Review* 10: 23–7.

Suruga, T. 1982. 'Functional Forms of Income Distributions: The Case of Yearly Income Groups in the Annual Report on the Family Income and Expenditure Survey (in Japanese)'. *Economic Studies Quarterly, Journal of the Japan Association of Economics and Econometrics* 33: 79–85.

Sutton, J. 1997. 'Gibrat's Legacy'. *Journal of Economic Literature* 35(1): 40–59.

Tabak, B. M., S. R. S. Souza, and S. M. Guerra. 2013. 'Assessing the Systemic Risk in the Brazilian Interbank Market'. Working Paper Series 318, Central Bank of Brazil.

Tachibanaki, T., T. Suruga, and N. Atoda. 1997. 'Estimations of Income Distribution Parameters for Individual Observations by Maximum Likelihood Method'. *Journal of the Japan Statistical Society* 27(2): 191–203.

Takayasu, H., A. H. Sato, and M. Takayasu. 1997. 'Stable Innite Variance Fluctuations in Randomly Amplied Langevin Systems'. *Physical Review Letters* 79 (6): 966–9.

Terai, M., Y. Fujiwara, K. Minami, and F. Shoji. 2016. 'Performance Analysis of the Graph Traversal Code using DebtRank Algorithm for Economic Simulation'. IPSJ SIG HPC Technical Report, Information Processing Society of Japan, Special Interest Groups, High Performance Computing.

The Institute of Statistical Mathematics. 2014. 'Web Decomp – Seasonal Adjustment & Time Series Analysis'. http://ssnt.ism.ac.jp/inets/inets_eng.html.

Timmer, M., A. A. Erumban, R. Gouma, B. Los, U. Temurshoev, G. J. de Vries, I. Arto, V. A. A. Genty, F. Neuwahl, J. Francois et al. 2012. 'The World Input-Output Database (WIOD): Contents, Sources and Methods'. Technical Report, Institue for International and Development Economics 2012.

Tobin, J. 1972. 'Ination and Unemployment'. *American Economic Review* 62(1): 1–18.

———. 1993. 'Price Flexibility and Output Stability: An Old Keynesian View'. *Journal of Economic Perspectives* 7: 45–65.

Traag, V. A. and J. Bruggeman. 2009. 'Community Detection in Networks with Positive and Negative Links'. *Physical Review E* 80: 036115.

Trust Companies Association of Japan, http://www.shintaku-kyokai.or.jp/html/trustbanks/e-1-2.html 2013.

Tzekina, I., K. Danthi, and D. N. Rockmore. 2008. 'Evolution of Community Structure in the World Trade Web'. *The European Physical Journal B* 63(4): 541–5.

Utsugi, A., K. Ino, and M. Oshikawa. 2004. 'Random Matrix Theory Analysis of Cross Correlations in Financial Markets'. *Physical Review E* 70: 026110.

Varian, H. R. 1992. *Microeconomic Analysis*, 3rd ed. New York: W. W. Norton.

Victoria-Feser, M. P. and E. Ronchetti. 1994. 'Robust Methods for Personal-Income Distribution Models'. *Canadian Journal of Statistics* 22(2): 247–58.

Vodenska, I., H. Aoyama, Y. Fujiwara, H. Iyetomi, and Y. Arai. 2016. 'Interdependencies and Causalities in Coupled nancial Networks'. *PLOS ONE* 11 (3): e0150994.

Weibull, W. 1939. *The Phenomenon of Rupture in Solids*. Stockholm: Generalstabens Litograska Anstalts Förlag.

———. 1939. *A Statistical Theory of the Strength of Materials*. Stockholm: Generalstabens Litograska Anstalts Frlag.

Wicksteed, P. H. 1932. *An Essay on the Co-ordination of the Laws of Distribution*, vol. reprint no. 12, London: London School of Economics, Generalstabens Litograska Anstalts Forlag, Stockholm 1894.

Wold, H. O. A. and P. Whittle. 1957. 'A Model Explaining the Pareto Distribution of Wealth'. *Econometrica, Journal of the Econometric Society* 591–5.

Yoshikawa, H. 2003. 'The Role of Demand in Macroeconomics'. *Japanese Economic Review* 54 (1): 1–27.

———. 2011. 'Stochastic Macro-Equilibrium and the Principle of Effective Demand'. *CIRJE Discussion Paper F Series, CIRJE-F-827*. http://www.cirje.e.u-tokyo.ac.jp/research/dp/2011/2011cf827.pdf.

———. 2014. 'Stochastic Macro-Equilibrium: A Microfoundation for the Keynesian Economics'. *Journal of Economic Interaction and Coordination*. DOI 10.1007/s11403-014-0142-4.

———, H. Aoyama, Y. Fujiwara, and H. Iyetomi. 2015. 'Deation/Ination Dynamics: Analysis Based on Micro Prices'. SSRN 2565599.

Yoshikawa, T., T. Iino, and H. Iyetomi. 2011. 'Market Structure as a Network with Positively and Negatively Weighted Links', in *Intelligent Decision Technologies*, edited by J. Watada, et al., 511–8. Berlin, Heidelberg: Springer.

———. 2012. 'Detection of Comoving Groups in a Financial Market', in *Intelligent Decision Technologies*, edited by J. Watada, et al., 481–6. Berlin Heidelberg: Springer.

———. 2012. 'Observation of Frustrated Correlation Structure in a Well-Developed Financial Market'. *Progress of Theoretical Physics Supplement* 194: 55–63.

———, Y. Arai, and H. Iyetomi. 2013. 'Comparative Study of Correlations in Financial Markets'. *Intelligent Decision Technologies, Frontiers in Artificial Intelligence and Applications* 255: 104–10.

Yule, G. U. 1925. 'A Mathematical Theory of Evolution Based on the Conclusions of Dr. J.C.Willis'. *Philosophical Transactions of the Royal Society of London B* 213: 21–87.

Yuta, K., N. Ono, and Y. Fujiwara. 2007. 'A Gap in the Community-Size Distribution of a Large-Scale Social Networking Site'. *Arxiv preprint physics/0701168*.

Zachary, W. W. 1977. 'An Information Flow Model for Conflict and Fission in Small Groups'. *Journal of Anthropological Research* 33(4): 452–73.

Zanette, D. and S. C. Manrubia. 1997. 'Role of Intermittency in Urban Development: A Model of Large-City Formation'. *Physical Review Letters* 79(3): 523–6.

Zwick, W. R. and W. F. Velicer. 1986. 'Comparison of Five Rules for Determining the Number of Components to Retain'. *Psychological Bulletin* 99: 432–42.

Index

Abenomics 223
Action 357
Actor, see Node
Adjacency list 227, 341
Adjacency matrix 228
Aggregate demand 118, 163
AI 357
AIC 54, 111
Akaike Information Criteria 54
Amplitude
    Average — 188
Angular frequency 198
Anti-Monopoly Act 82
AR 38, 152
ARCH 48
ARFIMA 46
ARIMA 38
ARMA 38
Arrow, Kenneth 2
Asset 321
Assortative mixing 240
Assortativity coefficient 242
Autocorrelation 152
Autocorrelation function 35
Autoregression model, see AR
Average 135, 146
Average mutual information 35

Bank loan 321
Bank of Japan 223
Bankruptcy 307

Basel Capital Accords 336
Beta
    — Distribution 62
    — Function 62
    — Type distribution 61
Binomial expansion 47
Bipartite network 321
Bisection method 249
Black Monday 33
Bohr 162
Boltzmann
    —'s constant 24
    — Ludwig 24
Bond, see Link
Bowtie structure 310
Box-counting dimension 45
Breadth-first search 233, 341
Brownian motion 41
    Fractional — 43
    Geometrical — 42
    Standard — 41
Bubble economy 188
Burr system 57, 58
Business cycle 163, 180, 191, 221

Cabinet Office 169
Cantor set 44
Capacity dimension 45
Causal relationship 272
CCDF 54
CDF 54

Center-of-mass motion 261
Central banks 214
Champernowne, David 6
Chemical potential 27
CHPCA 134, 137, 221, 266
City Banks (Japan) 323
Clockwise rotation 138
Clustering coefficient 245
Co-movement 207
Coarse graining 252
Collective motion 221, 261, 273
Combination dial lock 154
Common factor 169
Community 247, 262
    — Structure 268
Comovement 150
Comoving stocks 262
Complementary cumulative distribution function 54
Complete graph 280
Construction 215
Copula 104
    — Density 104
        Archimedean — 105, 107
        Clayton — 106
        Frank — 105
        Gumbel — 105, 116
        Production — 108, 110
        Survival — 107
Correlation
    Equal-time — 135
    coefficient 135, 146
Correlation matrix 135, 147, 260
    Complex — 266, 279
CPI, 210
    True Core — 214
CRD 94, 121
Credit networks 321
Crude oil price 217
Crystallization 251
Cumulative distribution function 54
Currency 155, 267

Dagum system 57
DCGPI 210
DebtRank 308, 324
Deep learning 357

Deflation 224
Degree 227
    — Correlation 243
    — Distribution 227, 240, 303
    — Sequence 238
    In- — 227
    Out- — 227
Depth-first search 228, 343
Detailed balance 88, 90
Digraph, *see* Graph, Directed
Dimension
    Fractal — 43
Distress 326, 336
Distribution
    Beta — 62
    Beta — of the first kind 64
    Beta — of the second kind 64
    Beta type — 61
    Bose-Einstein — 128
    Burr — 57, 65
    Burr III — 65
    Exponential — 68, 351
    Fermi-Dirac — 126
    Fisk — 68
    Gamma — 67
    Gaussian — 351
    Generalized beta — 62
    Generalized beta — of the first kind 62
    Generalized beta — of the second kind 63
    Generalized gamma — 64
    Inverse beta — of the first kind 63
    Log-normal — 66
    Lomax — 67
    Maxwell — 51
    Maxwell-Boltzmann — 27, 128
    Normal — 351
    Pareto — 53, 66
    Pearson — 64, 67
    Power — 66

    Power-law — 90
    Singh-Maddala — 65
    Type I Pareto — 66, 83
    Type II Pareto — 68
    Type IV Pareto — 65
    Uniform — 58
    Unit gamma — 63

Weibull — 67
Yule — 90
Dragon King 69
Drift 41
Driver node 296
Dugum system 59

Economic crisis 176, 273, 289, 294, 300
Economic Partnership Agreement 273
Economic Planning Agency 169
Economic shock 180, 186
Edge, see Link
Eigenvalue 150
Eigenvector 170
Einstein, Albert 134
Entrainment 180
    Frequency — 184
Entropy 24
    Boltzmann's — 25
    Gibbs' — 25
    Maximization of — 119
    Shannon's — 26
Equal a priori probability postulate 24
Equilibrium condition 131
Erdős–Rényi random graph, see Poisson random graph
Euclidean geometry 43
Euler
    — Constant 353
    — equation 120
Exchange rate 217

Feynman, Richard P. 5
Finished goods 165
Finite difference 31
Firm 53
    Number of employees of — 53
    Profit of — 53
    Total Asset of — 53
    — size 94
Fluctuation 27
Food 214
Fourier transform
    Discrete time — 139
Fréchet distribution 71
Fractal 43
    — dimension 43

Frequency entrainment 197
Frustrated correlation structure 263
Frustration 262

Gamma function 47
GARCH 48
Garibaldi, Ubaldo 4
GB2 63, 99, 116
Gibbs, Josiah Willard 10, 25
Gibrat
    — Regime 94
    —'s law 85, 88, 90, 94
Gibrat, Robert 6, 85
Grand canonical ensemble 126
Graph
    Adjacency list 227, 341
    Adjacency matrix 228
    Assortative mixing 240
    Bipartite — 227
    Breadth-first search 233, 341
    Clustering coefficient 245
    Configuration model 238
    definition 227
    Degree correlation 243
    Degree distribution 240
    Degree sequence 238
    Dense — 228
    Depth-first search 228, 343
    Directed — 227
    Distance 234
    Excess degree 238
    Modularity 242
    Poisson random — 236
    Random — 236
    Shortest path 234
    Sparse — 228
    Strongly connected component 231
    Undirected — 227
Great Recession 214
Greedy algorithm 249, 277
Group correlation 260

Hierarchical domain decomposition 252
Hierarchical scheme 249
High-frequency data 34
Hilbert transform 138, 195, 279

Hurst index 43

Iceland crisis 156
IMF 155
In-degree 267
Income elasticity 59
Incomplete gamma function 353
    The first-order — 78
Indices of industrial production 165
Induction 70
Inequality 53
Inflation 224
Inter-bank relationship 336
Inventory 165
Inventory stock 193
IPI 210
Ito process 41
Ito's lemma 42

Jaccard index 277, 283, 292
Japanese consumption tax 212
Job matching 119

K computer 310
Kaiser's selection rule 151
Kesten process 90
Keynes' principle of effective demand 170
Keynes, John Maynard 1
Keynesian theory 163
Kolmogorov-Smirnov test 88
Krugman, Paul 3
Kurtosis 55

Labor productivity 116
Labor wage process 90
Lag 134
Lag operator 47
Lagrangian multiplier 26
Lead 134
Lead-lag relation 267
Lehman crisis, *see* Sub-prime mortgage crisis
Lehman shock 333
Lending/borrowing 321
Limit-cycle 194, 202
Link 227
    Weight 227
Linked community 283, 292

Long-term memory 27, 46
Long-term memory process 35
Lorenz curve 92
Lost Decade 221
Lucas Jr., Robert E. 2

MA 38
Mantegna, Rosario N. 5
Market mode 261
Markov chain model 90
Matchability 296
Mathematica 249, 337, 345
Maximum bipartite matching 296
Maximum value 69
Mean 54
Mega Banks (Japan) 323
Merger 321
Minimum Spanning Tree, *see* MST
Mizuho Corporate Bank 328
Mode signal 136, 147, 172
Model
    AR(1) — 152
    Ehrenfest-Brillouin — 123
    Spring-charge — 250, 269
Modularity 242, 248, 268, 277, 280
    — Landscape 248
    Multiple — 278
Molecular dynamics 251
Molecule 357
Moment 54
Monetary base 216
Money stock 216
Monopoly 83
Monte Carlo simulation 74
Moving average 30
Moving average model, *see* MA
MST 324
Multiple edge 227
Multiplex network 278
NEEDS 98, 121
Negative weight 262
Neoclassical equilibrium theory 163
Network
    Equity-currency — 272
    Globally-coupled financial — 266
    Karate club — 250
    Power grid — 252

Stock correlation — 260
Synchronization — 268
Node 227
    Adjacent — 227
Non-self averaging 4
Non-stationarity 27
Non-stationary process 28
Nonlinearity 330
Normal distribution 55
NTAJ 85, 91

Oil 215
Old Keynesian view 177
Oligopoly 83
Open population model 90
Orbis 120
Order parameter 279, 287
Oscillator
    Kuramoto — 203
    Limit-cycle — 203
Out-degree 267
Overtime hours worked 216

Parallel analysis 151
Parallel edge 227
Pareto
    — Distribution 55
    — Index 66, 83, 93
    — Scale 66, 83
    —'s law 84
Pareto, Vilfredo 6, 83
Particle
    Distinguishable — 117, 128
    Indistinguishable — 129
PCA 134, 272
PDF 54
    Conditional — 85
    Joint — 88
Pearson
    — Distribution 55, 56
    — System 55–57
Penkov, Miroslav 1
Personal income 53
Phase
    Average — 188
Phase coherence 279
Phase locking 186, 197
    Partial — 186
Phase transition 357
Phillips curve 224
Poisson random graph 236
Polarization 261
Post-Lehman recession 165
Power spectrum 168
Power-law 53
PPI, *see* DCGPI
Pre-Lehman crisis 165
Price
    Consumer goods and services — 221
    Imported goods — 221
    Producer goods — 221
Probability
    Occurrence — 25
Probability density function 54
Production 165, 173
Production function 99
    CES — 102
    CES — 101
    Cobb-Douglas — 101, 163
Production network 253, 307
Productivity 97
    Labor — 98
Proverb 97
Quantum mechanics 357

R&D investment 165
Ramsey consumer 120
Random matrix theory 150
Raw moment 54
Real business cycle 163
Real Business Cycles Theory (RBC) 2
Reflection law 90
Reginal banks (in Japan) 321
Regression analysis 204
Return 32
    Log — 33
Ridge theory 101
RMT 156
Robustness 278
Romer, Paul 210, 225
Root 29
Rotational random shuffling 153
RRS 221
S&P 500, 264

Sagredo 356
Salviati 356
Samuelson, Paul 2
Sanwa Bank 331
Savings and loans 321
Scalas, Enrico 4
Scaling regime 94
SDR 155
Sea level pressure 154
Search theory 119
Seasonally-adjusted data 168
Sector
    Manufacturing — 132
    Non-manufacturing — 132
Self-affine 43, 45
Self-loop 227
Self-similarity 43, 44
Services 215
Shannon, Claude 26
Shipments 165, 173
Shock 38
    Common — 189, 191
    Demand — 165
    Individual — 189, 191
    Technological — 202
    Technology — 164
Short-term memory process 35
Simplicio 356
Simulated annealing 249, 262
Site, see Node
Skewness 55
Sklar's theorem 104
Slutsky effect 163
Sornette, Didier 4
Spacetime 357
Special Drawing Right, see SDR
Spectral decomposition 266
Stability of society 53
Standard deviation 55, 135, 146
Standardized time series 135, 146
Stanley, H. Eugene 5
Stationary
    Strongly — 29
    Weakly — 29
    — process 28
Statistical equilibrium 26
    — Theory of markets 117

Steepest descent path 248
Stiglitz, Joseph E. 3
Stirling formula 25
Stochastic macro-equlibrium 7
Stochastic model 11
Stochastic variable 54
Stock index 155
Stock market 261, 267
Strongly connected component 231
Structural balance theory 263
Structural controllability 295
Structure
    Balanced — 263
    Unbalanced — 263
Supercomputer 310
Supplier-customer network 253, 307
Surrogate variable 37
Synchronization 180, 191, 202, 269, 279

Temperature 27
    Negative — 118
Thermodynamics 24
Tie, see Link
Tokai Bank 331
Tokyo Mitsubishi UFJ Bank 328
Tokyo Stock Exchange 152
Trade liberalization 273
Trans-Pacific Partnership 273
Triangular relationship 263
TSE 260
TSR 94
Two-factor model 91

Unemployment rate 216
Unit root test 29
Universal statistics 123
Universality 83
Utility function 357

Value added 97, 278
Variance 55
Variation of information 278
VAT 212
Vertex, see Node
Visualization 250
Volatility 35, 37
Volatility clustering 35, 48, 50

Vulnerability of banks 332
Vulnerability of sector 335

Wage index 216

Wealth accumulation process 90
Wiener process 41

Yen-U.S. dollar exchange rate 221

Color Plates

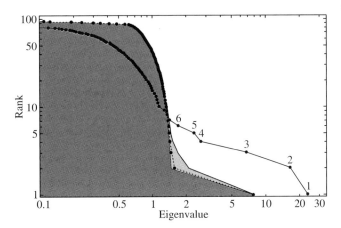

Fig. 5.10 Significant eigenvalues identified by CHPCA using the RRS method. The dot denoted by n shows the nth eigenvalue $\lambda^{(n)}$ (x-axis) and the eigenvalue rank (y-axis). The gray small dots and the lighter gray area show the average RRS and the 99% range. The six largest eigenvalues are clearly outside of their RRS ranges, and show significant relationships in the interdependent network (from Vodenska et al. (2016)).

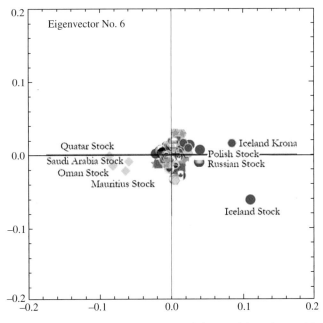

Fig. 5.13 The sixth eigenvector components of the world equity and foreign exchange markets (from Vodenska et al. (2016)).

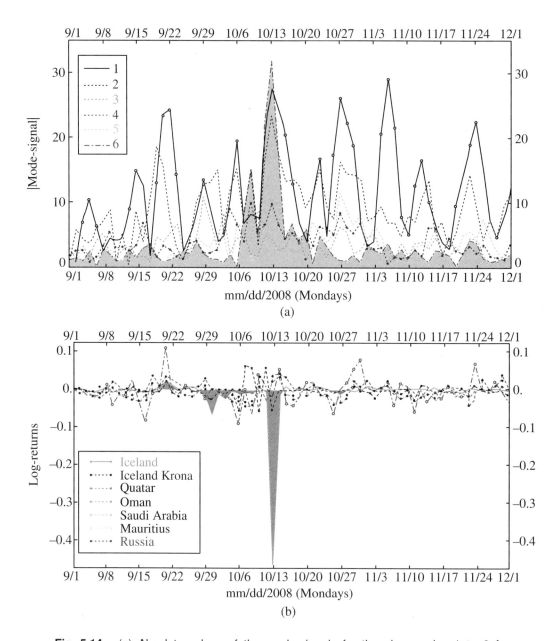

Fig. 5.14 (a) Absolute values of the mode-signals for the eigenmodes 1 to 6 from September 1st to December 1st of 2008. The red dots connected with red dash-dot lines show the sixth eigenmode. (b) Behavior of the log-returns of the time series that have large absolute values in the sixth eigenvector (from Vodenska et al. (2016)).

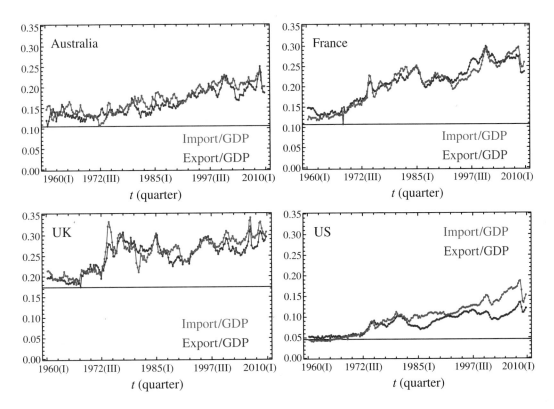

Fig. 6.33 Exports and imports relative to GDP for four countries: Australia, France, the United Kingdom and the United States. The ratios have increased for the past 20 years for all four countries. Trade data show that the imports (exports) relative to GDP is high except for the United States. These figures show that the importance of international trade has increased and therefore the interaction between countries is expected to have been strong (from Ikeda et al. (2013a)).

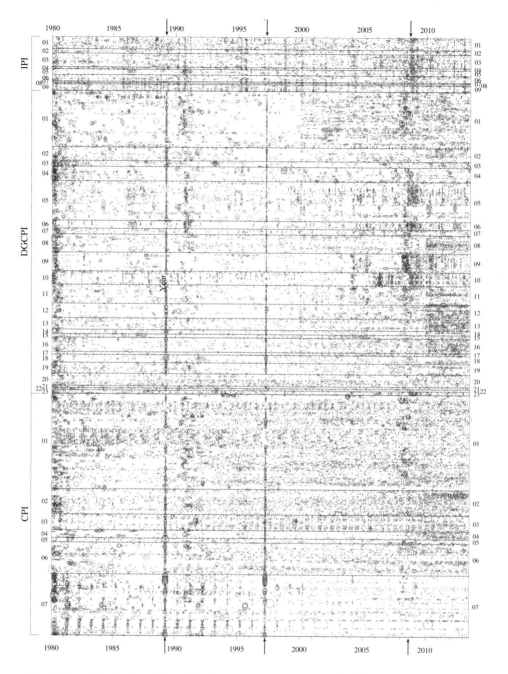

Fig. 7.1 Standardized logarithmic monthly growth rate $\widehat{r}_\alpha(t)$ of individual changes for $|\widehat{r}_\alpha(t)| > 1.0$. Three arrows correspond to the month when Japanese consumption tax (VAT) of 3% was introduced, the month it was raised to 5%, and the time when the Lehman Brothers declared bankruptcy during the subprime mortgage crisis, from left to right (from Yoshikawa et al. (2015)).

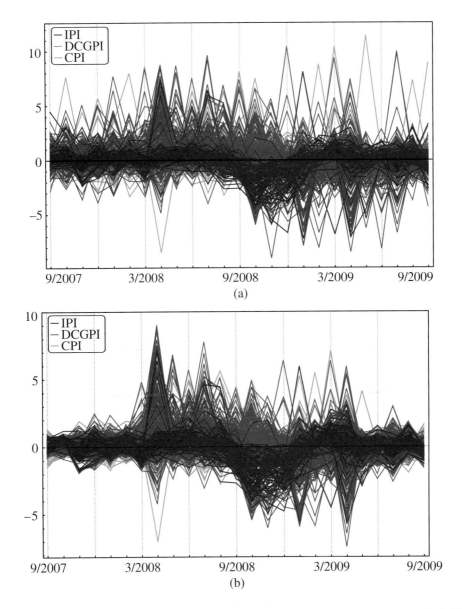

Fig. 7.4 The real part of the time series in and out of the Great Recession period. Plot (a) shows the original time series, and (b) the cleansed (from Yoshikawa et al.(2015)).

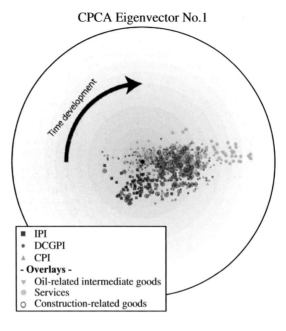

Fig. 7.5 The 1st eigenvector components in the complex plane (from Yoshikawa et al. (2015)).

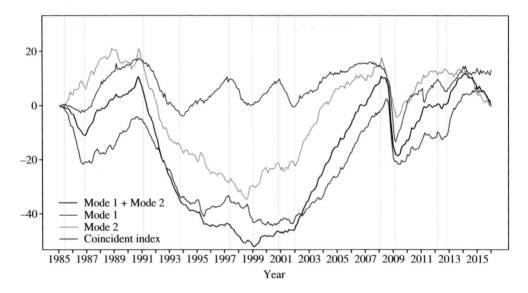

Fig. 7.11 Contributions of the first and second eigenmodes to the coincident index of business conditions compared with its standardized data, where all the results are temporally accumulated with the same initial value to facilitate comparison. The red and blue vertical lines indicate troughs and peaks in the Japanese business cycles, respectively. (From Kichikawa et al. (2016))

Fig. 8.16 Visualization of directed network of the production network in Japan, comprising a million firms as points. Colors of the points indicate industrial sectors of the firms. Note that upstream and downstream are depicted in the bottom and top sides respectively (from Fujita et al. (2016)).

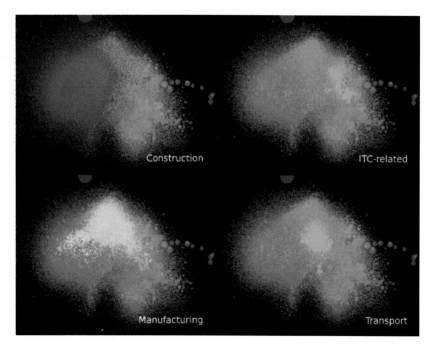

Fig. 8.17 Industrial sectors are identified for construction, ICT-related, manufacturing, and transport of automobiles (from Fujita et al. (2016)).

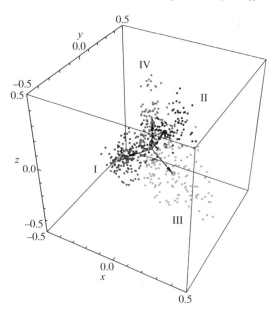

Fig. 8.20 Three-dimensional correlation state vectors of the stocks in the Tokyo Stock Exchange. The stocks in the community I are colored red, those in II are blue, those in III are green, and those in IV are brown. The arrows point to the positions of the center of gravity in individual communities.

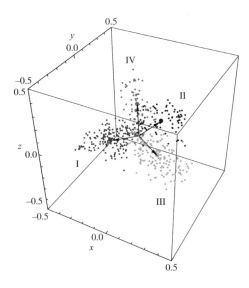

Fig. 8.22 Same as Fig. 8.20, but for the S&P 500.

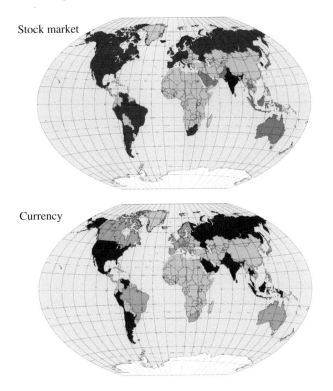

Fig. 8.23 Community structure of the financial network on a world map in the entire period, where communities dominated by stock markets are shown in red and orange, and those dominated mainly by currencies, in blue and green. (Adapted from Vodenska et al. (2016))

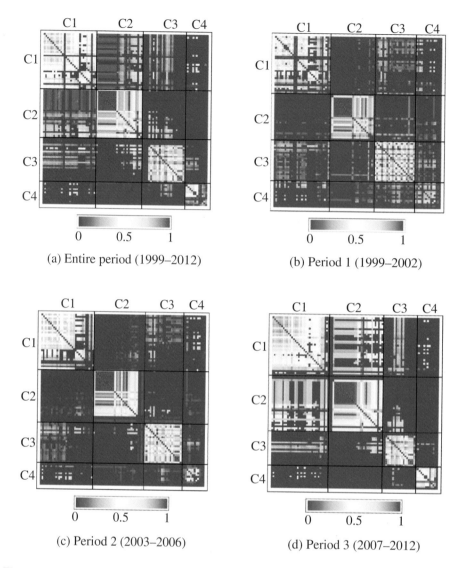

Fig. 8.24 Adjacency matrices of the financial network for the four periods, sorted according to the classification into four communities (C1, C2, C3, C4) of synchronizing nodes. (Adapted from Vodenska et al. (2016))

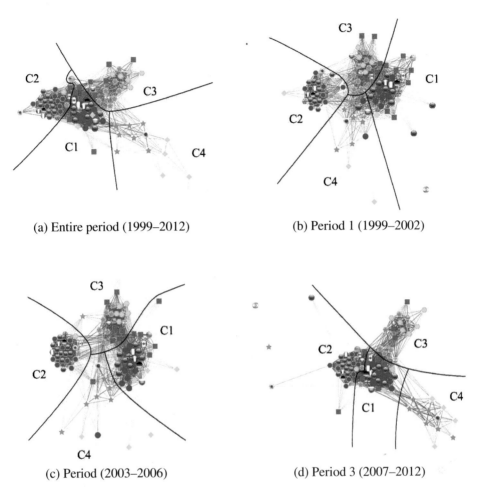

(a) Entire period (1999–2012)

(b) Period 1 (1999–2002)

(c) Period (2003–2006)

(d) Period 3 (2007–2012)

Fig. 8.25 Optimized layouts of the financial network in a spring–electrical model with boundaries separating the communities, corresponding to Fig. 8.24. (Adapted from Vodenska et al. (2016))

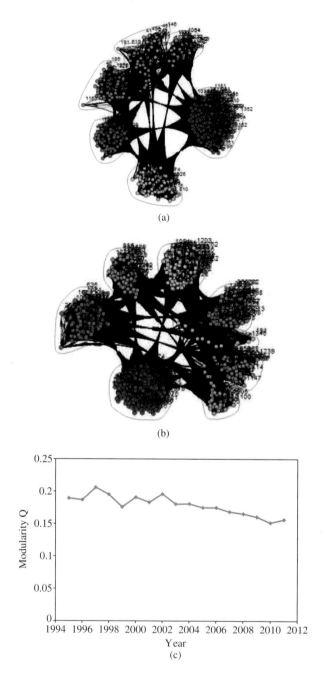

Fig. 8.27 The community structures for each time slice of the international trade network. Panels (a) and (b) lists examples of community structures for 1997 and 2009. There were seven communities for 1997, and eight for 2009. The temporal change of modularity Q is shown in panel (c). The value of the obtained modularity Q is about 2, which depends on the threshold of the weight of links $w_{A\alpha,B\beta}$ (from Ikeda et al. (2016)).

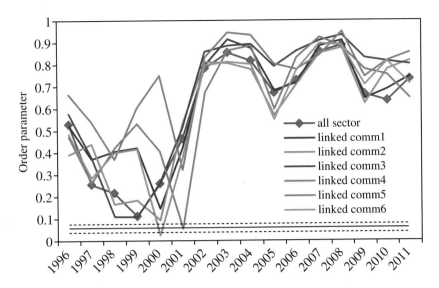

Fig. 8.31 The temporal change in amplitude for the order parameter $r(t)$ for the years 1996 to 2011. The phase coherence fell gradually in the late 1990s but increased sharply in 2001 and 2002. This temporal change might be related to the structural change in the international trade network discussed in Subsection ??. From 2002, the amplitudes for the order parameter $r(t)$ remain very high, although the years 2005 and 2009 are slightly lower (from Ikeda et al. (2016)).

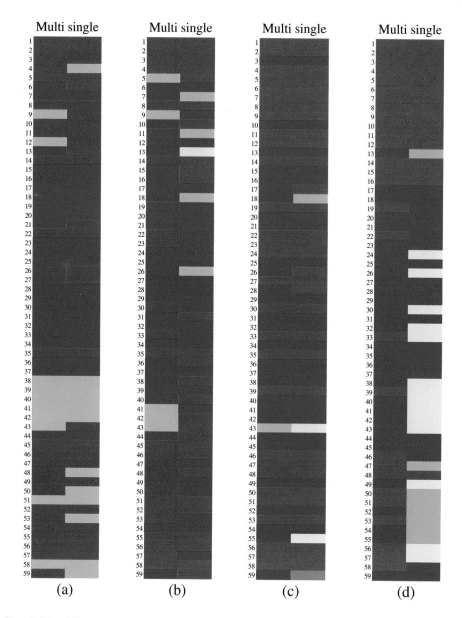

Fig. 8.35 We compare community structures between a multiplex network and a single-layer network for 2004, 2007, 2010, and 2013. The left and right column of each panel correspond to communities identified for a multiplex network and a single-layer network, respectively. Although the direction of links are ignored in the community analysis for a multiplex network, the two analysis agree reasonably well (from Ikeda et al. (2016)).

Color Plates

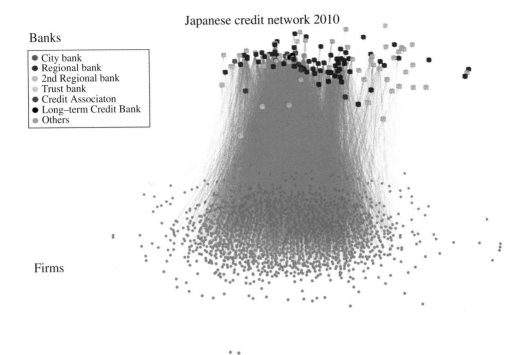

Fig. 9.10 Network formed by banks (upper layer) and firms (lower layer) in 2010, where edges are bank-loans unweighted by the amount (from Aoyama et al. (2013b)).

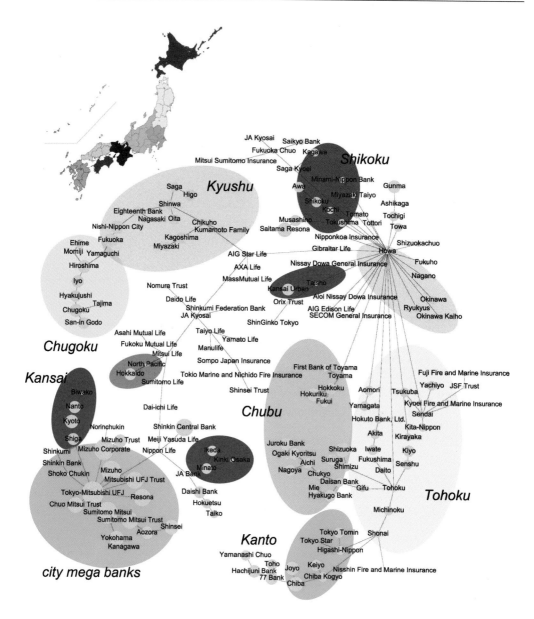

Fig. 9.11 Minimum spanning tree for the nodes of banks and the edges with weights of distances between banks. Distance of a pair of nodes is smaller if the banks have similar lending to firms. The node size is proportional to the logarithm of the bank's asset. Circles with colors are identified as geographical regions and city mega banks. (Upper-left) Japan and eight geographical regions; Hokkaido (red), Tohoku (yellow), Kanto including Tokyo (green), Chubu (cyan), Kansai including Osaka and Kyoto (blue), Chugoku (orange), Shikoku (purple), Kyushu including Okinawa (gray) (from Aoyama et al. (2013b).)

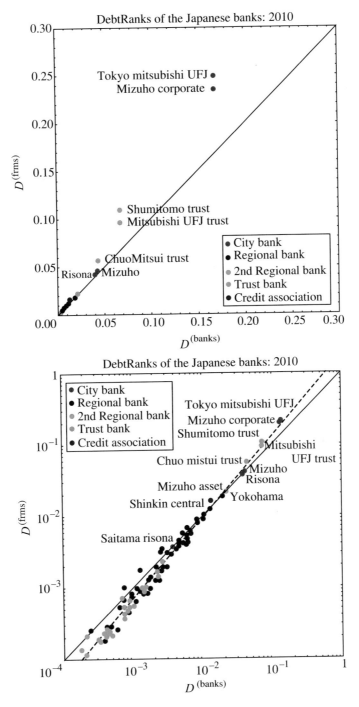

Fig. 9.14 The DebtRanks of the banks in 2010, the left in linear scale, the right in log scale (from Aoyama et al. (2013b)).

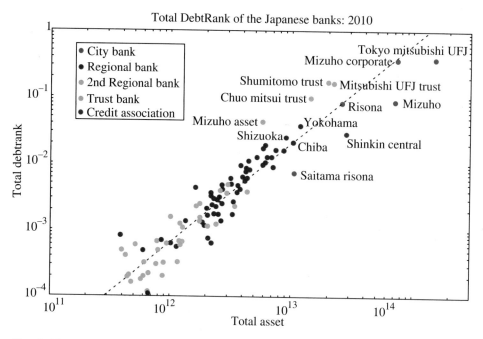

Fig. 9.15 Total asset vs. total DebtRank in 2010 (from Aoyama et al. (2013b)).

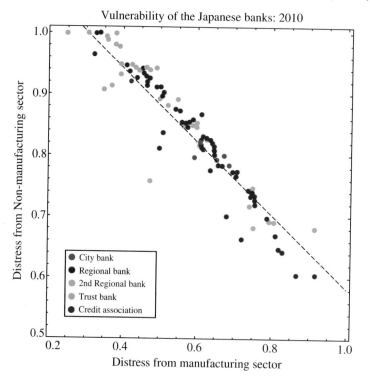

Fig. 9.17 Vulnerability of banks to distress in the manufacturing sectors and the distress in the non-manufacturing sectors (from Aoyama et al. (2013b)).

fs 995